Human Factors in Nuclear Safety

Human Factors in Nuclear Safety

EDITED BY

NEVILLE STANTON

UNIVERSITY OF SOUTHAMPTON

UK Taylor & Francis Ltd, 1 Gunpowder Square, London EC4A 3DE
USA Taylor & Francis Inc., 1900 Frost Road, Suite 101, Bristol, PA 19007

British Library Cataloguing in Publication Data

A catalogue record for this book is available from the British Library.

ISBN 0-7484 0166 0

Library of Congress Cataloging in Publication Data are available

Cover design by Christopher Gregory
Typeset in Times 10/12pt by Keyset Composition, Colchester
Printed in Great Britain by T.J. Press (Padstow) Ltd

Contents

Contributors

The names are listed in alphabetical order.

Ms Melanie Ashleigh
Department of Psychology
University of Southampton
Highfield
Southampton
SO17 1BJ
UK

Dr Christopher Baber
The Industrial Ergonomics Group
School of Manufacturing and Mechanical Engineering
University of Birmingham
Edgbaston
Birmingham
B15 2TT
UK

Professor Richard T. Booth
Director of Health and Safety Unit
Department of Mechanical and Electrical Engineering
Aston University
Aston Triangle
Birmingham
B4 7ET
UK

Professor Sue Cox
Centre for Hazard and Risk Management
Loughborough University of Technology
Loughborough
Leicestershire
LE11 3TU
UK

Professor Tom Cox
Centre for Organisational Health and Development
Department of Psychology
University of Nottingham
Nottingham
NG7 2RD
UK

Yves Dien
Direction des Etudes et Recherches
Département Etudes de Sûreté et de Fiabilité
Électricité de France
1 Avenue de Général de Gaulle
BP 408 92141
Clamart Cedex
FRANCE

Dr Marc Dorel
Laboratoire Travail et Cognition
Maison de la Recherche
Université de Toulouse-Le Mirail
5 allé Antonio Machado
31058 Toulouse Cedex
FRANCE

Yushi Fujita
Electrical and Control Engineering Department
Mitsubishi Atomic Power Industries Inc.
4-1, 2-Chrome
Shibakouen
Minato-ku
Tokyo 105
JAPAN

Dr Barry Kirwan
The Industrial Ergonomics Group
School of Manufacturing and Mechanical Engineering
University of Birmingham
Edgbaston
Birmingham
B15 2TT
UK

Dr Philip Marsden
Human Reliability Associates
1 School House
Higher Lane
Dalton
Wigan
Lancashire
WN8 7RP
UK

John M. O'Hara
Brookhaven National Laboratory
Building 130
Upton
New York 11973
USA

Dr Robert B. Stammers
Applied Psychology Group
Aston Business School
Aston University
Aston Triangle
Birmingham
B4 7ET
UK

Dr Neville Stanton
Department of Psychology
University of Southampton
Highfield
Southampton
SO17 1BJ
UK

Jody L. Toquam
Battelle Human Affairs Research Center

Dr Daniel L. Welch
4307 Harvard St
Silver Spring
Maryland
MD 20906-4461
USA

Preface

I believe that human factors has much to offer nuclear safety. The technical achievement in the nuclear industry is considerable, yet there is much to be discovered about human functioning in such environments. There has been some concern in recent years about safety. The incidents at Three Mile Island (in the USA) and Chernobyl (in the former USSR) are much cited in the press and technical literature. A recent near-incident at a nuclear utility in the UK has seemingly reinforced this concern. Whilst these nuclear power plants employ different technologies there is one common factor to these, and other, incidents: namely human beings. I am not laying the blame upon 'human error', a phrase that is all too common in nuclear and other types of incident. Rather, I am suggesting that the development of technological systems has not always paid heed to the social systems that are involved in their development, operation and maintenance. This book addresses the human aspect of nuclear power.

The idea for this book came to me when I was involved in the organisation of an international conference on nuclear safety in 1993 (a summary of the conference contributions can be found in *Nuclear Engineering International*, September 1993, pages 53–5). The conference emphasised the importance that the nuclear power community places upon human factors (human factors is the study of people in technological environments). In general, the nuclear industry is regarded as the Gold Standard in human factors research and development, but there is little room for complacency. Whilst much insight into the nature of human factors has been achieved and much practical advice can be given (as the contents of this book demonstrate) there is great scope for more research.

The focus of this book is upon Nuclear Safety, but many of the findings may be applied more widely to any complex technical environment. The text sets out to provide a balance between academic research and application of principles to the domain of nuclear power. To assist in the latter, guidelines and models have been included within individual chapters. In this way the book will appeal to both researchers and practitioners alike. I also hope that this book will leave the reader as enthusiastic about human factors and its contribution to nuclear safety as I am!

Thanks go to the people who made this book possible: the individual contributors (in order of appearance: Dan Welch, John O'Hara, Yves Dien, Chris Baber, Phil Marsden, Yushi Fujita, Jody Toquam, Melanie Ashleigh, Rob Stammers, Marc Dorel, Tom Cox, Sue Cox, Barry Kirwan and Richard Booth), the staff at Taylor & Francis (particularly Richard Steele and Christian Turner) and Maggie Stanton (for all her support and encouragement during this and other projects).

Neville Stanton
February 1996

CHAPTER ONE

The discipline of human factors?

NEVILLE STANTON

Department of Psychology, University of Southampton

1.1 What is Human Factors?

Most countries in the world have a national Human Factors/Ergonomics Society, for example, The Ergonomics Society in Great Britain (phone/fax from the UK 01509 234904) and the Human Factors and Ergonomics Society in the USA (phone from the UK 00 1 213 394 1811). According to Oborne (1982) initial interest in human-factors-related interests was conceived around the early 1900s, particularly problems related to increases in the production of munitions at factories during the First World War. The discipline was officially born at a multidisciplinary meeting held at the Admiralty in the UK on human work problems. We have an accurate recording of the date of this meeting, as can be expected of the armed services, and hence the birth of the discipline can be dated as 12 July 1949. The discipline was named at a subsequent meeting on 16 February 1950. To quote Oborne, 'The word ergonomics [cf. 'human factors'] was coined from the Greek: ergon – work, and nomos – natural laws.' (Oborne, 1982, p. 3.) The disciples of the discipline were initially concerned with human problems at work (on both sides of the Atlantic), in particular with the interaction between humans and machines and the resultant effect upon productivity. This focus has been broadened somewhat since then to include all aspects of the working environment. Figure 1.1 distinguishes human factors from psychology and engineering.

Human factors (HF) is concerned about the relationship between human and technology with regard to some goal-directed activity. Whereas engineering is concerned with improving technology in terms of mechanical and electronic design and psychology is concerned with understanding human functioning, HF is concerned with adapting technology and the environment to the capacities and limitations of humans with the overall objective of improving performance of the whole human–technology system. There are

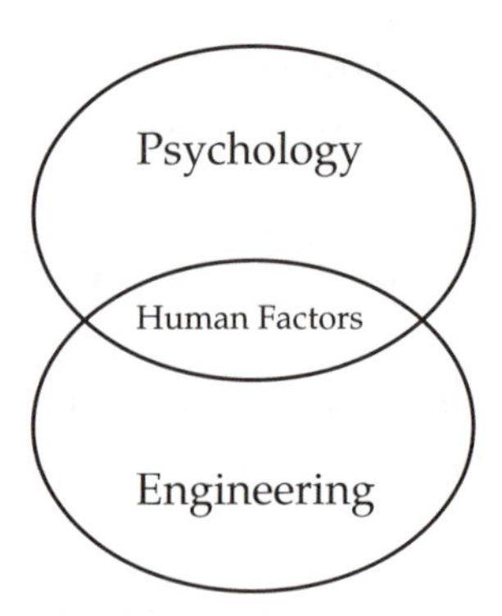

Figure 1.1 Distinction of human factors.

plenty of examples to show the consequences of the mismatch between human–technology systems: Zeebrugge, Three Mile Island, Chernobyl, Bhopal and Challenger. These disasters have an obvious and dramatic effect. Other forms of mismatch may have more insidious effects, e.g. reducing productivity and affecting employees' health. A systems viewpoint is taken in HF, to consider how human and technological systems can work together in a harmonious manner.

HF has emerged as a new discipline over the past 40 years in response to technological development. The role of HF is essential in the application of technological systems, as it focuses on the role of human involvement. The approach of HF is systematic, applying relevant and available information concerning human capabilities, characteristics, limitations, behaviour and motivation. In this sense, the problem space of HF can be defined as any system that requires human involvement. Thus the scope of HF covers the following:

- analysis, measurement, research and prediction of any human performance related to the operation, maintenance, or use of equipment or subsystem;
- research on the behavioural variables involved in the design of equipment, jobs and systems;
- the application of behavioural knowledge and methods to the development of equipment, jobs and systems;
- the analysis of jobs and systems to assist in the optimal allocation of personnel roles in system operation;
- research on the experiences and attitudes that equipment users and system personnel have with regard to equipment and systems;
- the study of the effects of equipment and system characteristics on personnel performance.

(from Miester, 1989).

Thus, the scope of HF is both broad and multi-disciplinary in nature. HF is a bridging discipline between psychology and engineering. As such it is able to offer input into most aspects of human-technology systems. Two general definitions of HF illustrate this point:

> The term 'human factors' is used here to cover a range of issues. They include the perceptual, mental and physical capabilities of people and the interactions of individuals with their job and working environments, the influence of equipment and system design on human performance and, above all, the organisational characteristics which influence safety related behaviour at work. *Source:* HSE (1989).

> [Human factors is . . .] the study of how humans accomplish work-related tasks in the context of human-machine system operation and how behavioural and non-behavioural variables affect that accomplishment. Human factors is also the application of behavioural principles to the design, development, testing and operation of equipment and systems. *Source:* Miester (1989).

These definitions highlight the need to consider the interactions between individuals, jobs, the working environment, the equipment, system design and the organisation. Given the continuity, and nature, of the interaction between different elements of complex systems, it is necessary to develop an integrated approach if the analysis is to be effective. The main objectives of HF are to improve the effectiveness and efficiency of the work system (i.e. the ease with which the symbiosis between humans and technology occurs) as well as reducing errors, fatigue and stress. Thus, it also aims to improve safety, comfort, job satisfaction and quality of working life (Sanders and McCormick, 1993).

Stanton and Baber (1991) offer four views of HF:

- a discipline which seeks to apply natural laws of human behaviour to the design of workplaces and equipment;
- a multidisciplinary approach to issues surrounding people at work;
- a discipline that seeks to maximise safety, efficiency and comfort by shaping the workplace or machine to physical and psychological capabilities of the operator;
- a concept, a way of looking at the world and thinking about how people work and how they cope.

Each of these views offers a subtly different perspective. The first suggests that 'natural laws' of human behaviour exist (e.g. Hick's law and Fitt's law), which can be applied to the design and evaluation of technological environments. Whilst such a view may produce important findings (which tend to be highly specific) it is dubious that such findings constitute immutable laws. This leads to the second viewpoint which draws on a potpourri of different subject matter. Alternatively the third viewpoint emphasises the need to design the job to fit the person. Problems with this approach arise from attempting to define the 'average person'. Finally the fourth viewpoint develops a notion of HF as an attitude: first it is necessary to recognise the need, then it is necessary to employ a body of knowledge and a set of skills to satisfy this need. Stanton and Baber favour the final view as it is distinctly different from the first three, it proposes HF as a philosophy rather than an 'add-on' approach; HF provides an overview of the human–technology system, rather than a discrete aspect of the system.

1.2 Human Factors Methodology

HF has developed approaches with which to combine different methods to study aspects of the workplace. In satisfying the need for improved HF we should consider what we mean by the overused clichés, 'user friendly' and 'user-centred design'. These phrases do not just mean asking the users what they want of a system, because often users may not be fully aware of the range of possibilities. Rather, this terminology highlights the need to design for the user, taking account of their capabilities and capacities. The HF engineer offers a structured methodological approach to the human aspects of system design. Traditionally HF assessments have been performed at the end of the design cycle, when a finished system can be evaluated. However, it has been noted that the resulting changes proposed may be substantial and costly. This has led to a call for HF to be more prominent in the earlier aspects of design.

HF engineers are particularly keen to be involved in user evaluations earlier on in the design process, where it is more cost effective to make changes. Prototyping is one means of achieving this. Developing prototypes enables the users to have a contribution towards the design process. In this way identified problems may be reduced well before the final implementation of the system. The contributions of the user do not only take the form of expressed likes and dislikes. Performance testing is also very useful. This could include both physiological measures (Gale and Christie, 1987) and psychological measures (e.g. speed of performance and the type of errors made). User participation may also provide the designers with a greater insight into the needs of the users. This can be particularly important when the users may not be able clearly to express what features they desire of the system, as they are likely to be unaware of all the possibilities that could be made available. It is important to remember that human factor specialists typically work alongside engineers and designers. The human factors specialists bring an understanding of human capabilities to the team, whilst the engineers bring a clear understanding of the machines' capabilities. These two approaches may mean that the final system is more 'usable' than would have been achieved if either the team components had worked alone.

1.3 Human Factors Applications

HF applications can be broadly divided into the following areas:

- specifying systems;
- designing systems;
- evaluating systems (does it meet own specifications?);
- assessing systems (does it meet HF specifications?).

This gives a field of activity comprising specification, design, evaluation and assessment. HF begins with an overall approach which considers the user(s), the task(s) and the technology involved, and aims to derive a solution which welds these aspects together. This is in contrast to the prevailing systems design view which considers each aspect as an isolated component. HF can provide a structured, objective investigation of the human aspects of system design, assessment and evaluation. The methods will be selected on the basis of the resources available. The resources will include time limits of project, funds, and the skills of the practitioners. Overall, HF can be viewed as an attitude to the specification, design, evaluation and assessment of work systems. It requires practitioners to recognise the need for HF, draw on the available body of knowledge and employ an appropriate range of methods and techniques to satisfy this need. It offers support to traditional design activity by permitting the structured and objective study of human behaviour in the workplace.

From the discussion it is apparent that HF has a useful contribution to offer. There is an awakening to this as the impending legislation demonstrates. The contribution comes in the form of a body of knowledge, methods and above all an attitude inherent in the HF approach. It is this attitude that provides HF with a novel perspective and defines its scope. HF has much to offer designers and engineers of technological systems, and this could be no more true than in the domain of the nuclear industry.

1.4 Human Factors in Nuclear Safety

There is an increasing awareness amongst the nuclear power community of the importance of considering human factors in the design, operation, maintenance and decommissioning of nuclear power plants. Woods *et al.* (1987) suggest nuclear power plants present a considerable challenge to the HF community, and their review concentrates upon control room design. Reason (1990a) reports that 92% of all significant events in nuclear utilities between 1983 and 1984 (based upon the analysis of 180 reports of INPO members) were man-made and, of these, only 8% were initiated by the control room operator. Thus, the scope of HF needs to consider all aspects of the human-technology system. The consideration of the human element of the system has been taken very seriously since the publication of the President's Commission's report on Three Mile Island (Kemeny, 1979) which brought serious HF problems to the forefront. In summary of the main findings the commission reports a series of 'human, institutional and mechanical failures'. It was concluded that the basic problems were people-related, i.e. the human aspects of the system that manufactures, operates and regulates nuclear power. Some reports have suggested 'operator error' as the prime cause of the event, but this shows a limited understanding of HF. The failings at TMI included:

- deficient training that left operators unprepared to handle serious accidents;
- inadequate and confusing operating procedures that could have led the operators to incorrect actions;
- design deficiencies in the control room, for example in the way in which information was presented and controls were laid out;
- serious managerial problems within the Nuclear Regulatory Commission.

None of the deficiencies accounts for the root cause of the incident as 'operator error', which is an all too familiar explanation in incidents involving human–technology systems. Reason (1987) in an analysis of the Chernobyl incident suggested two main factors of concern. The first factor relates to the cognitive difficulties of managing complex systems: people have difficulties in understanding the full effects of their actions on the whole of the system. The second factor relates to a syndrome called 'groupthink': small, cohesive and elite groups can become unswerving in their pursuit of an unsuitable course of action. Reason cautions against the rhetoric of 'it couldn't happen here' because, as he argues, one of the basic system elements (i.e. people) is common to all nuclear power systems. Most of the HF issues surrounding the problems identified in the analyses of TMI and Chernobyl can be found within this book. A committee was set up in the UK to advise the nuclear industry in the UK about how HF should be managed. The reports of this committee are the subject of the next section.

1.5 ACSNI Reports on Human Factors

The Advisory Committee on the Safety of Nuclear Installations (ACSNI) set up the Human Factors Study Group on 7 December 1987 to review current practice and make written reports on the important aspects of HF in nuclear risk and its reduction. They produced three reports covering training, human reliability assessment and safety culture. The third, and final, report was published in November 1992. The common theme of safety is apparent throughout, and the concept of safety culture is dominant.

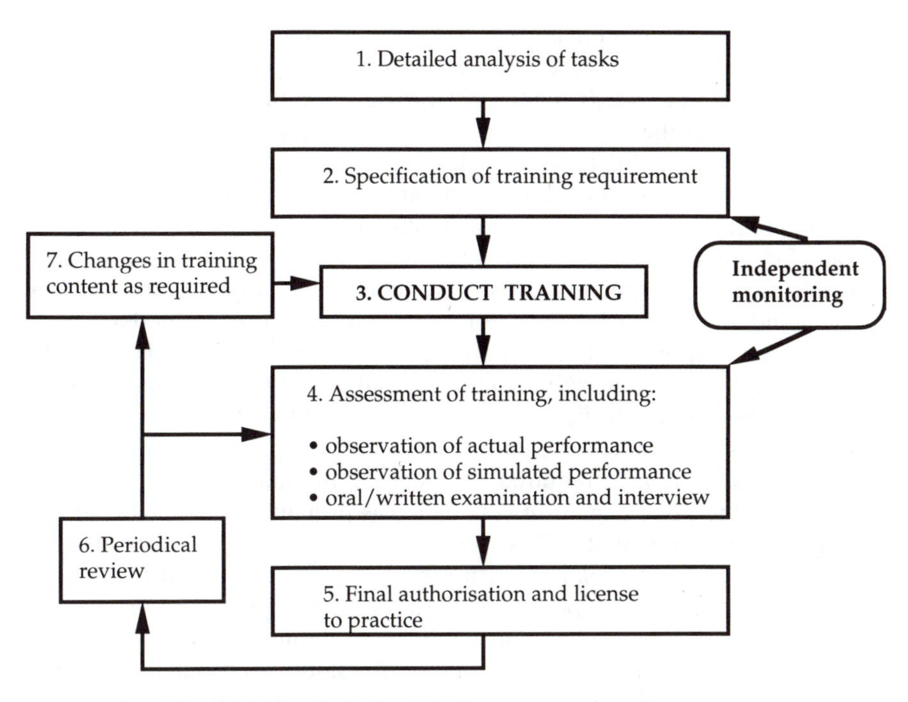

Figure 1.2 General framework for training.

The first report on 'Training and related matters' (HSC, 1990) identified safety culture as a key factor, proposing that this should be a high priority at the level of the Board of Directors within an organisation. Training issues covered include: initiation of training, monitoring, needs analysis, development of criteria and standards, training methods, training media and certification. A general framework for training was outlined, as illustrated in Figure 1.2.

As shown in Figure 1.2, the training begins with an analysis of the tasks to be trained and this leads to a training specification. Training conducted should include the use of simulation and the trainee's performance needs to be assessed in a variety of ways. If the trainees achieve a satisfactory level of performance, they require final authorisation granting a licence to practise. Regular formal assessments of performance should ensure that there is little opportunity for degradation in standards. The report also recommends that an independent body monitors training requirements and assessments to guarantee a minimum pass standard, but that this body should not be involved in the licensing of operators. The appendix to the report makes special note of the effects of stress upon operator performance. These effects are particularly important for tasks involving potential hazards. Stress may result from both work underload (e.g. boring and monotonous activities) as well as work overload. Other factors that appear to affect the level of perceived stress in the individual include the degree of: discretion that the individual has over the task, clarity of demands, feedback and support provided. Cox and Cox (Chapter 13) provide a comprehensive review of operator stress in the control room.

The report also highlights the need to consider training needs of all of the personnel, not just the control room operators, and to consider training as an ongoing process. For an overview of training issues see Stammers (Chapter 10) and for a review of simulator use, see Stanton (Chapter 7). It also suggests that selection of personnel ought to take their trainability into account (see Stanton and Ashleigh, Chapter 9, for a review of selection methods).

The second report on 'human reliability assessment' (HRA) (HSC, 1991) examines methods of determining human reliability in technological systems. The distinction between different types of 'aberrant behaviour' (i.e. errors and violations) owes much to Reason (1990b). The report acknowledges that the numerical estimates required for analysis of human error cannot be treated with the same degree of confidence as probabilistic risk assessments of engineering components. However, there do appear to be some benefits in considering human reliability in a systematic and structured manner. In brief, HRA techniques appear to share a generic approach of assigning a numerical value (of success and/or failure) to each human activity in a given scenario. Then by summing these numerical values, a probability estimate for the whole scenario may be derived. The study group point out three major caveats with probabilistic HRA: different techniques are appropriate for different types of errors, the resultant probabilities for scenarios are better for relative rather than absolute judgements, there is little point in identifying the errors unless there is an equally systematic approach for reducing the errors. A generic approach for HRA is shown in Figure 1.3.

As Figure 1.3 illustrates, there are six main stages common to most HRA techniques. In the first stage the problem to be analysed is defined. In the second and third stages, the task is analysed and the potential errors identified respectively. Quantification of the error likelihood occurs in the fourth stage. The consequences of the error are identified in the fifth stage. Finally, in the last stage, strategies for reducing the errors are determined. For a detailed discussion of HRA in the nuclear industry, see Kirwan (Chapter 14).

The final report on 'Organising for Safety' (HSC, 1992) again focuses on safety culture within organisations. The group define the concept of safety culture as '. . . the product of individual and group values, attitudes, competencies and patterns of behaviour . . .' and propose that 'for the organisation to be safe, every part of it must be concerned with safety.' In a safe organisation, resources are devoted to safety, all members of staff participate in safety, senior members of the organisation are regularly available, and there is a sensible balance between safety and production. The study group proposed a method for promoting safety culture in organisations which is illustrated in Figure 1.4.

As Figure 1.4 shows, steps 1 and 2 in this process require the organisation

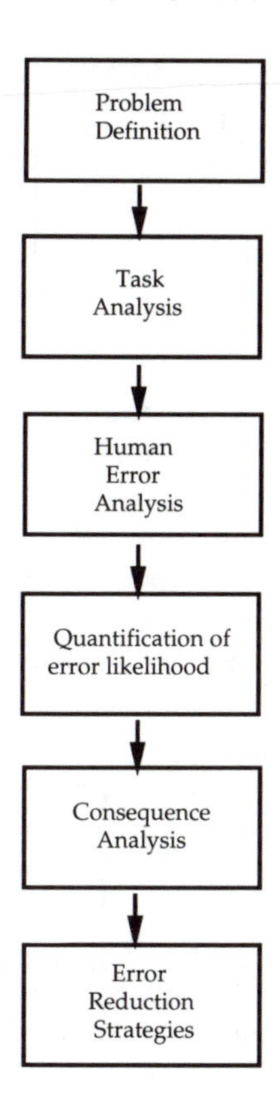

Figure 1.3 Generic HRA procedure.

to draw up a plan of action and allocate responsibilities to all of its senior managers. Steps 3 and 4 require that the current situation is assessed and the hazards which arise identified. Step 5 is to prioritise the hazards in terms of urgency and devise specific activities for the most important ones. Step 6 is to implement those activities in order to reduce the most urgent hazards. Finally, in step 7 the situation needs to be reassessed in light of the changes to ensure that the hazards have been brought under control. Then the whole process is set into motion once again. This cycle should be continually running, as organisations themselves are not static and risks are continually changing. A fuller discussion of the concept of safety culture and its measurement can be found in Booth (Chapter 15).

1.6 Structure of the Book

This book comprises 16 chapters and is divided into four main parts: organisational issues for design, interface design issues, personnel issues and

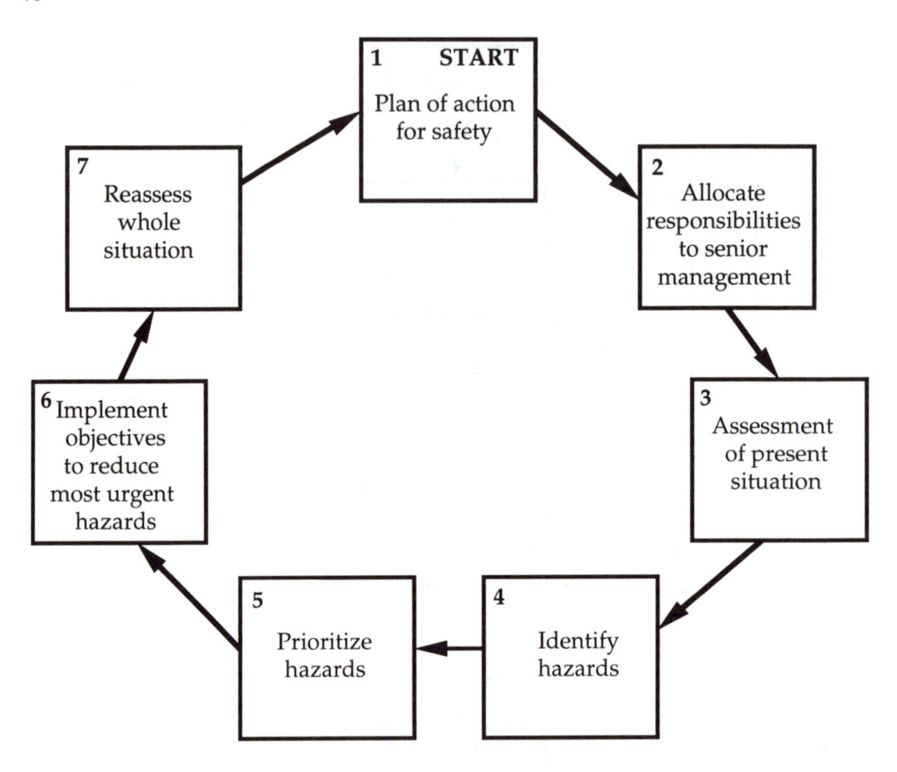

Figure 1.4 A process for promoting continuous improvement in safety culture.

organisational issues for safety. The logic for this organisation of the book is indicated in Figure 1.5. In addition there is an introductory chapter (this one) and a concluding chapter (Key Topics in Nuclear Safety) which draws the contributions together.

Part I (organisational issues for design) considers the process of developing nuclear power plant systems (Chapter 2) and ergonomic assessments in design (Chapter 3). Part II (interface design issues) considers human–computer interaction (Chapter 4), alarm design (Chapter 5), procedure design (Chapter 6) and simulators (Chapter 7). Part III (personnel issues) considers operator variability (Chapter 8), personnel selection (Chapter 9), training (Chapter 10), team processes (Chapter 11), shiftworking (Chapter 12) and stress (Chapter 13). Finally, Part IV (organisational issues for safety) considers human reliability analysis (Chapter 14) and safety culture (Chapter 15).

The reader is encouraged to discover human factors and its application to nuclear safety. The chapters are intended to strike a balance between offering a strong theoretical underpinning and practical advice that can be applied to the domain. As noted earlier, nuclear power plants offer a considerable challenge to human factors. The contributors to this book believe that human factors has an important part to play in nuclear safety.

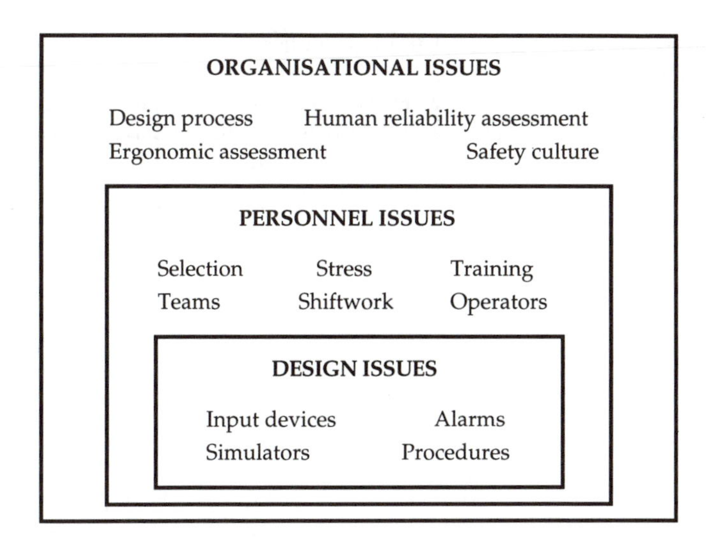

Figure 1.5 Design, personnel and organisational issues.

References

GALE, A. & CHRISTIE, B. (1987) *Psychophysiology and the Electronic Workplace*, Chichester: Wiley.

HSC (1990) Advisory Committee on the Safety of Nuclear Installations: study group on human factors. First Report: *Training and Related Matters*. London: HMSO.

(1991) Advisory Committee on the Safety of Nuclear Installations: study group on human factors. Second Report: *Human Reliability Assessment – a critical overview*. London: HMSO.

(1992) Advisory Committee on the Safety of Nuclear Installations: study group on human factors. Third Report: *Organising for Safety*. London: HMSO.

HSE (1989) *Human Factors in Industrial Safety*. London: HMSO.

KEMENY, J. (1979) *The Need for Change: The legacy of TMI*. Report of the President's Commission on the Accident at Three Mile Island: New York: Pergamon.

MIESTER, D. (1989) *Conceptual Aspects of Human Factors*. Baltimore: Johns Hopkins University Press.

OBORNE, D. J. (1982) *Ergonomics at Work*. Chichester: Wiley.

REASON, J. T. (1987) The Chernobyl errors. *Bulletin of The British Psychological Society*, **40**, 201–6.

(1990a) The contribution of latent human failures to the breakdown of complex systems. In Broadbent, D. E., Baddeley, A. and Reason, J. T. (eds), *Human Factors in Hazardous Situations*. Oxford: Clarendon Press.

(1990b) *Human Error*, Cambridge: Cambridge University Press.

SANDERS, M. S. & MCCORMICK, E. J. (1993) *Human Factors in Engineering and Design*, 7th edn. New York: McGraw-Hill.

STANTON, N. A. & BABER, C. (1991) Human factors in practice. *Journal of Health and Safety*, **6**, 5–12.

WOODS, D. D., O'BRIEN, J. F. & HANES, L. F. (1987) Human factors challenges in process control: the case of nuclear power plants. In Salvendy, G. (ed.). *The Handbook of Human Factors*. Chichester: Wiley.

PART ONE

Organisational Issues for Design

CHAPTER TWO

Human factors in the process of developing nuclear power plant systems

DANIEL L. WELCH[1] and JOHN M. O'HARA[2]

[1]*Carlow International Incorporated*
[2]*Brookhaven National Laboratory*

2.1 Introduction

By the end of the Second World War, the US Department of Defense (DoD) was becoming more and more aware that the accidents and operational difficulties involved with complex military systems (such as aircraft, ships, and armoured vehicles), often attributed to 'human error', were in fact the result of a complex interaction between the human operator, the design of the system, the procedures developed to employ the system, and the operating environment. The discipline and knowledge base of human factors engineering (HFE) was born, was nurtured within the military for a number of years, and eventually fledged to provide support to the design and development of non-military systems. The military, however, continued to define and refine the process by which the new discipline would be coordinated with other disciplines within the material acquisition process.

The new discipline was thrust into the arena of nuclear power on 28 March 1979. Following the Three Mile Island (TMI) incident there was general agreement that the cause of the accident was, in large part, related to the human element. Those problems were viewed as encompassing all areas of the system, including design, construction, operations, training, communication, and maintenance (Berghausen *et al.*, 1990).

In response to TMI, the US Nuclear Regulatory Commission (NRC) developed an action plan (NRC, 1980). Task I.D.1 of that plan specified that the NRC Office of Nuclear Reactor Regulation would order operating reactor licensees and applicants for operating licences to perform a sweeping review of their control rooms to identify and correct design deficiencies. This led to detailed control room design reviews (DCRDRs), which were a reactive attempt to correct problems that had been designed into nuclear power plant (NPP) systems. The effort continued throughout the 1980s and cost billions of dollars to upgrade systems to correct safety concerns.

In an effort to move from a reactive to a proactive posture, the NRC is currently developing a human factors engineering programme review model which will support the review of both the design process and final design of the HFE of advanced NPPs. This will attempt to entwine the process of human factors design into the larger effort of advanced plant design, in much the same way as the US military incorporates human factors effort into the material acquisition process.

This paper briefly presents the theoretical aspects of systems theory and the doctrinal aspects of the US military's development of a process for the integration of human factors into system design and development, and the NRC's current efforts to build on the military's experience to develop a review process that ensures that HFE is considered through the advanced NPP design and development process.

2.2 Human Factors Within the Military Systems Development Process

Applied general systems theory, embodied in the discipline of systems engineering, provides a broad approach to system design and development, based on a series of clearly defined developmental steps, each with clearly defined and attainable goals and with specific management processes to attain them.

A systems approach implies that all system components (hardware, software, personnel, support, doctrine, procedures, and training) are given adequate consideration in the developmental process. Department of Defense (DoD) policy identifies the human as an element of the total system (DoD, 1990). A basic assumption is that, rather than integrate personnel into the system after hardware and software design has been completed, the personnel element should receive serious attention from the very beginning of the design process. Indeed, in system elements where human performance is substantial or critical, human consideration should drive design, in order to achieve an optimum system configuration.

Systems engineering ensures the effective integration of human factors engineering (HFE) considerations by providing a structured approach to system development and a management structure which details the nature of that inclusion into the overall process. The systems approach is iterative, integrative, interdisciplinary and requirements-driven.

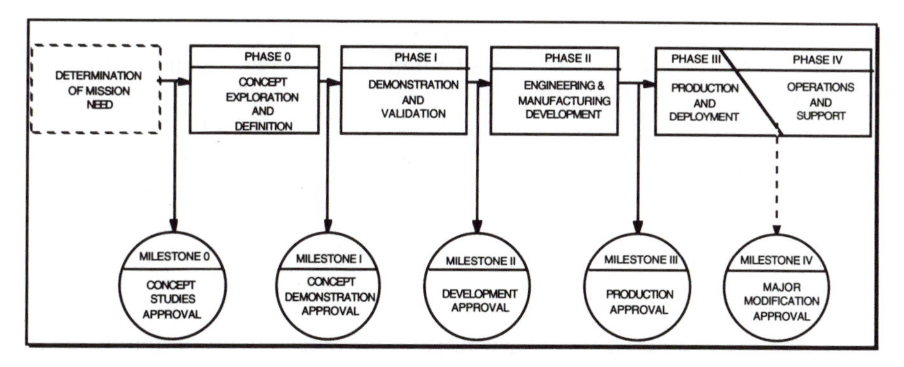

Figure 2.1 The phases and milestones of the military system development process.

2.2.1 The Military Systems Development Process

The military systems development process is normally divided into six phases to provide a basis for comprehensive management and the iterative and progressive decision making associated with programme maturation. The process phases and decision milestones are illustrated in Figure 2.1.

2.2.1.1 Pre-concept Activity – the Determination of Mission Needs

The system development process begins with the determination of mission needs. A mission need may be to establish a new operational capability or to improve an existing capability. It may also reflect a desire to exploit an opportunity that will result in significantly reduced ownership costs or improve the effectiveness of existing material.

Phase 0 – Concept exploration and definition The focus of this phase is on defining and evaluating the feasibility of alternative concepts to accomplish the system mission and providing the basis for assessing the relative merits of the concepts. Early life-cycle cost estimates of the competing alternatives are analysed relative to the value of the expected increase in operational capability. Trade-offs are made among cost, schedule and performance as a result of these analyses.

The objectives of Phase 0 are to explore various material alternatives to satisfy the documented mission need, define the most promising system concept(s), develop supporting analyses and information to support à decision to continue with development at Milestone I, and to develop a proposed acquisition strategy and initial programme objectives for cost, schedule, and performance for the most promising system concept(s).

Phase I: Demonstration and validation During this phase, multiple approaches and parallel technologies are investigated within the system concept(s). Prototyping, testing, and early operational assessment of critical systems, subsystems, and components are emphasised to identify and reduce risk and assess if the most promising design approach(es) will operate in the intended operational environment. Cost drivers and alternatives are identified and analysed. The cost of the design approach(es) is analysed as a function of risk and the expected increase in operational capability. Objectives of the phase are to:

- define the critical design characteristics and expected capabilities of the system concept(s);
- demonstrate that the technologies critical to the most promising concept(s) can be incorporated into the system design(s) with confidence;
- prove that the processes critical to the most promising system concept(s) are understood and attainable;
- develop the analyses and information needs to support a decision to continue development at Milestone II;
- establish a proposed development baseline containing refined programme cost, schedule, and performance objectives for the most promising design approach.

Phase II: Engineering and manufacturing development In this phase the most promising design approach developed in Phase I is translated into a stable, producible and cost-effective system design. The manufacturing or production process is validated and it is demonstrated through testing that the system capabilities meet contract specification requirements, satisfy the mission need, and meet minimum acceptable operational performance requirements. Planning for the next phase addresses design stability, production, industrial base capacity, configuration control, deployment, and support.

Phase III: Production and deployment The objective of this phase is to establish a stable, efficient production and support base, achieve an operational capability that satisfies the mission need, and conduct follow-on operational and production verification testing to confirm and monitor performance and quality and verify the correction of deficiencies.

Phase IV: Operations and support This phase overlaps with Phase III. It begins after initial systems have been fielded and continues until the system leaves the inventory. The purpose of the phase is to ensure that the fielded system continues to provide the capabilities required to meet the identified system need and to identify shortcomings or deficiencies that must be corrected to improve performance. In essence, it is only in this phase that reactive modifications are undertaken.

2.2.2 Integrating human factors into the development process

MIL-H-46855B, Human Engineering Requirements for Military Systems, Equipment and Facilities (DoD, 1979) establishes and defines the requirements for applying HFE to the development and acquisition of military systems, equipment, and facilities. It defines the objective of HFE within the system development process as:

> . . . to develop or improve the crew–equipment/software interface and to achieve required effectiveness of human performance during system operation/maintenance/control and to make economical demands upon personnel resources, skills, training and costs. The human engineering effort shall include, but not necessarily be limited to, active participation in the following three major interrelated areas of system development:
>
> - analysis
> - design and development
> - test and evaluation

Rather than simply apply static human factors guidance based on identified human capabilities and limitations, an effective HFE programme must begin with a study of the dynamics of the developing system. This yields the system understanding required to establish the nature and extent of human performance requirements for the system which will directly impact on the HFE design of the system. General requirements for HFE in the analysis area are described as follows:

> Starting with a mission analysis developed from a baseline scenario, the functions that must be performed by the system in achieving its mission objectives shall be identified and described. These functions shall be analyzed to determine the best allocation to personnel, equipment, software, or combinations thereof. Allocated functions are further dissected to define the specific tasks which must be performed to accomplish the functions. Each task is analyzed to determine the human performance parameters, the system/equipment/software capabilities, and the tactical/environmental conditions under which the tasks are conducted. Task parameters shall be quantified, where possible, and in a form permitting effectiveness studies of the crew-equipment/software interfaces in relation to the total system operation. The identification of human engineering high risk areas shall be initiated as part of the analysis.

While analysis provides the basis for HFE participation in system development, such efforts are only a means to an end and not an end in themselves. The desired end of HFE efforts in the material acquisition process is to *influence the design of the developing system*. Subsequent to, but largely overlapping and iterative with analysis efforts, general requirements for HFE effort in design are described as follows:

> Design and development of the system equipment, software, procedures, work environments and facilities associated with the system functions requiring personnel interaction shall include a human engineering effort that will convert the mission, system and task analyses data into detail design or development plans to create a personnel-system interface that will operate within human performance capabilities, meet system functional requirements, and accomplish mission objectives. The final developed design is the culmination of all of the initial planning, system analyses, criteria and requirements application, and engineering effort.

Finally, once design input has been achieved, it is necessary to test and evaluate the achieved system to ensure (1) that recommended HFE design elements were in fact incorporated into the system as required, and (2) that the HFE design elements in fact produce the desired results. General requirements for HFE in test and evaluation are described as follows:

> Test and evaluation shall be conducted to verify that design of equipment, software, facilities and environment meets human engineering and life support criteria and is compatible with the overall system requirements.

In terms of a staged approach to system development, the goals in Table 2.1 can be established for the HFE effort during the phases of the military system development process. The following sub-sections present a more detailed consideration of the HFE process in system development, as presented in Figure 2.2.

2.2.2.1 Programme Planning

The Integrated Program Summary (DoD, 1991), with its annexes, is the primary decision document used to facilitate top-level acquisition milestone decision making. It provides a comprehensive summary of plans, programme structure, status, assessment, and recommendations by the programme manager and the programme executive officer. Primary functions of the Integrated Program summary include:

- describing where the programme is going and how it will get there;
- summarising where the programme is versus where it should be;
- identifying programme risk areas and plans for closing risks; and
- providing the basis for establishing explicit programme cost, schedule, and performance (operational effectiveness and suitability) objectives and thresholds in the stand-alone acquisition programme baseline and programme-specific exit criteria for the next acquisition phase.

Table 2.1 Goals of HFE in the military material system development process phases

Pre-milestone 0 – Mission needs determination

- Define the basic operational requirement.
- Determine:
  - The mission and operational requirements
  - The functions required to accomplish each mission event
  - The performance requirements for each function
  - The initial allocation of function to hardware/software/human elements (including determination of the role of the human in the system)

Phase 0 – Concept exploration and definition

- Make trade-offs relative to how to put various elements together into a preliminary base-line system
- Assess alternative concepts
- Define:
  - The optimal design approach for accomplishing each hardware functional assignment (i.e. subsystem trade-off studies)
  - Preliminary operator, maintainer, and support task descriptions
  - Preliminary manning and training requirements definition
  - Preliminary information-flow and operational sequence diagrams

Phase I – Demonstration and validation

- Perform design trade-off studies to provide rationale for possible modification and improvement of the initial concepts
- Conduct system level design
- Perform:
  - Prototyping
  - Human–machine mock-up studies
  - Human–machine simulation studies
  - Fundamental (but applied) human factors studies
  - Time-line and link analyses
  - Refined task analysis and task description

Phase II – Engineering and manufacturing development

- Achieve and maintain close interaction between HFE and design engineering to successfully integrate hardware, software, and personnel subsystems
- Conduct subsystem/component level detailed design
- Provide specific design guidance
- Perform task analysis
- Perform hazards analysis
- Perform time-line analysis
- Perform link analysis
- Perform manning requirements analysis
- Perform training requirements analysis
- Provide training aids, equipment, and facilities design guidance
- Provide technical publications input and design guidance
- Provide HFE guidance to equipment procedure development

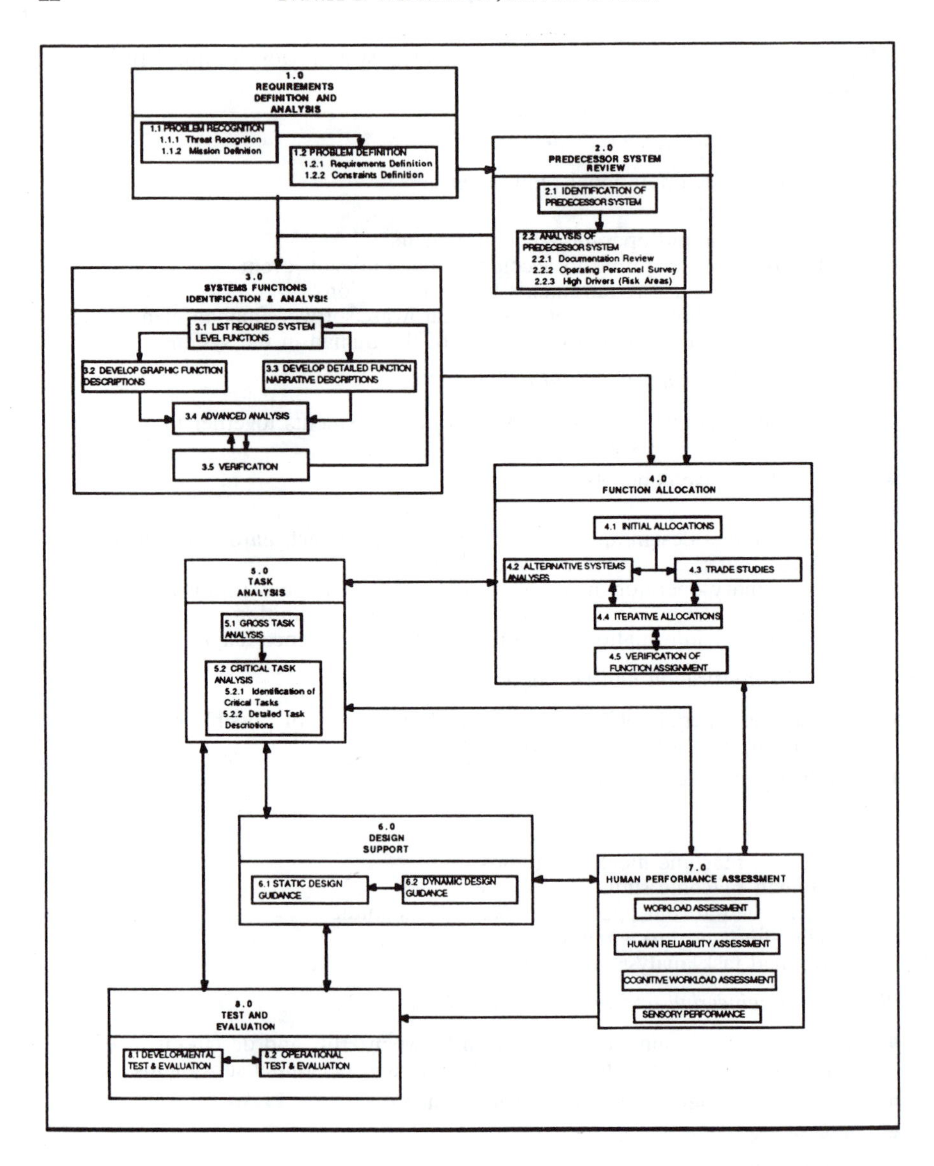

Figure 2.2 The HFE process in the military system development process.

Beyond the integrated programme summary, which is a highest DoD level document, there exist a number of discipline programme plans which define the contractor's approach to accomplishing the goals of the discipline. For example, the system engineering management plan (SEMP) is a concise top-level technical management plan for the integration of all systems engineering activities. The purpose of the SEMP is to make visible the organisation, direction and control mechanisms, and personnel for the attainment of cost, performance, and schedule objectives. The who, what, when, where, how, and why of the systems engineering activities, including relevant interfaces and engineering speciality areas, are clearly delineated. The SEMP, and indeed systems engineering itself, is generally composed of two components, systems engineering management and the systems engineering process. Both of these are addressed in the SEMP (Kockler *et al.*, 1990).

Since one objective of systems engineering is the integration of engineering specialities, the SEMP is generally supported by a number of engineering speciality plans. These include:

- integrated test plan (ITP);
- integrated logistics plan (ILP);
- reliability/maintainability plan (RMP);
- manufacturer's MANPRINT (manpower and personnel integration) management plan (MMMP).

The MMMP describes the management and process elements designed to implement the US Army's MANPRINT initiative (DoD, 1990; Westerman *et al.*, 1991), which seeks to integrate the following disciplines into a cohesive management effort:

- manpower;
- personnel;
- training;
- human factors engineering;
- systems safety engineering;
- health hazards analysis.

In turn, the MMMP incorporates the human engineering programme plan (HEPP) and the systems safety programme plan (SSPP). The HEPP includes detailed discussions of the following programmatic elements:

- HFE organisation;
- HFE in subcontractor efforts;
- HFE in system analysis;
- HFE in equipment detail design;
- HFE in equipment procedure development;
- derivation of personnel and training requirements;
- HFE in test and evaluation;
- HFE deliverable data products;
- time-phased schedule and level of effort.

2.2.2.2 *Requirements Definition and Analysis*

Within the military milieu, a problem is defined in terms of a threat. Threat recognition is the realisation that a problem or need exists and that the solution may be amenable to the systems approach. Mission definition is a statement of the objectives of the system required to counter the threat and the operational capability required to reach that objective. Requirements are expressions of the obligations the system must fulfil to successfully effect its mission and constraints are the temporal, scheduling, cost, technical and political factors which characterise the environment of the system development effort.

2.2.2.3 *Predecessor System Review*

As soon after mission definition as possible, problems and issues encountered in systems attempting to carry out the same or similar missions are identified and analysed so that special attention may be given to those problems and issues in the development of the current system, in order to avoid their repetition or, in the case of positive features, to ensure their retention.

2.2.2.4 *System Functions Identification and Analysis*

Functions may be defined as the most general, yet differentiable means whereby the system requirements are met, discharged, or satisfied. Function analysis is usually performed within the context of segments of the mission profile. A function is usually the primary reason for including a particular sub-system (equipment or crew member) in the design system. Functional operations or activities may include:

- detecting signals;
- measuring information;
- comparing one measurement with another;
- processing information;
- acting upon decisions to produce a desired condition or result on the system or environment.

System requirements determine system functions, and the function itself determines what equipment and human performance are necessary to carry out that function.

System requirements are analysed to identify those functions which must be performed to satisfy the objectives of each functional area. System requirements include, for each function: information, performance, decision, and support aspects. Each function is identified and described in terms of inputs, outputs, and interface requirements from the top down so that subfunctions are recognised as part of larger functional areas. Functions are arranged in a logical sequence so that any specified operational usage of the system can be traced in an end-to-end path.

2.2.2.5 *Function Allocation*

Functions can be assigned to humans, or to machines, or to combinations of humans and machines. The allocation of functions may be based on one or a combination of the following: (1) judgement and intuition based on similar experience; (2) imposition by the user or by management; and (3) critical analyses, experiments, and simulation studies involving tradeoffs of alternative configurations of humans and machines.

The function allocation process actually takes place throughout the design of a system. Usually, designers make the most substantial allocations early in the development of a system, and make minor changes later to finely tune the system. The allocation process is iterative. A designer originally allocates each function either to a human, machine or human–machine interaction. That allocation is then used as a working model until more is learned about system characteristics, when some changes to the original allocation should be made.

2.2.2.6 *Task Analysis*

Task analysis is the systematic study of the behavioural requirements of tasks, where a task is a group of activities that often occur in temporal proximity with the same displays and controls, and have a common purpose. Task analysis profiles the model of the human in the system.

Gross task analysis provides one of the bases for making design decisions; e.g. determining before hardware fabrication, to the extent practicable, whether system performance requirements can be met by combinations of anticipated equipment, software, and personnel, and assuring that human performance requirements do not exceed human capabilities. These analyses are also used as basic information for developing: preliminary manning levels; equipment procedures; and skill, training, and communication requirements of the system. Those gross tasks identified during human engineering analysis which are (1) related to end items of equipment to be operated or maintained by personnel and which require critical human performance, (2) reflect possible unsafe practices, or (3) are subject to promising improvements in operating efficiency, are further analysed through critical task analysis.

Critical task analysis provides a fine-grain profile of tasks with severe human performance requirements, high mission criticality, or safety implications, with emphasis on information requirements and flow, decision-making processes, and physical requirements. The combination of gross and critical task analyses provides the basis for all subsequent, dynamic HFE support to the development process.

2.2.2.7 *Design Support*

As stated previously, the end goal of an HFE programme in system development is to impact design. It is in the presentation of design guidance that such impact is achieved. Given a systems approach, HFE considerations may not be in all cases the sole or even primary areas of concern. However, in an effort to optimise system functioning, systems engineering attempts to integrate HFE principles and criteria, along with all other design requirements, during preliminary system and subsystem design, and in equipment detailed design.

Early in the design process, i.e. during preliminary system and subsystem design or prior to engineering and manufacturing development (Phase II), HFE is applied to designs represented by design criteria documents, performance specifications, and drawings and data (such as functional flow diagrams, system and subsystem schematic block diagrams, interface control drawings, and overall layout drawings). The preliminary system and subsystem configuration and arrangement should satisfy personnel–equipment/software performance requirements and comply with the static criteria specified in MIL-STD-1472.

During the detailed design of equipment, generally during engineering and manufacturing development (Phase II), the HFE inputs defined through analysis are converted into detail equipment design features. Experiments, tests and studies are conducted to resolve HFE and life-support problems specific to the system. HFE provisions are reflected by the detail design drawings for system and equipment to ensure that the final product can be efficiently, reliably, and safely operated and maintained. Design of work environment, crew stations, and facilities which affect human performance under normal, unusual and emergency conditions considers and reflects HFE guidance on such issues as: atmospheric conditions; adequate space for personnel, their movement and their equipment; effects of clothing and personal equipment; equipment-handling provisions; and provisions for minimising disorientation and maintaining situational awareness, among others. HFE principles and criteria are applied to the development of procedures for operating, maintaining, and otherwise using the system equipment. Finally, HFE provisions for equipment are evaluated for adequacy during design reviews.

2.2.2.8 *Human Performance Assessment*

In material development, human performance is a general term used to describe the ability of the person to accomplish task requirements for the human element of the man–machine system. A deficit in this 'ability' may be reflected in the depletion of attentional, cognitive, or response resources; inability to accomplish additional activities; errors; emotional stress; fatigue; or performance decrements. Human performance evaluation measures are generally obtained to evaluate the effects of different system designs (or operating conditions) on a human operator, or to quantify the effects of individual differences in the abilities or training of specific operators working with a given system.

Human performance assessment, depending on its specific objective at any given point in the development process, may involve workload assessment, cognitive workload assessment, error analysis, human reliability assessment, visual or auditory performance, response to fatigue/threat/emergency, etc.

The results of human performance assessment are employed both to provide information on the effectiveness of individual system alternatives for design purposes (i.e. for design guidance and developmental test and evaluation) and for the final (operational) test and evaluation of the operating system.

2.2.2.9 *Test and Evaluation*

Test and evaluation within a military framework are classically divided into developmental and operational aspects. Developmental test and evaluation programmes:

- identify potential operational and technological limitations of the alternative concepts and design options being pursued;
- support the identification of cost–performance trade-offs;
- support the identification and description of design risks;
- substantiate that contract technical performance and manufacturing process requirements have been achieved; and
- support the decision to certify the system ready for operational testing.

Developmental testing would normally occur during the earlier phases of development, especially during concept exploration, to help determine the most likely alternative concept, and during demonstration and validation, to perform system level trade-offs among system elements. Operational test and evaluation programmes:

- determine the operational effectiveness and suitability of a system under realistic conditions; and
- determine if the minimum acceptable operational performance requirements have been satisfied.

Operational test and evaluation normally occur at the end of engineering and manufacturing development, to demonstrate operational effectiveness of the system about to enter production.

2.2.3 Human Factors Design Process Beyond the Military

The standard, accepted practice within the disciplines of systems engineering and HFE is to apply HFE to the developing system as early in the design and development process as possible. This is expressed in a number of ways.

> During all phases of the life cycle of materiel systems, accepted principles of HFE will be used to integrate materiel development with personnel resources and capabilities. *Source:* (DoD, 1983).

> If human factors engineering is to have an impact on the system design process, the behavioral scientist must be called in at the beginning, while it is still possible to influence major design decisions. *Source:* (Chapanis, 1970).

The development of a major system is an involved process in any environment. In the arena of nuclear power, advanced control room (ACR) concepts are being developed in the commercial nuclear industry as part of future reactor design. The ACRs will use advanced human–system interface (HSI) technologies that may have significant implications for plant safety in that they will affect the operator's overall role (function) in the system, the method of information presentation, the ways in which the operator interacts with the system, and the requirements on the operator to understand and supervise an increasingly complex system. The NRC reviews the HFE aspects of advanced plant designs to ensure that they are designed to good human factors engineering principles and that operator performance and reliability are appropriately supported to ensure public health and safety.

The NRC staff is currently evaluating the HFE programmes submitted as part of the certification process for nuclear power plant (NPP) designs. The NRC has issued 10 CFR Part 52 (US Code of Federal Regulations, Part 52, 'Early Site Permits; Standard Design Certifications; and Combined Licenses for Nuclear Power Plants', Title 10, 'Energy') to encourage standardisation and to streamline the licensing process. Nuclear plant designers and vendors have begun the design of advanced standard plants, which are being submitted to the NRC for review and approval under Part 52. The term 'applicant' is used to refer to an organisation such as a nuclear plant vendor or utility that is applying to the US Nuclear Regulatory Commission for design certification or plant licensing. The licensing process of Part 52 consists of a final design approval by the NRC followed by a standard design certification that is issued as an NRC rule.

In the past, NRC evaluation of HFE acceptability was based on detailed plant design reviews. Since advanced HSI technology is continually changing, much of the design will not be completed before a design certification is issued for the advanced reactor designs currently under review. Thus, the NRC has concluded that it is necessary to perform HFE reviews of the design process as well as of the final design product.

There are other compelling reasons to perform HFE reviews during the design process. First, it is generally recognised in the nuclear industry that human factors issues and problems emerge throughout the NPP design process (O'Hara & Hall, 1991; Stubler *et al.*, 1991; O'Hara, 1993). For example, decisions regarding allocation of function can impact crew workload and performance reliability. Second, as was indicated in the introduction, when HFE issues are identified before the design is complete, solutions can be recommended in a more timely, cost-effective manner. Third, the evaluation of final designs is often based upon the quality of HFE analyses conducted during the design process, i.e. task analyses specify control and display requirements. Thus for several reasons a more proactive review approach, one which provides evaluations of design process as well as product, is supported.

The NRC, however, has not conducted a design process review as part of the reactor licensing process, so the evaluation criteria currently available provide little guidance for this type of evaluation. To provide criteria for the evaluation of a design process, as well as the final design implementation, the HFE Program Review Model (HFE PRM) was developed (O'Hara *et al.*, 1994).

This approach is also consistent with the recognition in the nuclear industry that human factors issues and problems emerge throughout the NPP design and evaluation process, and therefore human factors issues are best addressed with a comprehensive top-down programme (for example, see Beattie and Malcolm, 1991; O'Hara & Hall, 1991; Stubler *et al.*, 1991; O'Hara, 1993).

2.2.4 HFE PRM Development

The specific objectives of the HFE PRM development effort were:

- to develop a technical basis for the review of an applicant's HFE design process and final design implementation. The HFE PRM should be: (1) based upon currently accepted HFE practices, (2) well-defined, and (3) based on an approach which has been validated through its application to the development of complex, high-reliability systems.
- to identify the HFE elements in a plant/system development, design, and evaluation process that are necessary and sufficient requisites to successful integration of the human in complex systems;
- to identify the components of each HFE element that are key to a safety evaluation;
- to specify the review criteria by which HFE elements can be evaluated.

A review of current HFE guidance and practices was conducted to identify important human factors programme plan elements relevant to the technical basis of a design process review. Several types of documents were evaluated. First, the general literature providing the theoretical basis for systems engineering, as discussed in the previous section, was used as a basis. This material included the general HFE guidance which was generally applicable to the design and evaluation of complex systems. This basis was then tailored to NPP application by incorporating:

- NPP regulation – the regulatory basis for NPP review and related NRC literature, e.g. US Code of Federal Regulations (Parts 50, 52) and Nuclear Regulatory Commission (1980, 1981 (Appendix B)).
- NPP-specific HFE guidance – standards, guidance, and recommended practices developed in the NPP industry, e.g. the Institute of Electrical and Electronics Engineers (IEEE) STD 1023-1988 (IEEE, 1988), International Electrotechnical Commission (IEC) 964 (IEC, 1989), and EPRI Advanced Light Water Reactor Requirements (ALWR) Utility Requirements Document (EPRI, 1992).

From this review, an HFE development, design, and evaluation process was defined.

2.2.4.1 *General Description*

The HFE PRM defines the key HFE elements of NPP design and the general criteria by which these elements could be assessed. The HFE PRM represents a structured top-down approach to NPP HFE review and evaluation. That human factors is best addressed with a comprehensive top-down programme has been recognised in the industry (for example, see Beattie and Malcolm, 1991; EPRI, 1992). This section provides a brief overview of the HFE PRM (see O'Hara *et al.*, 1994 for a complete description).

The overall purpose of the HFE PRM review is to ensure that:

- HFE has been integrated into plant development and design.
- HFE components (e.g. HSIs, procedures, training) have been provided that make possible safe, efficient, and reliable performance of operation, maintenance, test, inspection, and surveillance tasks.
- The HFE reflects state-of-the-art human factors principles.
- The final design satisfies all specific regulatory requirements.

To accomplish these programmatic objectives, an adequate HFE programme plan is required which is implemented by a qualified HFE design team.

The HFE PRM evaluates the review process in ten elements reflecting four stages of design: planning, analysis, interface design, and evaluation. Each element consists of four sections:

- Background – a brief explanation of the rationale and purpose is provided for each element.
- Objective – the review objective(s) of the element is defined.
- Applicant submittals – materials to be provided for NRC review are listed. Generally three reports are identified: implementation plan, analysis results report, and design team review report.
- Review criteria – this section contains the acceptance criteria for design process products and for the final design review.

The HFE PRM consists of ten review elements:

Element 1 – HFE programme management The objective of this review is to ensure that the applicant has an HFE design team with the responsibility, authority, placement within the organisation, and composition to ensure that the design commitment to HFE is achieved. Also, the team should be guided by an HFE Program Plan to assure the proper development, execution, oversight, and documentation of the HFE programme. This plan should describe the technical programme elements assuring that all aspects of HSI are developed, designed, and evaluated based upon a structured top-down systems analysis using accepted HFE principles. Element 1 review topics include: general HFE programme goals and scope; HFE team and organisation; HFE process and procedures; HFE issues tracking; and technical HFE programme.

Element 2 – Operating experience review The main purpose of the operating experience review (OER) is to identify HFE-related safety issues. The issues and lessons learned from previous operating experience provide a basis for improving the plant design in a timely way; i.e. at the beginning of the design process. The objective of this review is to assure that the applicant has identified and analysed HFE-related problems and issues encountered in previous designs that are similar to the current design under review. In this way, negative features associated with predecessor designs may be avoided in the current design while positive features are retained. The OER should address the predecessor systems upon which the design is based, selected technological approaches (e.g. if touch screen interfaces are planned, HFE issues associated with their use should be reviewed), and NPP HFE issues (e.g. those identified in unresolved safety issues, generic safety issues, TMI issues, and NRC generic letters and information notices).

Element 3 – Functional requirements analysis and allocation This element involves two distinct review activities: functional requirements analysis and function allocation. Functional requirements analysis is the identification of those functions which must be performed to satisfy plant safety objectives, i.e. to prevent or mitigate the consequences of postulated accidents that could cause undue risk to the health and safety of the public. A functional requirements analysis is conducted to: (1) determine the objectives, performance requirements, and constraints of the design; (2)

define the functions which must be accomplished to meet the objectives and required performance; (3) define the relationships between functions and plant processes (e.g. plant configurations or success paths) responsible for performing the function; and (4) provide a framework for understanding the role of controllers (whether personnel or system) for controlling plant processes.

Function allocation is the analysis of the requirements for plant control and the assignment of control functions to (1) personnel (e.g. manual control), (2) system elements (e.g. automatic control), and (3) combinations of personnel and system elements (e.g. shared control and automatic systems with manual backup). Function allocation seeks to enhance overall plant safety and reliability by exploring the strengths of personnel and system elements, including improvements that can be achieved through the assignment of control to these elements with overlapping and redundant responsibilities. Function allocation should be based upon HFE principles using a structured and well-documented methodology that seeks to provide personnel with logical, coherent, and meaningful tasks.

The objective of this review is to assure that the applicant has defined the plant's safety function requirements and that the function allocations take advantage of human strengths and avoid allocating functions which would be negatively impacted by human limitations.

Element 4 – Task analysis Plant personnel perform tasks to accomplish their functional responsibilities. Task analysis is the evaluation of the performance demands on plant personnel to identify the task requirements for accomplishing the functions allocated to them (Drury *et al.* 1987). The objective of the review is to assure that the applicant's task analysis identifies the behavioural requirements of the tasks that the personnel subsystem is required to perform. The task analysis should form the basis for specifying the requirements for the displays, data processing and controls needed to carry out tasks. The analysis will help assure that human performance requirements do not exceed human capabilities, and provide important input for developing procedures, staffing, training, and communication requirements.

Element 5 – Staffing Plant staffing is an important consideration throughout the design process. Initial staffing levels may be established as design goals early in the design process based on experience with previous plants, customer requirements, initial analyses, and government regulations. However, staffing goals and assumptions should be examined for acceptability as the design of the plant proceeds. The objective of the staffing review is to ensure that the applicant has analysed the requirements for the number and qualifications of personnel in a systematic manner that includes a thorough understanding of task requirements and applicable regulatory requirements.

Element 6 – Human reliability analysis Human reliability analysis (HRA) seeks to evaluate the potential for and mechanisms of human error that may affect plant safety. Thus, it is an essential element in the achievement of the HFE design goal of providing operator interfaces that will minimise operator error and will provide for error detection and recovery capability. HRA has quantitative and qualitative aspects, both of which are useful for HFE purposes. HRA should be conducted as an integrated activity in support of both HFE/HSI design activities and probabilistic risk assessment (PRA) activities. The development of information to facilitate the understanding of causes and modes of human error is an important human factors activity. The HRA analyses should make use of descriptions and analyses of operator functions and tasks as well as the operational characteristics of HSI components. HRA can provide valuable insight into desirable characteristics of the HSI design. Consequently the HFE HSI design effort should provide special attention to those plant scenarios, critical human actions, and HSI components that have been identified by HRA/PRA analyses as being critical to plant safety and reliability. The objectives of the HRA review are to assure that the potential effects of human error on plant safety and reliability are analysed and that human actions that are important to plant risk are identified so they can be addressed in the design of the plant HFE.

Element 7 – Human – system interface design The selection of available HSIs and the design of new HSIs should be the result of a process which considers function/task requirements, operational considerations (e.g. the full-mission context within which the HSI will be used), and the crew's personal safety. The HSI should be designed using a structured methodology. The methodology should guide designers in the identification of what information and controls are required, the identification and selection of candidate HSI approaches, and the final design of HSIs. It should address the development and use of HFE guidelines and standards that are specific to the HSI design and provide guidance for resolving differences between different HFE guidelines. It should also address the use of analysis and evaluation methodologies for dealing with design issues. The availability of an HSI design methodology will help ensure standardisation and consistency in the application of HFE principles.

The objective of this review is to evaluate the process by which HSI design requirements are developed and HSI designs are selected and refined. The review should assure that the applicant has appropriately translated function and task requirements to the alarms, displays, controls, and aids that are available to the crew. The applicant should have systematically applied HFE principles and criteria (along with all other function/system/task design requirements) to the identification of HSI requirements, the selection and design of HSIs, and the resolution of HFE/HSI design problems and issues. The process and the rationale for the HSI design (including the results of trade-off studies, other types of analyses/evaluations, and the rationale for selection of design/evaluation tools) should be documented for review.

Element 8 – Procedure development While in the nuclear industry procedure development has historically been considered the responsibility of individual utilities, the rationale for including a procedure development element in the HFE PRM is that procedures are considered an essential component of the HSI design and should be a derivative of the same design process and analyses as the other components of the HSI (e.g. displays, controls, operator aids) and subject to the same evaluation processes. The objective of this review is to assure that the applicant's procedure development programme will result in procedures that support and guide human interaction with plant systems and control plant-related events and activities. Human engineering principles and criteria should be applied along with all other design requirements to develop procedures that are technically accurate, comprehensive, explicit, and easy to utilise.

Element 9 – Training programme development Training of plant personnel is an important factor in assuring safe and reliable operation of NPPs. Advanced nuclear power plants may pose demands on the knowledge, skills, and abilities of operational personnel that are different from those posed by traditional plants. These demands stem from differences in operator responsibilities resulting from advanced plant design features (e.g. passive systems and increased automation) and differences in operator task characteristics due to advances in HSI technologies. The objective of this review is to assure that the applicant establishes an approach for the development of personnel training that: incorporates the elements of a systems approach to training, evaluates the knowledge and skill requirements of personnel, coordinates training programme development with the other elements of the HFE design process, and implements the training in an effective manner that is consistent with human factors principles and practices.

Element 10 – Human factors verification and validation Verification and validation (V&V) evaluations seek to comprehensively determine that the design conforms to HFE design principles and that it enables plant personnel to successfully perform their tasks to achieve plant safety and other operational goals.

2.3 Conclusions

The HFE PRM has been developed to support a proactive approach to NPP HFE review by providing criteria that enable the review of the design process as well as final NPP designs. The overall approach is based in part on the DoD systems engineering process.

The HFE PRM is specified in a generic form and must, therefore, be tailored to the requirements of each specific review. Tailored versions of the model are currently being developed to support the staff reviews of the HFE programmes for the General Electric Advanced Boiling Water Reactor, Combustion Engineering System 80+, and Westinghouse AP600. While the HFE PRM was developed specifically to address the programmatic review of HSIs for advanced reactor designs, the principles can be applied to the evaluation of plant upgrades or to the evaluation of other complex human–machine systems.

2.4 Acknowledgements

This research is being sponsored in part by the US Nuclear Regulatory Commission. The views presented in this paper represent those of the authors alone, and not necessarily those of the NRC.

References

BEATTIE, J. & MALCOLM, J. (1991) Development of a human factors engineering program for the Canadian nuclear industry, *Proceedings of the Human Factors Society – 35th Annual Meeting 1991*. Santa Monica, CA: Human Factors Society.

BERGHAUSEN, P., DOOLEY, G. D. & SCHURMAN, D. L. (1990) Human factors in nuclear power: a perspective. *Nuclear News*, **33**(8), 46–8.

CHAPANIS, A. (1970) Human factors in systems engineering. In DeGreene, K. B. (ed.), *Systems Psychology*. New York: McGraw-Hill.

DEPARTMENT OF DEFENSE (1979) *Human Engineering Requirements for Military Systems, Equipment and Facilities* (MIL-H-46855B). Washington, DC: Office of Management and Budget.

(1983) *Human Factors Engineering Program* (AR 602-1). Washington, DC: Department of the Army.

(1990) *Manpower and Personnel Integration (MANPRINT) in the Materiel Acquisition Process* (AR602-2). Washington, DC: Office of Management and Budget.

(1991) *Defense Acquisition Management Policies and Procedures* (DODI 5000.2). Washington, DC: Office of Management and Budget.

DRURY, C., PARAMORE, B., VAN COTT, H., GREY, S. & CORLETT, E. (1987) Task analysis. In Salvendy, G. (ed.), *Handbook of Human Factors*. New York: Wiley-Interscience.

ELECTRIC POWER RESEARCH INSTITUTE (1992) *Advanced Light Water Reactor Utility Requirements Document*, Revision 4. Palo Alto, CA: Electric Power Research Institute.

INSTITUTE OF ELECTRICAL AND ELECTRONICS ENGINEERS (1988) *IEEE Guide to the Application of Human Factors Engineering to Systems, Equipment, and Facilities of Nuclear Power Generating Stations* (IEEE Std. 1023-1988). New York: IEEE.

INTERNATIONAL ELECTROTECHNICAL COMMISSION (IEC) (1989) *Design for Control Rooms of Nuclear Power Plants* (IEC-964). Geneva, Switzerland: Bureau Central de la Commission Electrotechnique Internationale.

KOCKLER, F., WITHERS, T., PODIACK, J. & GIERMAN, M. (1990) *Systems Engineering Management Guide* (AD/A 223 168). Fort Belvoir, VA: Defense Systems Management College.

NUCLEAR REGULATORY COMMISSION (1980) *Clarification of TMI Action Plan*

Requirements (NUREG-0737 and supplements). Washington, DC: US Nuclear Regulatory Commission.

(1981) *Guidelines for Control Room Design Reviews* (NUREG-0700). Washington, DC: US Nuclear Regulatory Commission.

(1984) *Standard Review Plan, Revision 1* (NUREG-0800). Washington, DC: US Nuclear Regulatory Commission.

O'HARA, J. (1993) The effects of advanced technology systems on human performance and reliability. In *Proceedings of the Topical Meeting on Nuclear Plant Instrumentation, Control, and Man–Machine Interface Technologies*. La Grange Park, ILL: American Nuclear Society.

O'HARA, J. & HALL, R. (1991) Advanced control rooms and crew performance issues: implications for human reliability, *Conference Record of the 1991 IEEE Nuclear Science Symposium*. Washington, DC: IEEE.

O'HARA, J., HIGGINS, J., STUBLER, W., GOODMAN, C., ECKENRODE, R., BONGNARRA, J. & GALLETTI, G. (1994) *Human Factors Engineering Program Review Model* (NUREG-0711). Washington, DC: US Nuclear Regulatory Commission.

STUBLER, W., ROTH, E. & MUMAW, R. (1991) Evaluation issues for computer-based control rooms, *Proceedings of the Human Factors Society – 35th Annual Meeting, 1991*. Santa Monica, CA: Human Factors Society.

US CODE OF FEDERAL REGULATIONS, Part 50, *Domestic Licensing of Production and Utilization Facilities*, Title 10, *Energy*. Washington, DC: US Government Printing Office (revised periodically).

Part 52, *Early Site Permits; Standard Design Certifications; and Combined Licenses for Nuclear Power Plants*, Title 10, *Energy*. Washington, DC: US Government Printing Office (revised periodically).

WESTERMAN, D. P., MALONE, T. B., HEASLY, C. C., KIRKPATRICK, M., BAKER, C. C. & PERSE, R. M. (1991) *HFE/MANPRINT IDEA: Integrated Decision/Engineering Aid*. Fairfax, VA: Carlow Associates.

CHAPTER THREE

Ergonomic assessment of a nuclear power plant control room: requirements in a design context

YVES DIEN
Électricité de France, Clamart

3.1 Introduction

The control room is the nerve centre of a nuclear power plant in that it is the place where most process commands are initiated and where most information on the process is received. In order to ensure a high level of safety and satisfactory operation of the plant, it is necessary for the control room to be perfectly tailored to the requirements of all its users. The Three Mile Island accident is just one example of how the improper adaptation of the control room to operating needs could result in unacceptable consequences. It can be seen that matching of the control room to its users must be an ongoing concern throughout the design process if one is to achieve valid control room ergonomics. This chapter examines the methodologies for ergonomic assessment as part of the design process for NPP control rooms. For the practical aspects of the assessment, standards either already published or at press can be referred to:

- ISO 11064: *Ergonomic Design of Control Centres*;
- IEC 964: *Design for Control Rooms of Nuclear Power Plants*; Supplementary standard to the IEC 964: *Nuclear Power Plants – Main Control Room, Verification and Validation of Design*;
- NUREG-0700: *Guidelines for Control Room Design Reviews*; Rev. 1: *Human–System Interface Design Review Guideline*.

3.2 Different Types of Assessment

The three major types of assessment are:

1. Verification, which checks first that the functional specifications comply with the design principles, and then that as it is built, the control room matches the specifications. This verification therefore takes place in two stages, at two different phases of the design process.
2. Functional or technical assessment, which checks that the plant can be controlled under any operational circumstances within the process, with all the required performance in terms of safety, production, and availability. The functional assessment can be undertaken at any time in the design process, using the appropriate tools. Despite this, it should be noted that the relevance of the results is greater downline in the design process.
3. Ergonomic assessment, which checks the match between design choices and ergonomic requirements, i.e. user requirements in any control situation. This ergonomic assessment can (and must) be carried out at different phases in the design process. It is a complement to the verification and functional assessment. Downline in the design process, care should be taken to perform the ergonomic assessment after the other two types of assessment in order to be sure that the 'object' assessed from the ergonomic point of view does indeed correspond to the initial design principles.

3.3 Clarification Prior to the Ergonomic Assessment

The ergonomic assessment of the control room requires certain conditions to be fulfilled before it is carried out.

3.3.1 Knowledge of the Initial Design Targets

The design choices with respect to the technology used and the organisation provided for the work are based on the expected results in terms of benefits and/or progress (relative to other control rooms).

Knowing the initial design targets (i.e. expected benefits of the various choices) makes it possible to define part of the scope of the assessment: one role of the assessment is to help designers eventually validate (or invalidate) their initial choices.

3.3.2 Benchmark and Threshold

Assessing a control room means making a value judgement about it. The characteristics defining the value of the control room depend on the objectives of the assessment. Without going so far as to counter or c each other, these characteristics can be different. Thus, for the tech. assessment, the objective is to check the existence and proper operation o. all the control functions for the process, whereas the objective of the ergonomic assessment is to check that the control room is used as efficiently as possible and under the best possible conditions.

In order to make a value judgement (= assess), it is necessary to determine the benchmark(s) to which the results obtained for the control room will be compared. There are three types of benchmark, bearing in mind that safety and production aspects remain the primary considerations:

1. design choices: the purpose of the assessment is to see to what degree the results expected are actually achieved;
2. another control room: the purpose of the assessment is to see to what degree the new control room is better than another;
3. user satisfaction: the purpose of the assessment is to see to what degree the conditions of use of the control room allow for 'control comfort'.

Once the benchmark(s) has/have been chosen for each 'element under assessment', it is necessary to define the thresholds of acceptability or refusal in order to be able to cast judgement. These thresholds can concern the performance of the process, dysfunction or errors observed, or opinions expressed. They set the acceptable margins for the differences found between the results observed (during the assessment of the control room) and the results of the benchmark(s).

3.3.3 Results of an Ergonomic Assessment

The objective of an ergonomic assessment is to check that the design leads up to a control room from which the process can be controlled in all operational situations while providing suitable conditions for the users. The result of the assessment will be either validation (from the ergonomic point of view) of the technical design choices, or recommendations for improvement of the design.

The results of the assessment will chiefly concern the adaptation or evolution of design choices in accordance with observed difficulties in use. Nevertheless, in order to reply to the general objective of suitable functioning of the user/control room system, some recommendations may concern training.

3.3.4 Different Types of Users

Since we are dealing with control rooms, the type of users that immediately springs to mind is the operators controlling the process. However, the control room is a place where other categories of people are occasionally required to carry out tasks: these people include management, maintenance staff, cleaning staff, and visitors. Their presence in the control room is not permanent, but the control room must be designed so that even 'secondary users' can perform their work and/or not hinder the main users (i.e. the operators).

The ergonomic assessment must therefore also take account of secondary users.

3.4 Steps in the Design Project and Ergonomic Assessment

Among the various steps (theoretically present) in the design project of a control room, those described hereunder are the most relevant for the ergonomic assessment and therefore for consideration of the user in the design process.

- Choice of technical options. The characteristics of the instrumentation and control (I&C) system and equipment are (pre)selected during this phase. These choices are often based on operating feedback (capitalisation of previously acquired knowledge) and on the state of the art of technologies at the time the choices are made. The ergonomic assessment either i) validates or calls for re-examination of all or part of the choices, or ii) postpones the choice until the time when all the elements required for fully cognisant decision-making have been brought together.
- Preliminary specifications. All or part of the functions and characteristics of the control room are described at the outcome of this step. All or part of the functions or characteristics can be judged at this moment, by virtue of ergonomic assessment.
- Construction of the control room on the site. Even if this phase is beyond the design process, assessment of the application on the site gives a real image of the control room, and the results obtained can be used as benchmarking for the design of subsequent control rooms.

The choice of the time(s) to carry out ergonomic assessment(s) depends on:

- the potentialities of the results, i.e. the possibility for going back on the design choices (demands of deep changes in design choices have more chance to be taken into account during the specifications step rather than when the control room is on site);
- the expected relevance of the results, which will be inversely proportional to the 'distance' between the function or characteristic assessed and the function or characteristic actually achieved (design choices are evolving all along the design process);
- the time taken to achieve the required result.

It should be noted that the more 'upline' the assessment is in the process:

- the shorter the result lead times will be (to read and to analyse specifications is quicker than to built a prototype, to design collecting data tool(s), to perform tests and to analyse results);
- the more localised and specific to a particular point of the design the answers will be, for the interactions between the functions are still relatively unintegrated;
- the less 'immovable' the characteristics of the control room will be, which results in a twofold phenomenon:
  - — it is easier to go back on the design in order to change it (there is more room for manoeuvre for proposing modifications);
  - — the degree of certainty on the relevance of the functions or characteristics assessed is low (are they those of the final control room?).

Lastly, the various means or 'tools' for carrying out an assessment are not available at all stages of the design process.

3.5 Means Available for Assessment

An ergonomic assessment can use the following means:

- *Expert judgement* This does not necessitate the implementation of external physical resources, but requires knowledge of the results of similar or assimilable applications and of the principle of the 'operation' of human beings in occupational situations. In other terms, expert judgement is backed up by references, i.e. standards, scientific results, results of previous assessments, etc.

  This method is rapid (short lead time before results are obtained) and inexpensive to implement. On the other hand, its use is governed by the existence of references, with all the problems of transfer and generalisation of the results.
- *Mock-up* This is a tool which reproduces or simulates specific parts (physical or functional) of the control room. A faithful representation of the future application is not considered to be fundamental.

  The use of mock-ups allows for an experimental approach for assessment of specific functions or characteristics (in as far as they are representative of the future control room). The results obtained are relevant but localised, i.e. they are centred on the functions or characteristics assessed but show interactions on only a feeble scale.
- *Prototype* This is a tool which simulates the (quasi) totality of the future control room, in terms of both physical and functional aspects. The

difference relative to the future control room generally lies in the method of production.

As a means of assessment, a prototype represents a substantial investment in terms of time and money, but it allows for testing the dynamic integration of the functions and characteristics of the future control rooms.

- *Control room on site* On-site assessment gives access to all the interactions, but by virtue of that very point, the difficulty of allocating the result of the observation to a particular characteristic or function is augmented. In addition, at this stage it is practically impossible to go back on the design, particularly as far as equipment is concerned.

3.6 Effect of Means of Assessment in the Design Project

The use of resources like mock-ups or prototypes implies the addition of two additional stages in most design projects:

1 mocking-up (breadboarding); and
2 prototyping.

These two stages which are inserted into the specification phases must be taken into account in the general schedule of the project, particularly with respect to the time aspect. Breadboarding and prototyping 'clock up time' in two ways:

1 the construction period itself, i.e. the time of construction of the resource, testing of design alternatives, and analysis of results. This period can be scheduled;
2 the period for integration of the results into the design project. This period can hardly be scheduled, for it depends on the number and amplitude of the results obtained.

3.7 Choice of the Assessment Resource

Not all resources are available at a given moment in the design process. Expert judgement may be expressed very early on in the design process, as soon as the preliminary options have been drafted. Use of a mock-up can be envisaged only once the preliminary specifications have been drawn up, and a prototype can be envisaged only at the end of the general specification phase. As for the on-site control room study, it is possible only at the end of the design project. The possibility of using a new resource does not mean the previous resources are abandoned: for example, expert judgement remains a rapid means of obtaining results at the specification stage.

When choosing a resource, it must also be remembered that:

- The complexity of the assessment in practical terms increases as the resources become more sophisticated (from expert judgement to on-site assessment).
- The assessment's room for manoeuvre, i.e. the possibility of going back on the design, diminishes as the resources become more sophisticated, for the use of sophisticated resources is generally linked to the chronology of the design process.

Finally, it is seen that each resource has its advantages and disadvantages. The question of choosing a (single) resource to help in design implies that resources could be mutually exclusive. In reality, each resource has a specific role, and its answers are adapted to the stage of the design process when it is 'usable'.

3.8 Methodological Aspects of an Assessment Procedure

The development of an assessment method must take account of the specific features of the means of assessment used. In all cases, the method must be structured and defined beforehand if it is to be efficacious:

- structured, in that each element in the assessment (points to be analysed, how to analyse them, etc.) is situated relative to the other elements;
- defined beforehand, in that the assessment method as a whole is defined in detail before the assessment is carried out.

Consequently, the preparation of the assessment takes time, and must be taken into account in the schedule.

With regard to the main users of the control room, the procedure described below is particularly well adapted when the means of assessment is a prototype. Nevertheless, each step described can be taken into account whatever is the means of assessment used and can be followed after adaptation.

Figure 3.1 summarises the overall evaluation procedure.

3.8.1 Method of Assessment for the Main Users

3.8.1.1 Preliminaries

The preliminaries for the assessment are the collection and organisation of practical elements that will be conducive to the smooth performance of the assessment.

To start with, it is necessary to collect the available documentation of use for the assessment. This concerns both assessment reports (on previous

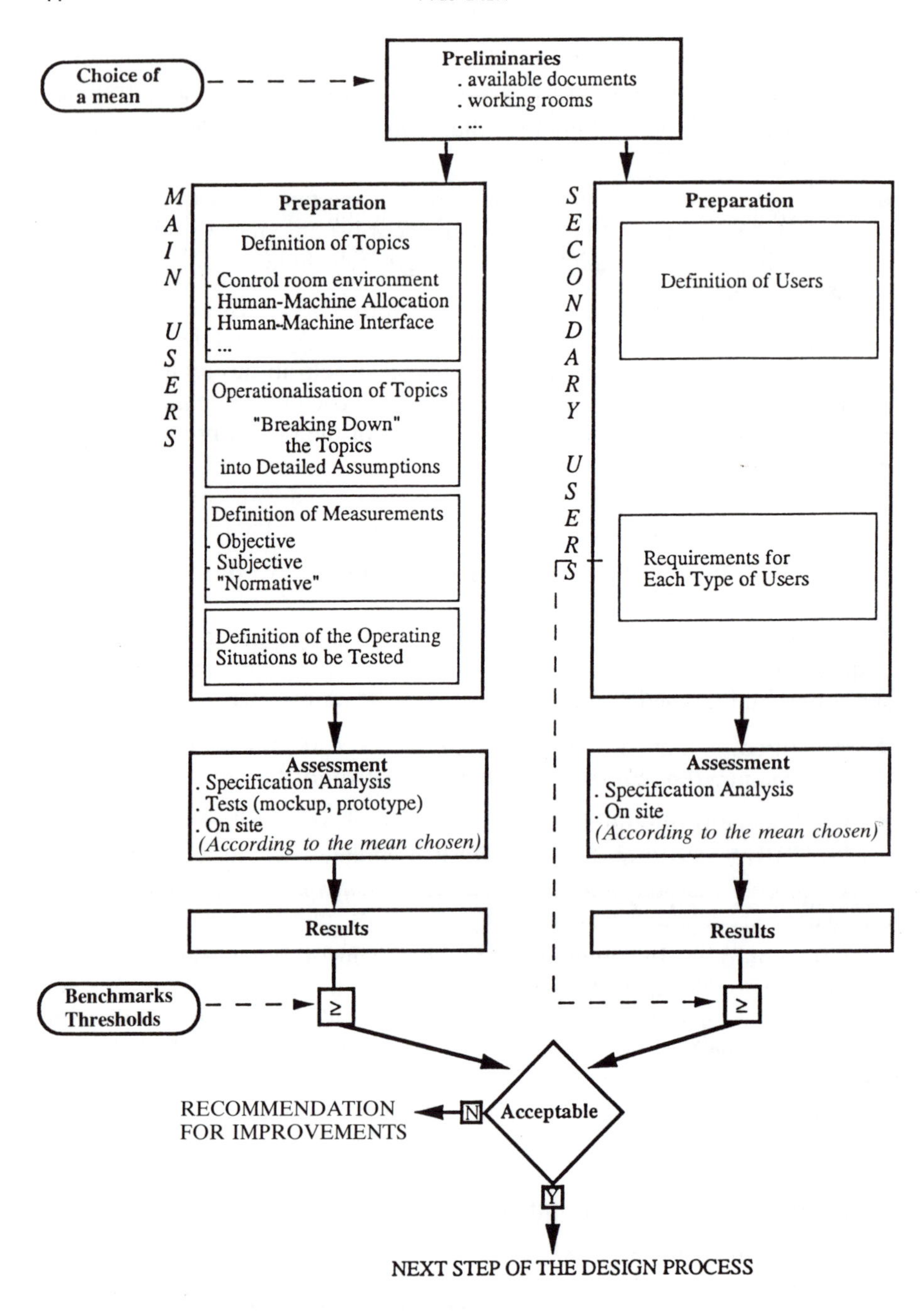

Figure 3.1 Evaluation process flow chart.

assessments or similar control rooms), process operation descriptions, control room drawings, control procedures, descriptions of the training courses envisaged, and standards. Of course it stands to reason that, depending on the time in the assessment process, certain documents will not be available.

It is also necessary to set up an assessment team. The size of the team will chiefly depend on the expected results, the means of assessment used, and the budget allocated for the assessment, but the team should consist of, at least, a human factors specialist, a control specialist, and an I&C specialist.

If the means of assessment is sophisticated (mock-up, prototype, control room), provision must be made for the tools for collecting the data (at least in budgetary terms) and for premises (for performing the assessment, work space for the assessment team, for sorting the data collected, etc.).

The schedule for the assessment must also be drawn up in order to incorporate the time in the design project, especially if the results affect the subsequent steps.

Lastly, the benchmarks and thresholds must be defined so that it is possible to compare the results during the assessment.

Some of the resources set up can also be used for assessing the control room for secondary users.

3.8.1.2 *Steps in the Preparation of the Assessment*

The assessment of the control room for operators must be envisaged through control activity in its entirety. Control is a complex activity carried out by a crew of operators that has its own specific characteristics at the collective level (personal relationships amongst individuals, mutual trust and familiarity, etc.) and at the individual level (physiological aspects, cognitive aspects, etc.).

The basis for the method is a breakdown of the activity into topics for assessment. Activity as a whole can be reconstituted by means of a synthesis of the results.

Topics for Assessment

- *Physical and environmental aspects of the control room* This general topic covers concerns relative to the following:
  - — The design of work stations, i.e. the size of the work station, but also the layout of dialogue tools (controls, information support). It is checked that operators can work comfortably and efficiently.
  - — The architecture of the control room, i.e. the location of the work stations, displays (e.g. wall mimic board), and access. It is checked that movements, communications, and obtaining of information within the control room are easy.

— The atmosphere within the control room, i.e. illumination, auditive, thermal, and air-quality aspects. It is checked that the operators can accomplish their tasks with an acceptable level of comfort.
— In the case of a computerised control room, in addition to all the above, particular attention is to be given to 'visual aspects': visual fatigue, readability, light and colour constraints, reflections, etc.).

- *Human–machine allocation* This question covers the way in which tasks are shared between the operator and the automatic system. It is checked whether the breakdown enables 'parties' to implement their own capabilities and skills.
- *Human–machine interface* This question involves the coherence and reciprocal compatibility of the various control dialogues and modes for presentation of information. It also concerns the resources present alongside the work station (are they adequate, and are they used?). It is checked that the operators have the means for accessing the relevant information and controls quickly, comfortably, and without confusion. (In the case of a computerised control room, it is checked that work station management actions, i.e. the actions necessary for access to the controls and information, do not outweigh the control actions in relative terms.)
- *Work procedures* This question concerns the control procedures to be used by the operators. It is checked that the procedures are effectively of assistance in control.
- *Cognitive aspects* This question concerns the human functions linked to the processing of information: perception, memory, attention (= situation awareness), learning, etc. It is checked that these functions can be carried out efficiently in the control room.
- Organisational aspects: this question must be examined from two levels:
  — 'micro-organisation', i.e. individual activities within the group (distribution of tasks between operators, co-ordinators, etc.);
  — 'macro-organisation', i.e. the formal organisation adopted (structuring of crews, length of shifts, rostering, training, career evolution, etc.).

It is checked that the system set up (or planned) is sufficiently flexible to provide the necessary potential for adaptation to the imponderables of the situations (in terms of organisational situations and process situations as well).

Operationalisation of topics Each topic represents too broad a breakdown to be assessed directly, and so must be broken down again into points to be analysed in order to assess the adaptation of the control room to operators and process control needs. These points can be more or less detailed.

Moreover, they are adapted to the characteristics of the control room. Some examples of points to be analysed for these topics are:[1]

- *Design of the work-station* The location of the various controls makes them easy to see and reach in the work situation, ...
- *Sharing of human–machine tasks* The automatic systems are designed so that the operator can monitor the evolution of the process and over-ride manually if necessary, ...
- *Human–machine interface* The coding of information is relevant and coherent, the number of information devices is adequate, the access to controls is easy, several devices are available to perform an action, ...
- *Work procedures* The procedures guide operators for process control, prescriptions within procedures are readable (visibility) and understandable (legibility), ...
- *Cognitive aspects* Obtaining information becomes more economic with time, ...
- *Organisational aspects* The operators can modify the prescribed sharing of tasks in accordance with the situation and their workload, ...

Definition of measurements Measurements are the data to be collected during the assessment and they must be chosen in accordance with their 'sensitivity' to the points of detail to be assessed. For example, to assess the obtaining of information, during the assessment the number and type of items of information used for a given phase of control are collected.

The measurements may be:

- objective, i.e. refer to identifiable elements of the activity [behaviour (posture, eye movements, . . .) and activity related to the process control (information taken into account, operation performed, . . .)] and/or status of the process (parameters values, alarms occurrence, . . .);
- subjective, i.e. be related to opinions or beliefs of people (usually operators) involved in the tests carried out for assessment (mock-up, prototype, on site);
- normative, i.e. be derived from standards, scientific results, or the results of previous assessments (in order to compare the assessment results and the benchmarks results).

Measurements are not mutually exclusive and their choice depends both on the points to be assessed and the means of assessment used.

Once the measurements have been defined, the tools for making them must be built.

Definition of control situations The control situations are envisaged on the basis of the points to be analysed. Some points may be satisfied with all types of situations; others necessitate particular process conditions (e.g. accidental situation to test emergency procedures).

3.8.1.3 *Assessment*

The performance of the assessment depends on the assessment tool.

Expert judgement is not very efficient with respect to cognitive and organisational aspects. Assessments involving expert judgement are based on comparisons with existing results by applying task analysis methods, i.e. which will be assessed if the prescribed work can be performed via the control room. Expert judgement does not require tests and so future users (or representatives of future users) are usually poorly involved in such an assessment.

Using a mock-up generally makes it possible to assess a single topic of specific aspect at a time, and the mock-up is based on trial methods (with excellent control of input variables). A mock-up is well adapted for assessing issues related to the human–machine interface.

Using a prototype makes for a realistic approach to the control room. The assessment is based on methods for analysing the real activity of the operators (i.e. ergonomic methods), as in the case of the on-site control room assessment.

In as far as possible, assessment with mock-ups and prototypes is done with the participation of the future operators of the control room or staff representative of the future operators. The use of operators in the assessment process requires the following precautions:

- forecast logistics during the preparation of the assessment (requesting operators to participate, accommodation, etc.);
- forecast training on the characteristics of the control room prior to the assessment;
- pay attention to ensuring that the operators involved did not participate in the design so that they are not put in the position of having to judge themselves.

3.8.2 **Assessment Method for Secondary Users**

The assessment for secondary users is more 'lightweight' than for the main users.

3.8.2.1 *Definition of Secondary Users*

Secondary users are the persons whose activity brings them into the control room occasionally. They are:

- auxiliary operators;

- members of auxiliary control services;
- management;
- maintenance people;
- cleaning people;
- visitors.

3.8.2.2 *Definition of Requirements*

The requirements for each type of secondary user are defined. The requirements of secondary users cover the aspects of information (obtaining information on the process, and oral or written transmission of information to the operators or other secondary users) and access to equipment (for maintenance, instrumentation and control engineers, and cleaning personnel) to perform their tasks.

3.8.2.3 *Assessment*

The purpose of the assessment is to check that the control room meets the requirements of secondary users *without* disturbing the main users.

The assessment can be based on expert judgement. On-site assessment increases the knowledge base for subsequent assessments.

3.8.3 Design Feedback

The result of analyses of the assessment is a fully reasoned and substantiated opinion on the control-room characteristics deemed acceptable or unacceptable from the point of view of ergonomics.

Recommendations for improvements must be given for the characteristics deemed unacceptable.

3.9 Conclusions

The procedure presented here has its strengths and weaknesses. The structuring and prior preparation of the method can constitute a trammel on the assessment which prevents the detection of a problem or generic aspect which was not incorporated in the preparation. *It is therefore necessary for the assessment to remain open* to possibilities of readjustment in order to complement the aspects for which no assessment tool was envisaged initially. Similarly, the prior definition of measurements and situations of the process can create an implicit 'route map' for the analysis of results, introducing the risk of inducing a halo effect on interpretation. The assessment topics are somewhat heterogeneous in that some of them are specific (e.g. work station design) whereas others are transversal (e.g. cognitive aspects of the use of the interface; work procedures and collective aspects). In spite of everything, the breakdown into topics makes it possible to demarcate and detail what the assessment has to deal with by providing a global approach for evaluation of use of a control room. When evaluating on site or with a prototype (i.e. having a wide set of detailed topics analysed), a way to 'rebuild' global reality in terms relevant and usable for designers, is to reassemble the topics in 'design-oriented objects'. These objects are for instance 'alarms' or 'displays'. Every result connected to alarms or displays and coming for different topics as human–machine allocation, human–machine interface, organisational aspects and so on, are gathered and analysed in inter-relation in order to give results close to designer culture. Furthermore, prior structuration allows for organisation of the technical and human resources to be implemented for the assessment. Finally, this procedure is a means of introducing the question of human factors into the design project, even if the assessment aspect is to be no more than a complement to the human factors introduced at the moment of design choices and decisions.

Notes

1 The points to be analysed are drafted in the affirmative, and the aim of assessment is to judge whether the affirmation is true.

Bibliography

Dien, Y., Kasbi, C., Lewkowitch, A. & Montmayeul, R. (1991) Méthode d'évaluation ergonomique d'une salle de commande de processus continu. *E.D.F Bulletin de la Direction des Études et Recherches, Série A*, **2**, 9–18.

Kirwan, B. & Ainsworth, L. K. (eds) (1992) *A Guide to Task Analysis*. London: Taylor & Francis.

Leplat, J. (ed.) (1992) *L'analyse du travail en psychologie ergonomique, Tome 1*. Toulouse: Octares Éditions.

PART TWO

Interface Design Issues

CHAPTER FOUR

Human factors of human–computer interaction techniques in the control room

CHRISTOPHER BABER

Industrial Ergonomics Group, University of Birmingham

4.1 Introduction

In control rooms built during the 1950s, information presented to operators was an analogue of plant functioning, for instance, movement of a needle on a dial represented a change in temperature or pressure. Each measured variable from the plant could be assigned its own indicator, which resulted in large panels of hard-wired displays. Many of the measured variables could be acted upon from within the control room through switches or levers. Panels were thus laid out so that controls and displays were correctly positioned, both in terms of physical proximity between a control and its associated display and in terms of the physical operation of a control.

With the rise of computerisation in control rooms, from the early 1960s, came a change in the way control room staff interacted with plant. Much has been written on the change in display formats and design, but surprisingly little attention has been focused on the changes relating to the physical activity of the operator. One can imagine that this is due, in part, to the need to design efficient means of displaying process data on visual display units (VDUs) and, in part, to the belief that operator activity has become far more cognitive than physical. On both counts, consideration of physical activity could be assumed to be of little value. However, as more computers are used in control rooms, there is an increasing variety of interaction techniques which are available; no longer do systems designers need to think solely in terms of a keyboard for computer control, but can consider touchscreens, mice, trackballs amongst other options. Selection of input devices is often made without use of available empirical literature, with designers basing their choice on personal preference or anecdotal evidence (Carey, 1985). Thus, it would seem useful to provide some contemporary perspective on the potential uses of input devices, with reference to their usability, performance and operation.

4.1.1 Changes in Operators' Work Roles in Response to Computers

With the shift towards computerisation came changes in the work activity of control room operators. The most obvious change has been in terms of work role from an operating engineer, who would trim process states, to a process manager, who needs to understand not only the intricacies of process activity but also the functioning of automatic control systems. In a pre-computerised control room, the operators' jobs would involve interrogating hard-wired displays, usually by walking up and down the control room, taking readings from specific items, and physically adjusting controls where necessary. This would require not only a knowledge of the process, but also an awareness of the physical layout of the controls and their relationship with plant items. The complexities of controlling nuclear power stations, coupled with reductions in the size of control rooms, has tended to render wall-mounted displays unrealistic. Furthermore, there has been an increase in the range of uses of computer technology, from system control and data acquisition (SCADA) through to 'intelligent' control and operator support. Thus, the operator in a modern plant is more likely to sit at a control desk than to move around the control room. This means that exploration of process state and activity will be conducted through the paging of information on VDUs rather than by physically moving around hard-wired displays.

There has been much effort in the human factors community to define appropriate means by which operators can receive all relevant information to aid their work. Computers have been seen as an optimal means of providing access to a range of information sources. Their strength lies not only in their ability to manipulate data, but also in the flexibility with which they allow operators to manipulate information. With the introduction of computers come changes in the way information is accessed and, more importantly, acted upon. Baber (1991a) notes that operators perform three types of activity in relation to the information displayed on VDUs: they can scan the information, or search for and retrieve additional information; they can manipulate the information; they can perform control actions in response to the information. If we consider these actions performed on a hard-wired display and on a computer, there are several points to note.

On a hard-wired display, information search will take the form of physically scanning indicators, perhaps looking for patterns of display activity, or will take the form of consulting some other information source, e.g. print outs of previous states. The point is that search can be time-consuming and require the operator to determine which links to make before the search, i.e. to decide what to look for. On a computer, search can be made across a number of sources, links can be generated on an *ad hoc* basis. Scanning will take the form of paging through different displays, i.e. will require a physical action to call up items of information. Thus, one principal difference, relating to the size of the display area, concerns the manner in which information is searched for – visually, on wall-mounted display, and physically on computer screens. One could suggest that the very transience of computer-displayed information is both its advantage and disadvantage: on the one hand, it allows for disparate information sources to be linked together in a fast search, which may aid diagnosis; on the other hand, it may encourage links to be made randomly. On a hard-wired display, where the operator would need to walk to the appropriate indicator, errors could be made in selecting the wrong display. On a computer, where the operator can select a page of information or an object, errors can still be made in terms of selecting the wrong source of information. It is proposed that some interaction techniques may be more prone than others to such error.

It is difficult to conceive of data manipulation on hard-wired displays; often such activity took the form of laborious form-filling and manual calculation. The computer provides a means for not only manipulating current information, but also predicting future trends and conducting 'what-if' scenario testing. In this type of activity, one might conceive of errors as arising from either inappropriate selection of information source or performing inappropriate manipulation on the information. While one might suggest that this latter problem can be circumvented using some intelligence built into the software, this may not always be possible.

While control actions traditionally are few in number (Umbers, 1979), one can see quite marked differences between performing a control action on a physical control, i.e. turning a knob to close a valve, and through a computer interface, i.e. selecting a valve icon. Having said this, systems exist which require operators to select and manipulate a knob icon, i.e. by placing the cursor on the knob and moving the cursor to turn the knob. While this may represent a consistent mapping between traditional and computerised operation, it is a particularly poor design for an interface, combining both poor use of the computer input device and requiring operators to perform a redundant task. This example also raises the issue of whether 'compatibility', which played so useful a role in the design of traditional control rooms, can be applied to computer-based systems.

4.1.2 Legacies of Three Mile Island

The Three Mile Island (TMI) incident resulted in many recommendations for the nuclear industry in the US, which have had an impact across the world (Kemeny, 1979; Perrow, 1984). As the incident at TMI unfolded, some of the key information was hidden, either by the mass of other information or by physical objects, and some of the principal controls were difficult to reach, in terms of physical location. With the development of computer-based control, one can foresee a shift in the loci of these problems. Rather than having information physically hidden, information will be 'conceptually hidden', by which I mean that the many pages through which an operator needs to scroll to find a relevant piece of information could lead to a situation in which finding information is prone to error; either in terms of 'finding' the wrong information or in terms of being unable to find the desired information. Rather than having controls which are physically inaccessible, the placement of controls on screen-based mimics or in menus will again lead to difficulty in finding the correct control, and the current iconography of display screens may lead to errors in selection.

Referring again to the TMI incident, there were problems with the spatial relationship between controls and their displays in the TMI control room, in particular the 'mirror imaging' of panels in the control room. This is a problem of compatibility in the general layout of the controls. In computer-based operation, the problems relating to information retrieval and control manipulation will be compounded by the simple fact that a small set of control actions, i.e. positioning a cursor and pressing a button, will be required for all control-room activity.

4.2 Compatibility

In this section we will consider types compatibility (movement; spatial; operational; conceptual; modality), and their relationship with the design of computer input devices. The section will consider the potential for human error when the principles of compatibility are violated in computer-based systems. In its simplest form 'compatibility' can be defined as the relationship between a person's expectations of how an object could be used and how the object should be used, i.e. how it is designed to be used. To take a simple example, consider a lever-operated press. In order to raise the press, one should pull the lever down and in order to lower the press, one pushes the lever up. While this accords well with basic engineering, there is a possibility that, under stress, the operator could convert an intention to lift the press up into the physical action of lifting the lever up – which would bring the press crashing down. This is precisely what actually happened in a steel forging mill. In this instance, the operator fails to translate from intention to action, defining the adverb 'up' not in terms of the movement of the press but in terms of the movement of his arm. This point can be further illustrated by a standard human factors experiment.

People are presented with a row of lights and corresponding sets of switches. Performance is superior in configurations in which the layout of the switches mirrors that of the lights, i.e. where a row of switches is positioned under a row of lights, as opposed to having the switches positioned randomly.

This rather obvious finding again relates to the issue of translation; in configuration two, the respondents were required to identify the light using a code, search for the switch with the corresponding code and press the switch, whereas in configuration one all that was required was to press the switch below the light.

These examples illustrate the principles of movement and spatial compatibility, which we can define as

- movement compatibility: when the action required to move an object corresponds to the user's intention;
  - corollary 1 If the movement is opposite to intention or if it requires translation, then error is possible.
  - corollary 2 If the movement needs to be fully completed before the user can detect its outcome, then it is not possible to correct errors until the operation is fully completed.
- spatial compatibility: when a control and the object it controls (or a display of the object's state) are positioned proximally.
  - corollary 1 Proximity can be defined spatially by placing the control adjacent to the object.
  - corollary 2 Proximity can be defined spatially by placing the controls in a similar arrangement to the objects.

These principles are well documented in the ergonomics literature (the reader is referred to Sanders and McCormick, 1992, p. 302 for a full discussion). An interesting point to note, with reference to corollary 2 of 'movement compatibility', is that operation in a nuclear control room takes place across several time-scales. This means that feedback from control activity is typically obtained some time after the action.

The issue of movement compatibility for input devices would appear to be relatively straightforward. Many of the available input devices can be used to move a cursor on the VDU. In this case, movement of the device will result in a corresponding movement of the cursor. Providing the orientation of the device is correct, one should be able to move the cursor with relatively little difficulty in the appropriate direction. While this may appear self-evident, the issue of movement compatibility for cursor control raises an interesting point. In non-computerised control rooms, feedback from the operation of a control was initially kinaesthetic (i.e. if you pressed a button, you felt it move) and then visual, in the form of changes on a display. Rarely would the visual feedback be simultaneous with control operation; an exception to this would be for fine tuning of variables where a display would function rather like the old 'dials' on radios with a needle moving through the wavebands. In such a case, one would most likely perform the control action at a rate commensurate with using the visual feedback. However, the introduction of a pointing device means that operators are effectively having to 'aim' the cursor at an object, 'steer' the cursor towards the object and then select the object. In other words, while we can optimise the movement compatibility of these devices, there is still a question as to whether operators ought to be performing these additional activities; one could argue that, not only does computer-based operation restrict the operator's 'window' on plant activity, it also erects a barrier to task performance by requiring a whole host of trivial activities to be performed in association with the operator's 'real' goals. My favourite example of this is the use of 'virtual control knobs' on displays – in order to 'turn' the knob, pictured on the VDU, the operator moves the cursor to the bottom of the knob and then drags the image to a new position. This manages to be both similar to and significantly different from the use of a traditional knob; the difference lies almost entirely in the introduction of trivial activity.

A second point to note concerning movement compatibility relates to the speed of cursor movement. Much attention has been given to the setting to control : display (C : D) ratios to match the speed of cursor movement to that of device movement, principally to reduce the impression of lag between movement although it is possible to incorporate some acceleration in the cursor. Tränkle and Deutschmann (1991) have shown that C : D ratio may not be as important as previously believed, and argue that movement is far more likely to be influenced by factors such as target size, distance moved and, to a lesser extent, screen size. This would suggest that, for input devices, movement compatibility may be translated, i.e. rather than being a simple matter of using an input device to move the cursor around the screen, the input device is used to move the cursor to specific targets. This distinction may appear, at first sight, fatuous, but leads to an important point concerning the planning and control of actions using input devices. If the use of an input device to move a cursor could be related solely to movement compatibility, then all one would need to do would be to ensure appropriate C : D ratio and device orientation, but if the use of the device requires planning and control then consideration needs to be given the points at which this cognitive activity could break down, e.g. would different devices lead to over-shooting of targets, would different devices lead to different ways of moving a cursor around the screen, would different devices impose different levels of workload on the user?

I have heard of people who were able to learn to use inverted mice, i.e. they placed the mouse with its lead facing away from the computer (assuming the lead was the mouse's tail) and learned that movement of the mouse to the left would lead to a movement of the cursor to the right. Other anecdotes tell of people who, when told to use the mouse to move the cursor, placed the mouse on the computer screen. Naturally, the stories illustrate an extreme of human behaviour (which the reader will no doubt be able to supply with a fitting label). However, this raises an interesting question of how does the design of an input device *suggest* its use? One could answer this question with reference to an everyday object, such as a door handle. A flat plate on a door is clearly intended to be pushed, a door knob, on the other hand, would appear to be designed to be grasped and pulled. However, while the flat plate is unambiguous, in that there is only one operation available, the door knob has a degree of ambiguity, in that one could also grasp and push. One could also consider the act of turning off a water-tap (faucet) in the same context; while the tap can be considered in terms of movement compatibility, relating off to clockwise rotation for instance, when the tap is open and water is flowing, how can one determine which way to turn it? This illustrates a further type of compatibility which, for the sake of consistency, will be termed 'operational compatibility'.

- operational compatibility: when the design of an object permits an unambiguous set of actions;
  - — corollary 1 If the object can be acted upon in different ways, then error is possible.
  - — corollary 2 If the movement needs to be fully completed before the user can detect its outcome, then it might not be possible to correct the error.

For some computer input devices, as noted above, users might have an initial difficulty in interpreting the correct set of actions. While this problem can be addressed by training, one needs to bear in mind that during periods of high workload or stress, people could revert to pre-training assumptions and perhaps respond to the operational compatibility of a device. As an example, consider a trackball which is designed such that the ball is mounted in a holder which also houses a button at the top of the housing. Movement of the cursor will result from finger-tip control of the trackball; how will the button be pressed? One could suggest that the button is positioned such that its operational compatibility invites pressing using a finger, but the fingers are already engaged in operating the ball and if moved they could shift the cursor. Alternatively, users could attempt to press the button using their thumb but this would require even greater contortion of the hand.

Another example concerns the use of multibutton mice. A company, which invited us to review its product range, had developed a sophisticated image analysis system and used a three-button mouse as an input device. Interaction took the form of selection from menus, or selection of objects, or classification of objects. In the first two activities, the right-hand mouse button was used to click objects (the other buttons had no effect), while in the third activity, the left-hand button signalled 'accept', the middle button signalled 'undecided', and the right-hand button signalled 'reject'. There are obvious problems of operational compatibility in this example (notwithstanding the fact the original design actually used different combinations of button presses at different points in the program), with the principal one being the definition of the right-hand button; in the first two activities, one could propose the right-hand button signalled a positive selection, whereas in the third activity, it signalled a negative selection. If the users persisted with a notion of right-hand equals positive selection, then they would erroneously select inappropriate information.

With reference to spatial compatibility, it can be difficult to define how an input device can have spatial compatibility with the object(s) it is being used to control. One can draw a distinction between direct-pointing devices, such as touchscreens and light-pens, in which the users will physically touch a specific object, and indirect-pointing devices, such as mice, trackballs, joysticks, in which the user moves a cursor to an object and then perform some selection action, such as a button press. In the former, spatial compatibility is simply a matter of permitting an object to act as its own button, i.e. an icon of a valve on a touchscreen could be touched to change its state. However, this relates to the design of the display rather than the use of the device. In the latter case, one might propose that object selection requires a degree of translation (or at least, imposes some additional task on the operator). A point worth considering is, given the drive to design and implement multi-purpose controls, is there a possibility that errors could arise due to lack of spatial compatibility? The answer to this question would seem to be yes, as is shown by an example from Moray (1988).

A computer system, used to control the engines on a ship, had been developed and trialled to the enthusiastic response of operators. The basic control desk was configured in a very similar way to the control desks of nuclear power stations, with a central VDU reserved primarily for overview information (although it could, of course, be reconfigured) and two adjacent VDUs, which could be reconfigured to display a range of information. In front of the VDUs were sets of function keys to permit system control. In one 'test incident', an operator was confronted with an overheating starboard engine. The operator correctly determined that the course of action required was to throttle back on the starboard engine, and called up the 'gearbox information' page on the left-hand VDU. Using the controls under this VDU, the operator attempted to perform the appropriate control actions. After about thirty seconds of repeated button pressing to no avail, the operator contacted the system engineer to report a fault. Of course, the left-hand buttons were dedicated to the control of the port engine (which the operator had throttled back), and the starboard engine controls were to be found on the right-hand side.

The potential for violation of spatial compatibility offered by reconfigurable displays is clearly illustrated by the example. One could propose design solutions which allowed controls to be reconfigurable as well as displays, but that could only exacerbate matters. Alternatively, one could attempt to label displays to indicate what the operator was looking at. However, as this example illustrates, people are likely to read information they are looking for (which may not always be what they are looking at). Alternatively, one could attempt to design the interaction so that specific control actions are performed on specific objects, i.e. in this example, having the engine throttle incorporated into the display could permit the operator to select that object on that display and act upon it. In this latter case, spatial compatibility would be achieved by having the control adjacent to the object.

The corollaries to several of the principles of compatibility outlined above imply a need for adequate provision of outcome information, or feedback. Obviously, it is important for the user to be informed of the success or otherwise of an action. This type of feedback incorporates a number of levels, and one needs to consider the compatibility of each of these levels with the task in hand in order to provide sufficient information without distraction. This introduces the final types of compatibility to be considered: conceptual and modality. Suppose the input device used was a speech recogniser, and the task was to enter specific commands, such as opening and closing valves. In this instance, feedback needs to be provided as to the activity of the input device, i.e. whether the words had been recognised correctly, and the task activity, i.e. whether the correct valve had been operated on in the correct fashion. One could decide to provide separate feedback for each type of information. However, performance has been shown to be far better if the feedback can be incorporated into the primary task display (Baber *et al.*, 1992). In this instance, one can retain some conceptual compatibility between the performance of the task and the information received concerning that performance.

The issue of modality compatibility is relatively new, but of interest to this paper. By way of illustration, consider a display of variation of temperature against time. This display could be presented using a graph, with a line representing temperature. However, it could also be represented as a series of discrete spoken values. In the first instance, the visual modality is used to represent changes in a spatial relationship, in the second instance, the auditory modality is used to represent this relationship. This suggests that some types of information should be presented visually, while others can be represented verbally. The principal issue here is how one should define the information. For example, returning to the speech example; one could propose that issuing a spoken command ought to have verbal feedback, and closing a valve ought to have visual feedback, and yet the study found that presenting verbal feedback, in the form of text, had a detrimental effect. In this instance, the information required related to the performance of the task, rather than the use of the input device. One could suggest from this that successful implementation of computer technology will not place an obstacle between operator and primary task performance, i.e. the primary task will not be computer operation, but process control.

- conceptual compatibility: when the design of an information permits unambiguous interpretation by users;
  - — corollary 1 If the information cannot be interpreted, then error is possible.
  - — corollary 2 If the information can be interpreted in different ways, then error is possible.

— corollary 3 If the information detracts from the primary task, then error is possible.

- modality compatibility: when information is presented in a format suitable for its interpretation, and when actions are performed using appropriate responses.
  — corollary 1 If information is presented to a modality already used, there might be interference in processing.
  — corollary 2 If the information is presented to an inappropriate modality, then error is possible.

From this discussion, one can see that, although principles of compatibility relate to earlier design principles, one needs to reconsider them in terms of computer input devices. Computer input devices permit users to manipulate objects on computer screens. This means that the manipulation will be dependent upon the design of the objects and the operational characteristics of the devices. It also means, of course, that users no longer operate directly on plant items. While the 'knobs and dials' control room provided a means for operators to act analogously on the plant, the computer control room entails operation on analogues of analogues, e.g. VDU displays are full of pictures of valves which represent the valve control on wall-mounted mimics, which represent the valve on plant. Given this level of removal, it is worth considering the relationship between actions performed with an input device and knowledge of the plant. This point will be returned to in Section 4.5.3. In the following section, the performance aspects of input devices which are usually used in common rooms will be considered.

4.3 Performance

Initially, computer applications in control rooms relied on the standard alphanumeric keyboard which was supplied with the computer. This required operators to type strings of letters and numbers to identify plant items, to issue commands, to request information, and to manipulate this information. Given the problems associated with command languages in general (Shneiderman, 1992) and the fact that operators were seldom trained typists, efforts were made to minimise the use of text entry, e.g. through the design and development of menu systems or through the use of function keyboards. With the development of more sophisticated graphics, it became possible to move a cursor around displays to select objects. This was performed using cursor keys on older systems, but these have quickly been superseded by other pointing devices. Contemporary computer systems in control rooms employ 'direct manipulation' interfaces, i.e. it is possible to select and manipulate objects on the VDU without recourse to entering commands.

As this potted history implies, there are several generations of computer system in operation, with new systems in development. This means that there exist a number of different approaches to operation of computers in control rooms. Future technology allows even greater opportunity to manipulate and interact with plant objects (virtual reality control rooms are even being discussed as a serious option). All of which makes it important to determine whether the devices chosen are suitable for the task in hand, and whether their selection has been unduly influenced by factors other than performance, e.g. cost or space requirements. In this section, we consider performance in terms of three metrics: speed, accuracy, and error. Accuracy and error have been separated because there are some types of error which are not directly related to the accuracy of performance.

4.3.1 Speed

There are two main activities involving input devices in which speed is an issue: cursor movement and text and data entry.

While one might feel that speed at which a cursor can be moved is not an important issue in the use of input devices in the control room, it can be proposed that devices which lead to long performance times will be disliked. There is a consistent finding in the literature that cursor keys are the slowest form of cursor movement and are unanimously disliked. Having said this, cursor keys are also the fastest form of cursor movement for closely spaced objects (Card *et al.*, 1978). Thus, one must be careful of blanket statements concerning input device performance, without making reference to task requirements. This is particularly true when it comes to performance times. Moving the cursor is a task which is rarely performed for its own sake; usually cursor movement is associated with either selecting an object, or moving the cursor to a specific place on the screen, or moving an object, or drawing. In the process control room, one would anticipate that object selection would be one of the most common actions, with moving objects and drawing relatively infrequently performed as yet. Thus, we will focus our attention on the selection of objects.

Given a small number of objects to select and act upon, the function keyboard (in which each key is labelled with the name of an object) represents a fast technique (Goodwin, 1975). However, there is clearly a problem of space, as the number of objects increases. Thus, it should be possible to perform object selection on relatively small devices, either through the device directly controlling a display or through a reconfigurable device. The latter can be characterised by the traditional alphanumeric keyboard, discussed above, or through automatic speech recognition (see Section 6.4).

Providing the objects on a display are of a sufficient size and spacing, touchscreens would appear to be the fastest input device for object selection and the one which users tend to prefer (Usher, 1982; Haller *et al.*, 1984; Thomas and Milan, 1987). The principal reason for this speed advantage might well relate to the fact that the touchscreen does not, in fact, require the use of a cursor; the user simply positions a finger on the desired object. In effect the finger acts as a cursor, but the main issue seems to be that touchscreens do not require the additional task of cursor manipulation. Rather than having to monitor the position of the cursor relative to the object, the user touches the object.

From a review of computer input devices in general, Milner (1988) notes that 'The empirical support for the ubiquitous mouse input device is poor in the literature.' Having said this, more recent research has found the mouse useful for a specific range of tasks. MacKenzie *et al.* (1991) have shown mice to be superior to trackballs in terms of both pointing at, and dragging, objects. We would suggest that the advantage of the mouse for dragging objects relates to its design and operation. In order to drag something across the screen, one must both 'grab' the object (usually by holding down a button, which effectively renders the object as a cursor) and move it (as one would move the cursor). The position of the button on the mouse allows this operation to be performed relatively easily; one can press the button with the index finger while retaining control of the mouse through the thumb and ring finger. For the trackball, control is through the fingers (usually the first two), which means button pressing might have to be performed with the thumb, throwing the hand off balance as it were and rendering the task difficult.

4.3.2 Accuracy

While touchscreens have been found to yield fast object selection times, their accuracy has been questioned by some researchers (e.g. Ball *et al.*, 1980). The studies in which touchscreen accuracy is low tend to employ targets smaller than the recommended size, and to use technology which suffers from parallax (where the point at which the user's finger touches the screen does not correspond to the object's position, due primarily to the relationship between the curvature of the screen and the distance the touch-sensitive area is above the screen). As far as mice and trackballs are concerned, some studies find in favour of the mouse (MacKenzie *et al.*, 1991) and some find in favour of the trackball (Ritchie and Turner, 1975), while some find no difference (Haller *et al.*, 1984). A point to note is that these studies use different tasks and one might suggest an interaction between task and input device in terms of accuracy. However, there would seem to be reasonable evidence to suggest a compromise such that, given sufficient object size and that objects are not spaced too widely apart, there will be similar performance from these devices.

4.3.3 Error

There are two principal types of error one could anticipate with the use of input devices: the first concerns object selection, the second concerns cursor control and positioning. The first error type relates to the design and layout of displayed information, and would result in operators selecting the wrong object. This could be exacerbated by objects having a similar appearance, although attempting to produce distinct objects would be impossible. Usually discrimination is performed on the basis of additional labelling or coding of objects. The second type of error relates to the actual control of the input device. We have noted that some forms of touchscreen technology could be prone to problems of parallax (although this usually applies only to older, infrared models or to products at the lower end of the price range). Mice and trackballs, on the other hand, can be prone to target overshoot, especially with unskilled users.

4.4 Operational Considerations

4.4.1 Desk-space Requirements

The input device which has minimal requirement for desk-space is the touchscreen, which is mounted over the VDU. With reference to the process control domain, Usher (1982) has been instrumental in introducing touchscreens into the control room, and the coal-fired power station at Didcot, Oxfordshire is currently using this technology. For many designers, the touchscreen is considered too expensive and cheaper alternatives are sought. Among these, the mouse and the trackball would appear to be the most popular (with the trackball being by far the more common in the UK). Informal discussion and observation of current control room design suggests that the use of trackball is based primarily on its limited space requirements and its lack of trailing lead during operation. The mouse requires an area of desk space, preferably with a pad (this is, of course, very important for use with an optical mouse).

4.4.2 Durability

Trackballs have been in use on the reactor desk at WYLFA nuclear power station since 1982, and have, as far as I know, withstood constant use. Trackballs are also the devices used at THORP reprocessing plant at Sellafield. A previous implementation of trackballs for accessing grid diagrams (at the CEGB National Control Centre), proved somewhat problematic due to slippage and interference between ball and casing, caused by a build-up of dust and grease. However, this problem can be overcome with regular maintenance. As one might expect, from the similarities in their engineering, mice exhibit similar characteristics of durability to trackballs (Commerford, 1984). Having said this, it is likely that mice require more regular servicing than trackballs due to the fact that they are moved around the desk surface and can pick up dust and debris, whereas trackballs are relatively well protected by their casing. Depending on the technology used, touchscreens are also fairly durable. Their main problem seems to relate to the fact that VDUs become covered with fingerprints, which obviously require regular cleaning.

Thus, as far as durability is concerned the input devices considered in this section would appear to be able to meet the rigorous requirements of the nuclear industry, and can be found in control rooms around the world.

4.5 Task-compatibility

4.5.1 Task Activity

A number of authors decompose task activity with input devices into a set of generic tasks (Foley *et al.*, 1984; Carey, 1985), and we have produced a simple classification for the purposes of this chapter. As Table 4.1 illustrates, some devices are better suited to some generic tasks than others.

Naturally, the question is, how best to select a device suited to the range of generic tasks required, while also considering its performance characteristics. It is proposed that this table, together with the previous discussion, will provide the reader with an idea of performance of a few input devices.

4.5.2 Physical Operation

4.5.2.1 Musculoskeletal Problems

The touchscreen requires users to move their hand to points on the screen. If the screen is mounted perpendicular to the user, this may result in strain from prolonged use. An alternative might be to tilt and lower the screen towards the desk surface, although this presents a problem relating to glare from overhead lighting. Having said this, strain from prolonged use would be apparent only if the user had to make a great many selections per hour, e.g. as might be found in data entry, which is unlikely. In the system installed at Didcot, the touchscreen is used for object selection while a keyboard is used for data entry.

Many writers have suggested that mice are trouble free, as far as musculoskeletal problems are concerned, suggesting that working with the mouse allows users to change posture easily (Engelbart, 1973; Foley *et al.*, 1984; Dix *et al.*, 1993). These assumptions have passed into computer-lore with little or no questioning. However, in recent surveys of office staff who are using windows-based software, we have noted a high incidence of reports of musculoskeletal problems relating to the right wrist and shoulder; in many of these cases the mouse is positioned at a distance from the keyboard (Noyes *et al.*, 1994). It would appear that prolonged mouse usage can lead to

Table 4.1 Characteristics of input devices

Input device	Select object	Act on object	Specify action	Enter alphanumerics
Keyboard	Type code	Type cmd	Type cmd	Type text/numbers
Function keys	Press object key	Press cmd key	Press cmd key	—
Cursor keys	Move cursor to object and confirm	Select cmd from menu	Select cmd from menu	—
Trackball	Move cursor to object and confirm	Select cmd from menu	Select cmd from menu	—
Mouse	Move cursor to object and confirm	Select cmd from menu	Select cmd from menu	—
Touchscreen	Touch object	Select cmd from menu	Select cmd from menu	(Type using keyboard overlay)

problems, especially where the space in which the mouse can be operated is limited.

For large movements, across the screen or to large targets, it is possible to move the mouse using full arm movement, but for small movements, either across small distances or to small targets, more precise control will be had from using wrist movements, often with the fourth and fifth fingers resting on the table surface. As screens become both larger and more cluttered, with an increase in small targets, there may well be a mismatch between these different task requirements. As desk space is often at a premium in the control room, this may become an important issue.

The mouse is often controlled using the forearm/wrist muscle grouping (Whitefield, 1983). It is probably for this reason that users rest the heel of the hand or their forearm on the desk while using the mouse. However, some evidence is beginning to accumulate in the literature relating extensive mouse use to specific injuries, principally related to ulnar deviation resulting from holding the mouse and pressing buttons (Davie *et al.*, 1991; Franco *et al.*, 1992). These factors could be addressed by using pads large enough to rest the hand on, in order to reduce pressure, and through training operators in optimal mouse handling, or by altering the design to encourage users to rest their hands on the mouse.

There does not appear to be a great deal of information relating to the musculoskeletal problems of trackball use. One might anticipate some similarity in problems arising from mouse use, particularly as the trackball is often operated with the forearm and wrist resting on the desk. One might also anticipate some problems in the fingers following prolonged periods of fine control, in which the hand may need to be held in tension to maintain control of the cursor. The solution to these problems would seem to be provision of adequate support, such as matting (which is noticeably lacking in many control rooms) and encouragement of breaks during use, i.e. preventing operators from 'resting' their hands on the trackball during use.

4.5.2.2 *Control of Action*

There is an assumption among many of the people to whom I spoke when preparing this paper that, under stress, the trackball is less prone to error than the mouse. I would suggest, from the discussion below, that this assumption is probably erroneous. For pointing activity, as we can see in Figure 4.1, the path taken by the cursor is very different when using a mouse or a trackball. This suggests that while mouse movement is based on an initial ballistic swing about the wrist or forearm, followed by fingertip correction, the trackball is based on a series of discrete, fingertip corrections. This would lead to both slower positioning time and a tendency to spend a significant portion of time 'hovering' over the target in final correction phase for the trackball. One could further argue that, as the control of movement is strongly influenced by factors such as workload and stress (Fleischer and Becker, 1986), this pattern would either be exacerbated by high workload or that additional attention would need to be allocated to the control of the trackball during periods of high workload.

Norman (1988) has suggested a simple, seven-stage model of human activity which can be paraphrased for the purposes of this chapter as follows:

- form goal;
- form intention;
- specify action;
- execute action;
- perceive results of action;
- interpret results of action;
- evaluate outcome.

Thus, prior to any action, the operator will both formulate a goal and plan a set of actions to achieve that goal. Rasmussen and Goodstein (1988) suggest that this stage will be influenced by the type of system; the task; the user interface; the knowledge, skills and abilities of the operator. The type of system and the task will obviously depend on the process state and the type of reactor. The knowledge, skills and abilities of the operator will depend on training and experience. We are interested in the influence the user interface might have on forming goals and intentions. While the current emphasis in control room ergonomics is on display design for situational awareness, this will aid only the formation of goals. At the level of action, the operator will specify actions in terms of the opportunities provided by the interface. In traditional control rooms, one could suggest that actions had equal potential, as all controls were accessible. In computer systems, with controls arranged on different pages in hierarchies, there will be variable potential for different controls. In both cases, the operator will need to determine which control to use in order to perform a specific action but, as

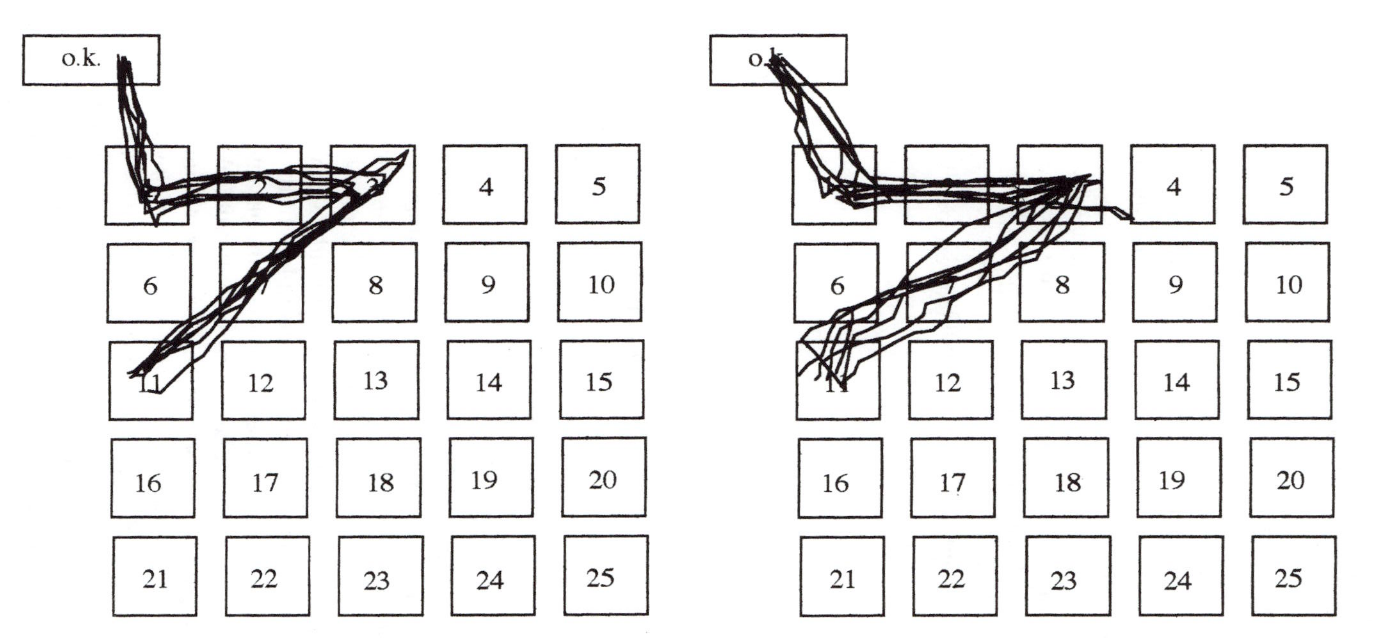

Figure 4.1 Patterns of cursor movement with (a) mouse and (b) trackball. In this task, six people were required to move a cursor through a series of numbered squares. The paths of the cursor for movement sequence 1-3-11 are presented for (a) mouse and (b) trackball. The paths show greater variability for the trackball than the mouse. This is reflected in the fact that performance is significantly faster when using the mouse for this task. Presumably this difference in variability in cursor path is due to problems relating to the control of the trackball in its housing.

we have already noted, there will be differences in the time required to search for the different controls. This would suggest that action planning will be influenced by the interface.

Hesse and Hahn (1994) show that interface layout also has a bearing on the execution of actions, and the subsequent stages of monitoring the effects of the action. Thus, different layouts of interfaces, and different ways in which an operator can interact with them, may require different forms of action planning and execution. We have noted that mice and trackballs do not use identical control movements, and that cursor tracking may be a problem. We have also noted use of 'virtual controls', i.e. knobs or sliders which require the cursor to be placed on screen objects and then dragged to operate the control. While these designs may fulfil a requirement to represent the controls in a fashion to which operators can relate, they would seem to introduce additional action planning and execution tasks in the primary control task.

4.5.3 Strategy

As Barker *et al.* (1990) demonstrated with reference to mice, performance would appear to be strongly related to visual monitoring of cursor position. Thus, one would predict the input devices which use cursors have an impact on the strategy used to inspect information. The point to note here is that, although we can define and modify movement compatibility for cursor positioning, is this being conducted at the expense of primary task activity, i.e. does the control and operation of a cursor constitute a task in itself?

If we are to believe the notion of direct manipulation, which is fashionable in interface design circles, the act of physically selecting an object on the screen requires little effort – it is 'direct'. However, the fact that cursor movement requires visual monitoring suggests that there is more to object selection than this suggests. At a simple level, this implies problems in terms of actually finding the cursor, with operators scanning the display for the cursor's current location, but at a deeper level, one might ask whether the activity of moving the cursor will narrow the operator's focus on the display. We know that operators can become decoupled from ongoing process activity, leading them to narrow their focus on plant parameters. One might feel that the visual monitoring of cursor movement would lead to operators' attention being directed to the vicinity of the cursor, and that the act of moving the cursor may be sufficient to interrupt or otherwise impact upon thinking. Moray (1988) points out that observers of visual displays typically move their visual attention to different sources at a rate of twice per second, although obviously this figure should be considered relative to the characteristics of the display, the signal-to-noise ratio and the task demands. If this rate is exceeded, one might suggest that the individual is overloaded. One could propose that the monitoring of cursor positioning would act to increase workload, by requiring greater-than-average shifts in visual attention. From this, one might expect a specific strategy to follow the cursor, and to act within the space defined by the cursor's position.

We have been investigating the interaction between devices used for object selection and strategies people use for tasks such as fault diagnosis. For instance, a study comparing user performance on a simulated fault-finding task with either a mouse or a touchscreen to select nodes for testing showed a significant interaction between device and strategy. The touchscreen led to users systematically working through the nodes, following recognised fault-detection strategies, while the mouse, in contrast, led to users searching within specific areas. Although the data are by no means unequivocal, one could suggest that the presence of the cursor in the mouse condition, led to users narrowing their field of search. However, there were no differences in terms of overall performance time. Baber (1991b) compared speech with function keys for fault diagnosis tasks, and found that the function keys led to users following the arrangement of keys. With speech, users issued fewer commands, took longer to rectify faults and missed more faults than with the function keys. The results seem to indicate a relationship between strategy and type of pointing device, such that with some devices users employ a strategy which is more focused on the arrangement of information in the display. While this section is based on simple laboratory studies, the findings raise an interesting question concerning spatial compatibility, viz., to what extent will the design of information on a screen encourage certain patterns of use, and how will this interact with the use of specific input devices?

4.5.4 Process Feel

The final point in this section concerns the impact of computerisation on process feel. Speaking to operators with experience of pre-computerised control rooms, one is struck by their symbiotic relationship with the plant via the control panels. The position of needles in dials, the orientation of levers or knobs on a section of the panel could provide valuable information on plant status. At the level of control operation, the feel of control movement and the knowledge engaged in turning a control through a specified range could be used in order to define a suitable level without having recourse to the scale around the control. Further, some controls would be operated within specific ranges, with 'out of range' being signalled by constraints on control movement. The operators' experience and knowledge were expressed as much through physical action as through cognition. While it is not credible to advocate a return to these traditional control rooms, it is a moot point as to how much process feel can be obtained from computerised control rooms. We are familiar with the argument that the computer effectively shrinks the window on the process and plant for an operator, but to what extent does the restriction of control action to typed commands or cursor movement have a bearing on operator activity? One could consider the possibility of incorporating a variety of feedback signals to operators via their controls, e.g. through kinaesthetic feedback in terms of the operation of an input device or through simple auditory cues relating to specific actions. In this way, one might be able to present some sense of process feel through computer control – but this leads into the next section.

4.6 A Look to the Future

The following sections consider current trends and developments in input device technology, and relate these to possible advanced control room concepts. The point is not to 'sell' the technology, so much as to raise an awareness of the possible and the practicable.

4.6.1 Remote Operation

It is a moot point whether future control rooms will seek to centralise operations further, with fewer, more powerful workstations, or whether future control rooms will dispense with work stations altogether, relying on large, computer-driven back-panels. There are several prototype systems which demonstrate the latter concept. Given a move away from control desks, one needs to ask how can operators still interact with the plant? The answer to this lies in the technologies of gesture and speech recognition, discussed in Sections 4.6.3 and 4.6.4 below.

4.6.2 Mobile Communications

Technology is developing to allow more flexible communication between the control rooms and other operational staff. Developments in computer technology will permit more flexible work arrangements, which in turn require mobile communication between plant and control room. While one can imagine radios as a possible solution, hand-held computers are already capable of transmitting both sound and text, and one can foresee such devices generating high-quality images in the near future. Thus, one needs to consider how people will interact through these devices.

4.6.3 Gesture

Gesture recognition is a generic term used to cover applications in which physical activity on the part of the user results in changes in the state of the computer. In some systems, the gestures could be physical, e.g. pointing to objects, while in other systems, the gestures could be mediated by some hand-held object, such as a pen. The recognition of physical gestures is more difficult than pen-based gestures because there are more parameters to consider, i.e. the path of the hand and the movement of each finger, rather than simply the path of a pen.

4.6.3.1 *Physical Gesture Recognition*

At the crudest level, one could attach some form of light- or sound-emitting device to an operator's finger, and ask them to direct the output towards specific objects. This would function somewhat like a remote touchscreen. Alternatively, one could attempt to track to movement of specific parts of the hand, by optically tracking of fingertip movement or by analysis of video images. More commonly, people employ glove-based devices (Sturman and Zeltzer, 1994). Some recent developments involve the use of eye-movement as a means of selecting objects (Istance and Howarth, 1994).

At present there is little in the way of empirical comparison between these different devices. One could imagine them as extensions of current direct manipulation practices, especially when considering the use of glove-based devices in virtual reality environments. It would be fair to say that current technology makes possible demonstrator systems, but not implementable technology. Thus, this particular form of technology, while representing a means of selecting and acting on objects in the future, is not immediately viable. Furthermore, one might feel that the redesign of control rooms required to exploit this technology will not warrant consideration, and that the technology will be more suitable to the constrained, physical space of aircraft cockpits. However, given the advances in miniature displays, there are systems which can be used for maintenance work on site, providing access to computers in the control room. The devices may well require some form of gesture-based interaction.

4.6.3.2 *Pen-based Gestures*

Pen-based computing has, in the last three years or so, moved quickly into commercial fact, with a number of products on the market which support either pen-based gestures, i.e. pointing to objects and entering commands in the form of 'squiggles', or handwriting recognition. Basically, pen-based systems provide the user with electronic paper, which can be written, drawn or scribbled on, and with a means by which objects can be selected and manipulated. Given that research has found pen-based gestures easier to learn and use than keystrokes (Wolf, 1992), one might anticipate an explosion in the technology very soon. However, such optimism needs to be tempered with an awareness of the limitation of the technology. Frankish *et al.* (1994) note that unconstrained writing often leads to unacceptably high levels of error, and that such activity often arises from users having a poor understanding of how the device 'recognises' handwriting. Thus, pen-based systems would appear to be most suited to tasks which do not require extensive handwriting, but for conventional pen-and-paper tasks, such as memos, sketchpads, note-taking and diaries. One could imagine applications relating to general 'house-keeping' functions in the control room and to maintenance work on plant, using pen-based systems.

4.6.4 Speech

With automatic speech recognition (ASR), an operator can identify objects (by naming them) and enter commands and information without the requirement of being in close proximity to the work station. Thus, ASR offers a highly flexible interaction technology, permitting maximum operator mobility. There has been much interest in the technology in military circles, and several applications in industry. The US Electrical Power Research Institute (EPRI) commissioned an investigation (EPRI, 1986) into the potential benefits to be obtained from the use of speech technology in control rooms. In the preamble to this report, the authors propose,

> Voice devices now on the market could boost efficiency and reduce error and fatigue . . . The potential for . . . voice technologies is greatest . . . where workloads are heavy and responses must often be immediate.

Given that this report was compiled at a time when ASR performance was notably inferior to current standards, one could suggest that the proposal is even more viable today. The report concludes that ASR has the potential to be a useful medium for supervisory control. This conclusion is by no means unequivocal with later research both supporting (Baber and Usher, 1989) and questioning it (Mitchell and Forren, 1987).

There have been some studies investigating the performance of ASR relative to other input devices. However, many of these studies are flawed in that their experimental design fail to control for different performance characteristics of the technologies (Baber, 1993). Having said this, ASR would appear to offer some benefits as a means of selecting objects (Damper, 1994), and for entering commands strings which are either long or complex (Welch, 1977; Baber, 1991a).

4.7 Conclusions

The aim of this chapter was to make the reader aware of the effects of different input devices on control room performance, and to raise questions of how the physical side of operator activity is affected by the tools with which the operator is provided. The chapter covers some of the pros and cons of different devices, and provides some indication of selection criteria for system designers. In this chapter, we have seen the importance of considering input devices in system design and some of the consequences of poor design. We have seen that some of the most common input devices are not necessarily the most efficient, although they do seem to represent reasonable compromise solutions when bearing in mind other constraints. We have glimpsed into the near future to see what type of technology could be employed in the next decade of control room designs. Prediction is always a risky business, but I would anticipate a rapid growth in the use of pen-based systems for the nuclear industry, both for maintenance work and in the control room, and some semi-serious flirtation with speech, although it is unlikely that working applications of ASR-driven control rooms will be realised for some time.

References

BABER, C. (1991a) *Speech Technology in Control Room Systems: a Human Factors Perspective*. Chichester: Ellis Horwood.

(1991b) Human factors aspects of automatic speech recognition in control room environments, *IEE Colloquium on Systems and Applications of Man–Machine Interaction using Speech I/O*. London: March, IEE Digest 1991/066.

(1993) 'Novel' techniques for HCI in the control room of the future, *Proceedings of the Human Factors of Nuclear Safety*. London: IBC.

BABER, C. & USHER, D. M. (1989) *ASR in Grid Control Rooms: part 1. Feasibility Study*, Bristol: CEGB Report TD/STM/89/10024/N.

BABER, C., USHER, D. M., STAMMERS, R. B. & TAYLOR, R. G. (1992) Feedback requirements for ASR in the process control room. *International Journal of Man Machine Studies*, **37**(6), 703–19.

BALL, R. G., NEWTON, R. S. & WHITEFIELD, D. (1980) Development of an off-shore high-resolution, direct touch input device: the RSRE touchpad. *Displays*.

BARKER, D., CAREY, M. S. & TAYLOR, R. G. (1990) Factors underlying mouse pointing performance. In Lovesey, E. J. (ed.), *Contemporary Ergonomics 1990*. London: Taylor & Francis.

CARD, S. K., ENGLISH, W. K. & BURR, B. J. (1978) Evaluation of mouse, rate-controlled isometric joystick, step keys and text keys for text selection on a CRT. *Ergonomics*, **21**, 601–13.

CAREY, M. (1985) *The Selection of Computer Input Devices for Process Control*. Stevenage: DTI, Warren Springs Lab. Report no. 101/41/4.

COMMERFORD, R. (1984) Pointing devices innovations enhance user-machine interfaces. *Electronic Design News*, **26**, 54–65.

DAMPER, R. I. (1994) Speech as an interface medium: how can it best be used? In Baber, C. and Noyes, J. M. (eds), *Interactive Speech Technology*. London: Taylor & Francis.

DAVIE, C., KATIFI, H., RIDLEY, A. & SWASH, M. (1991) 'Mouse'-trap or personal computer palsy, *Lancet*, **338**, 832.

DIX, A., FINLAY, J., ABOWD, G. & BEALE, R. (1993). *Human–Computer Interaction*. New York: Prentice Hall.

ENGELBART, D. (1973) Design considerations for knowledge workshop terminals, *Proceedings of NCC AFIPS*, **42**, 9–12.

EPRI (1986) *Speech Recognition and Synthesis for Electric Utility Dispatch Control Centres*. Palo Alto, CA: Electric Power Research Institute Report EL-4491/report 2473-1.

Fleischer, A. G. & Becker, G. (1986) Free hand-movements during the performance of a complex task. *Ergonomics*, **29**(1), 49–63.

Foley, J. D., Wallace, V. L. & Chan, P. (1984) The human factors of computer graphics interaction techniques. *IEEE Computer Graphics and Applications*, **4**(11), 13–48.

Franco, G., Castelli, C. & Gatti, C. (1992) Tenosinovite posturale da uso incongruo di un dispositivo di puntamento. *Medicina del Lavoro*, **83**, 352–5.

Frankish, C. F., Morgan, P. & Noyes, J. (1994) Pen computing in context. *Proceedings of Designing Future Interaction*. Warwick University, April.

Goodwin, N. C. (1975) Cursor positioning on an electronic display using light pen, light gun or keyboard for three basic tasks. *Human Factors*, **17**, 289–95.

Haller, R., Mutschler, H. & Voss, M. (1984) Comparison of input devices for correction of typing errors in office systems, *Proceedings INTERACT '84*. Amsterdam: Elsevier.

Hesse, F. W. & Hahn, C. (1994) The impact of interface-induced handling requirements on action generation in technical system control. *Behaviour and Information Technology*, **13**(3), 228–38.

Istance, H. & Howarth, P. A. (1994) Input device emulation using an eye-tracker for interaction with graphical user interfaces, *Proceedings of Designing Future Interaction*, Warwick University, April.

Kemeny, J. (1979) *The Need for Change: the Legacy of Three Mile Island*. Washington, D.C.: Government Printing Office.

MacKenzie, I. S., Sellen, A. & Buxton, W. (1991) A comparison of input devices for elemental pointing and dragging tasks, *Proceedings CHI'91*, 161–6.

Milner, N. P. (1988) A review of human performance and preferences with different input devices to computer systems. In Jones, D. M. and Winder, R. (eds), *People and Computers IV*. Cambridge: Cambridge University Press.

Mitchell, C. M. & Forren, J. F. (1987) Multimodal user input to supervisory control systems: voice augmented keyboard. *IEEE SMC-17* (4), 594–607.

Moray, N. (1988) Mental workload since 1979. In Oborne, D. J. (ed.), *International Review of Ergonomics 2*, pp. 123–50. London: Taylor & Francis.

Norman, D. A. (1988). *The Psychology of Everyday Things*. New York: Basic Books.

Noyes, J., Baber, C. & Steel, A. (1994) Ergonomic surveys of VDTs and Manufacturing Workstations: a Case Study, *Ergonomics Society and IOSH: Working Together to Meet the Challenge of the New EC Regulations*, Bath, October.

Perrow, C. (1984) *Normal Accidents: Living with High Risk Technologies*. New York: Basic Books.

Rasmussen, J. & Goodstein, L. P. (1988) Information Technology and Work. In Helander, M. (ed.), *Handbook of Human–Computer Interaction*. Amsterdam: North Holland.

Ritchie, G. & Turner, A. (1975) Input devices for interactive graphics. *International Journal of Man–Machine Studies*, **7**, 639–60.

Sanders, M. S. & McCormick, E. J. (1992) *Human Factors in Engineering and Design*. New York: McGraw-Hill.

SHNEIDERMAN, B. (1992) *Designing the User Interface: Strategies for Effective Human–Computer Interaction*. Reading, MA: Addison-Wesley.

STURMAN, D. J. & ZELTZER, D. (1994) A survey of glove-based input. *IEEE Computer Graphics and Applications*, **14**(1), 30–9.

THOMAS, C. M. & MILAN, S. (1987) Which input device should be used with interactive video? *Proceedings INTERACT '87*. Amsterdam: Elsevier.

TRÄNKLE, U. & DEUTSCHMANN, D. (1991) Factors influencing speed and precision of cursor positioning using a mouse. *Ergonomics*, **34**(2), 161–74.

UMBERS, I. (1979) Models of the process operator. *International Journal of Man-Machine Studies*, **11**, 263–77.

USHER, D. M. (1982) A touch sensitive VDU compared with a computer aided keypad, *Displays*, **5**, 157–61.

WELCH, J. R. (1977) *Automatic Data Entry Analysis*. Rome, NY: Rome Air Development Centre Report RADC-TR-77-306.

WHITEFIELD, A. (1983) Pointing as an input technique for human-computer interaction. *Future Input Techniques for Man–Machine Interaction*. London: IEE Colloquium, April.

WOLF, C. G. (1992) A comparative study of gestural, keyboard and mouse interfaces. *Behaviour and Information Technology*, **11**, 13–23.

CHAPTER FIVE

Operator reactions to alarms: fundamental similarities and situational differences

NEVILLE STANTON

Department of Psychology, University of Southampton

5.1 Introduction

The general aims of the survey were to obtain information to enable an assessment of control desk engineers' (CDEs) reactions to their alarm system. This information was sought to explore the differences in alarm handling in different industries. This is a valid area for human factors research, as Lees (1974) wrote:

> Alarm systems are often the least satisfactory aspects of process control system design. There are a number of reasons for this, including lack of clear design philosophy, confusion between alarms and statuses, use of too many alarms, etc. Yet with the relative growth in the monitoring function of the operator, and indeed of the control system, the alarm system becomes increasingly important. This is therefore another field in which there is much scope for work.
>
> *Source:* Lees (1974, p. 418).

In particular, the survey was conducted to find out what differences exist in the way in which alarm information is used in different industries. By comparing different industries, it was hoped that both differences and similarities would become apparent. This begs the question: 'is nuclear power a special case?' In many ways the nuclear industry is special: it is at the forefront of technological development, it has (potentially) high risks associated with non-contingent events (e.g. release of radioactive material), it is complex and requires a high degree of training for operators. In other ways nuclear power is similar to other human supervisory control environments: the process is monitored via a central control room, large amounts of information are presented which the operator is required to assimilate, workload is highly variable, operators are required to monitor and interact with systems processes, a layer of devices and automatic systems protect the plant (and sometimes impede the progress of the operator, etc.). Given these apparent similarities and differences, it was hypothesised by the author that there may be some fundamental similarities in the use of alarm systems across industries, as well as industry-specific differences. Specific objectives of the survey were:

- to elicit the CDEs' definitions of the term 'alarm';
- to examine the CDEs' alarm-handling activities;
- to get information on problems with the alarm system.

An 'alarm-handling questionnaire' was constructed in line with the three main objectives. The design was in accordance with an approach proposed by Youngman (1982). Content analysis was employed for analysing the questionnaire data (Oppenheim, 1992). The main phases of this technique are: data reduction, classification, and inference. The data were reduced by transcribing the open-ended items and compiling the closed data. The open-ended items were then classified by looking for repetition in responses. With all of the responses collated, inferences were drawn which are presented with each of the studies and again in the conclusions. All of the phases were carried out by the author, although the classification system was checked by colleagues.

The sites that responded to the questionnaire were a chemical manufacturer, a confectionery company and a nuclear power generation company (all in the UK). All sites were similar in the respect that they essentially had complex plant that was monitored from a central control room by skilled operators. The purpose of this chapter is to consider each of the sites with respect to the objectives of the questionnaire and to examine which elements were common to all sites and which were not.

The results of the surveys are presented within the chapter, with conclusions comparing the sites at the end.

5.2 The Questionnaire Study

The respondents from the nuclear utility were 11 self-selected RDEs (Reactor Desk Engineers) from a Magnox power station in the UK. Their central control room experience ranged from a few months to over 18 years. The respondents from the confectionery company were eight self-selected CDEs from a process plant in the UK. Their central control room experience ranged from 7 to 38 years. All respondents were assured that their answers were to remain confidential, and therefore numbers rather than names were assigned to individual questionnaires. The respondents from the chemical manufacturer were three teams of CDEs (five CDEs in each team) from a process plant in the UK. Their central control room experience ranged from

1 to 5 years. The respondents were requested to complete the questionnaire by the assistant fire, safety and security officer. Unfortunately, due to some misunderstanding the questionnaires were answered by a whole team together, rather than individually. This probably led to some censorship of thoughts. Also complete anonymity was not assured, because the assistant fire, safety and security officer wished to 'check' the questionnaires before they left the establishment.

5.2.1 Definition

The questionnaire sought to gain the plant operator's definition of the alarm system(s) with which they are familiar by asking three questions:

1 What did they think the alarm system was designed for?
2 What information did they get from alarms?
3 How did they use alarms in their daily activities?

Content analysis of their responses resulted in four major classifications:

1 to gain attention;
2 to provide warning;
3 to support monitor and control actions;
4 to provide factual information.

To aid clarity, the frequency with which these answers were given is presented in a bar chart in Figure 5.1.

The answers suggested that the majority of RDEs from the nuclear utility believe that the alarm system:

- was designed to attract their attention;
- provides warning information;
- is used in accordance with monitoring and control activities.

The majority of CDEs from the confectionery company believed that the alarm system:

- was designed to attract their attention and to warn of faults;
- provided warning information about the nature and location of faults;
- provided information to initiate fault-finding activities.

Figure 5.2 illustrates the frequency of responses.

All of the CDEs from the chemical manufacturer believed that the alarm system:

- was designed to alert them to abnormal conditions;
- provided information on the nature and location of the problem;
- was used to support prediction and control of the plant state.

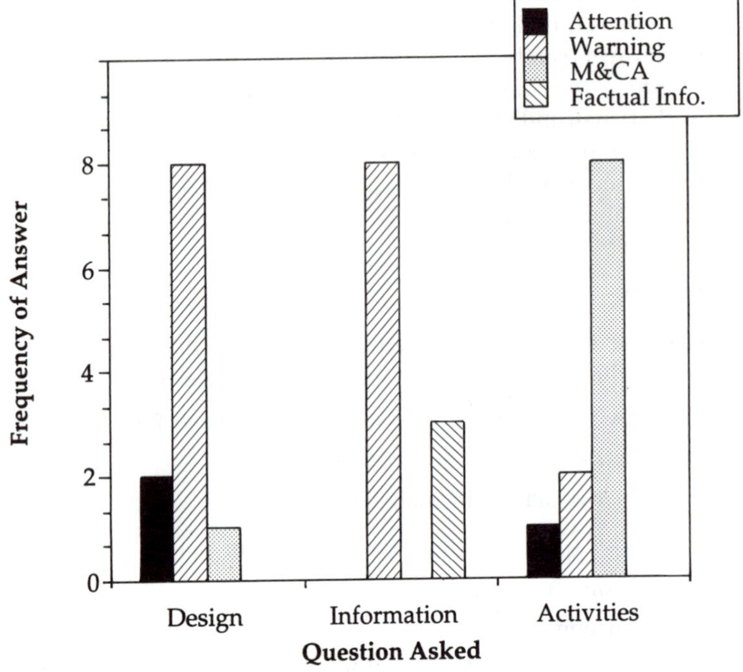

Figure 5.1 Frequency of answer to items 2, 3 and 4 from the nuclear utility.

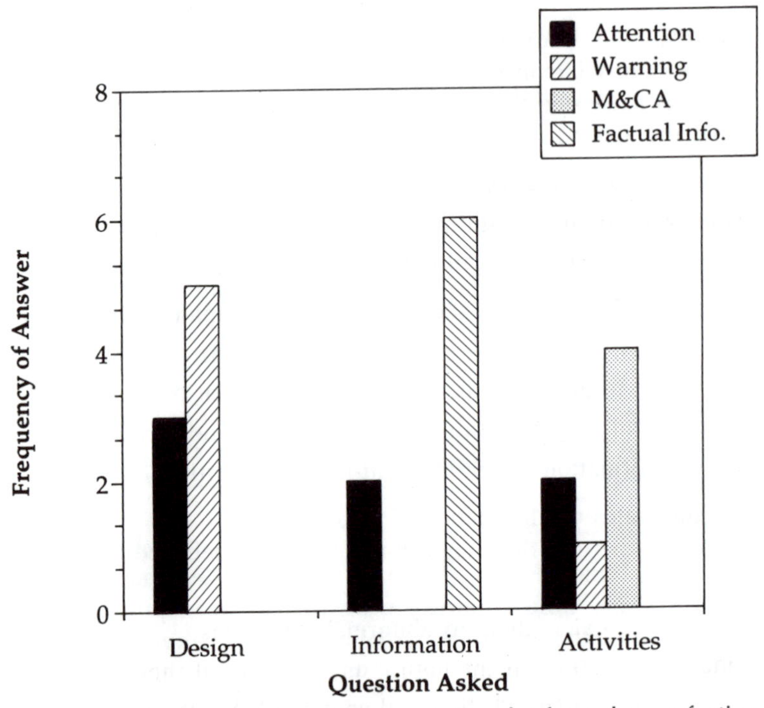

Figure 5.2 Frequency of answer to items 2, 3 and 4 from the confectionery company.

This unanimity in the responses from the CDEs at the chemical manufacturer was probably due to the fact that teams, rather than individuals, were responding.

Overall, these responses suggest that there are some slight differences in the perception of the purpose of the alarm system at different sites.

5.2.2 Alarm Handling

This section of the questionnaire attempted to explore CDEs' alarm-handling activities in greater detail. It was hoped to draw a distinction between what they do in routine situations, and what they do in critical situations.

5.2.2.1 *Priorities*

This section sought to discover what the operational priorities of the CDEs were, and to see if there was any consensus among the respondents. Item 5 asked respondents if it was more important to them to rectify an alarm that had already occurred or to prevent an alarm from occurring. Most responses went for the latter option, suggesting that the majority of this sample feel that it is more important to prevent rather than cure an alarm. This is illustrated in Figure 5.3.

Item 7 sought information on how regularly plant operators deliberately scanned for alarms if unprompted. Figure 5.4 shows that there is quite a spread in opinion to this question, and this may largely depend on individual plant operators' preferred working practices. The respondents from the chemical manufacturer reported that they never scanned for alarms, and always waited for the audible cue before attending to the alarm panel.

The next item was concerned with the operational goals of the plant operators in their daily activities. What is their order of priority, and where does alarm handling fit into their task goals? They were requested to rank each of the following items in order of importance to them. The results show some consistency, and the overall rankings for the nuclear utility, the confectionery company and the chemical manufacturer are shown in Figures 5.5, 5.6 and 5.7 respectively.

This may reflect a very strict training in which this order of priority has been drilled in to each member of the team across a range of industries.

5.2.2.2 *Modelling CDE Behaviour*

Many theoretical models have been developed to explain control-room behaviour, but very little empirical evidence has been presented in their support, outside of the experimental laboratory. This questionnaire aimed to get some insight into plant operators' activities by the use of critical incident technique: to ask them what they did in response to routine alarms,

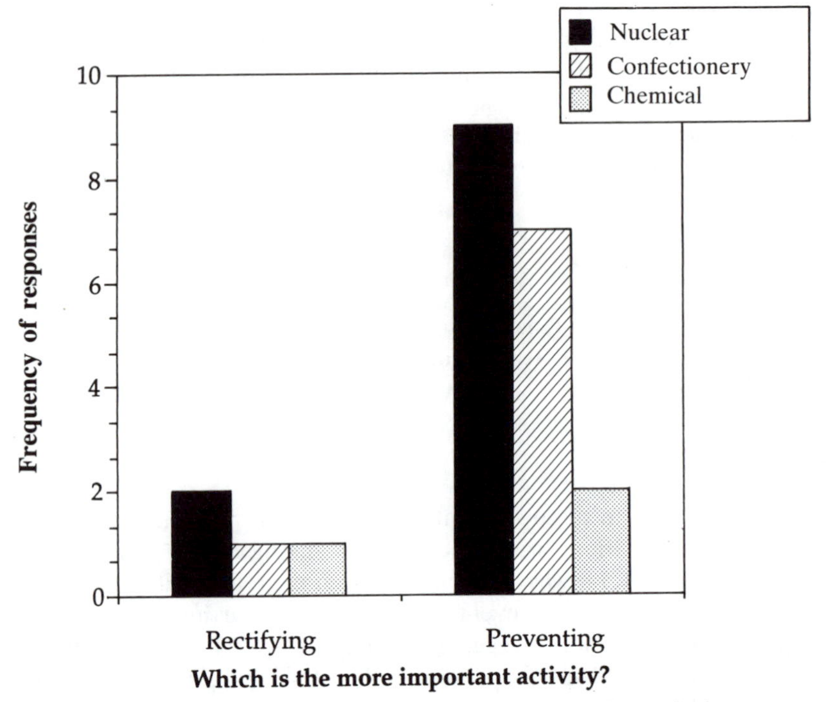

Figure 5.3 The importance of preventing alarms or rectifying faults.

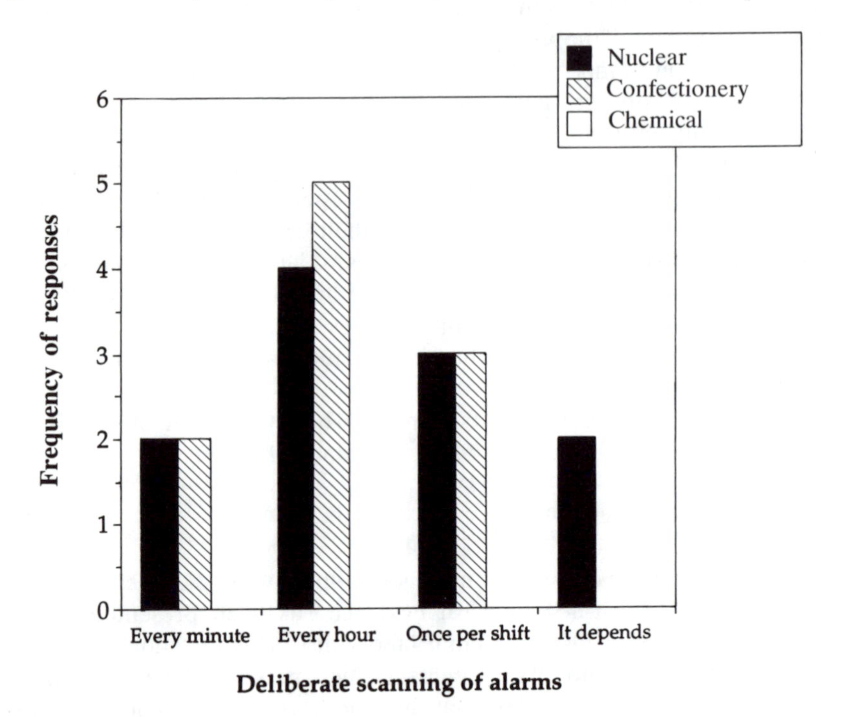

Figure 5.4 The frequency with which CDEs scan for alarms.

1. Ensuring safety
2. Following operational procedures
3. Meeting loading targets
4. Increasing efficiency
5. Reducing the number of alarms.

Figure 5.5 CDEs order of priority at the nuclear utility.

1. Ensuring safety
2. Following operational procedures
3. Meeting loading targets
3. Increasing efficiency
3. Reducing the number of alarms.

Figure 5.6 CDEs order of priority at the confectionery company.

1. Ensuring safety
2. Following operational procedures
3. Increasing efficiency
4. Meeting loading targets
5. Reducing the number of alarms
6. Maintaining morale.

Figure 5.7 CDEs order of priority at the chemical manufacturer.

and what they did in response to critical events. The following model was constructed from the answers to item 11 and further supported by the answers to item 18 which highlights the difference between routine incidents involving alarms and critical incidents involving alarms. The responses are shown in Figure 5.8 (the nuclear utility), 5.9 (the confectionery company) and 5.10 (the chemical manufacturer).

In more critical situations additional activities are performed. The CDE will 'try to find out why the alarm has initiated through investigative procedures'. This is the main activity that distinguishes critical from routine alarm handling shown in Figure 5.8. In all other respects the activities may be classified under the same headings.

Additionally they report that in critical incidents involving alarms they perform an additional activity of 'investigating why the alarm went off' before attempting to rectify the fault. This is the main activity that distinguishes critical from routine alarm handling shown in Figure 5.9.

However, if the alarm is associated with a 'critical' situation, additional activities are carried out. These are to 'check the plant, inform the supervisor and initiate the appropriate recovery sequence'. This is the main activity that distinguishes critical from routine alarm handling shown in Figure 5.10.

Further investigation into the alarm handling was promoted by items 6 and 12. Item 6 asked plant operators to give the approximate percentage of time spent in each of the stages whilst handling alarms. The stages presented in Figure 5.13 were derived hypothetically from the literature and through

1. Observe the alarm.
2. Accept the alarm.
3. Decide if it's important.
4. Take the necessary action to correct the fault condition.
5. Monitor the situation closely until stable again and alarms are reset.

Figure 5.8 Alarm handling at the nuclear utility.

1. Observe the alarm.
2. Press the 'acknowledge' button.
3. Decide on the importance of the alarm.
4. Find the fault.
5. Rectify the fault.
6. Monitor the panel.

Figure 5.9 Alarm handling at the confectionery company.

1. Hear the alarm.
2. Mute the audible alarm.
3. Display the alarm.
4. Accept the alarm.
5. Assess the situation.
6. Take the appropriate action.
7. Monitor alarm condition until out of alarm status.

Figure 5.10 Alarm handling at the chemical manufacturer.

discussion with colleagues. This is presented below in Figure 5.11. Recent research into models of alarm handling have come to question this simplistic stage-based model, to propose a process model of human alarm handling. This development is presented in Section 5.4: A model of alarm handling.

It is interesting to note that the largest percentage of effort is spent in the assessment stage (see Figure 5.12).

Item 12 relates to one stage of alarm handling, namely diagnosis. Plant operators were asked how they diagnose faults based on alarm information. As Figure 5.13 illustrates, this is mainly based on past experience, but sometimes also includes the pattern of alarms and order of occurrence. The suggestion that past experience plays the major part in diagnosis places an important emphasis on the training of plant operators to present to them a wide range of conditions, some of which may only be encountered very infrequently.

Bainbridge (1984) suggested that past experience can play a major part in diagnosis, but that this could also be a potential source of error, as it may be misleading. It puts an emphasis on encountering a wide range of plant conditions if it is to be a successful diagnosis strategy.

Item 16 asked plant operators what aspects of the information presented to them in the CCR hindered the diagnosis of the cause of the alarm. The data as presented in Figure 5.14 may appear conflicting at first sight, i.e. the

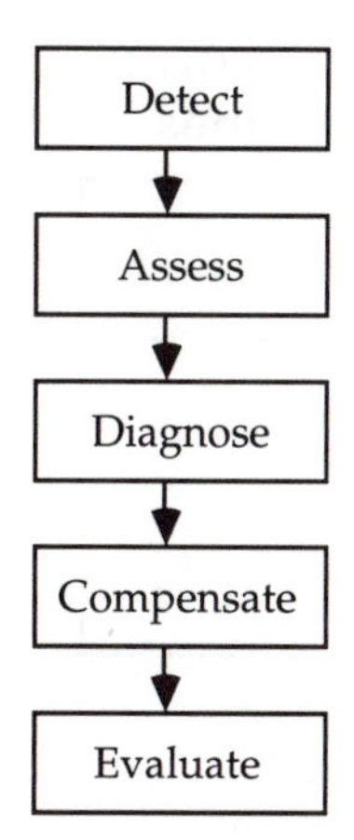

Figure 5.11 A 'stage' model of alarm handling.

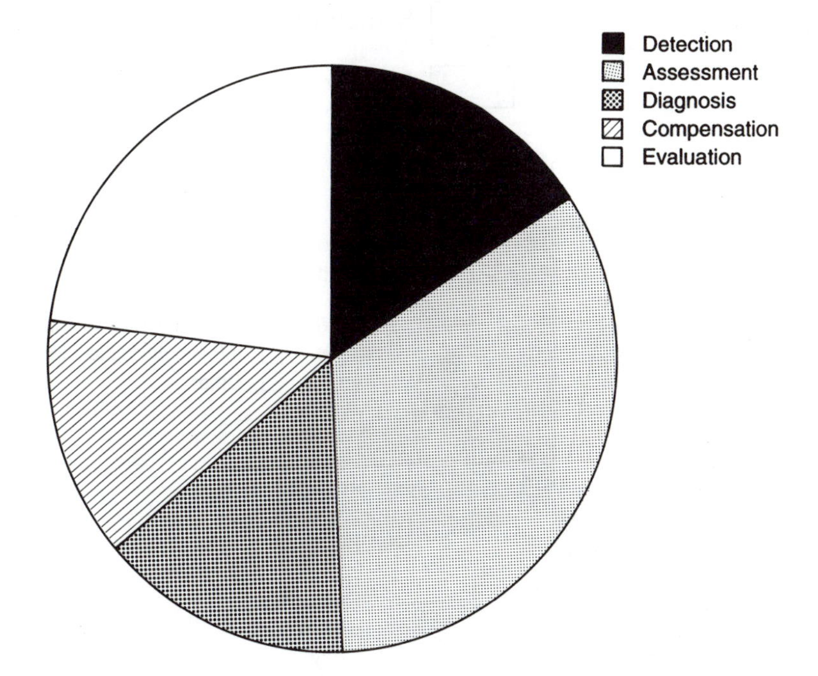

Figure 5.12 Percentage of perceived distribution of effort in stages whilst alarm handling.

responses were both too much and too little information. However, when considered with other responses (difficulties in finding and interpreting information), it may be inferred that there are problems with the appropriateness of the information presented. For instance there may be too much of the wrong sort and too little of the right sort, making the process of

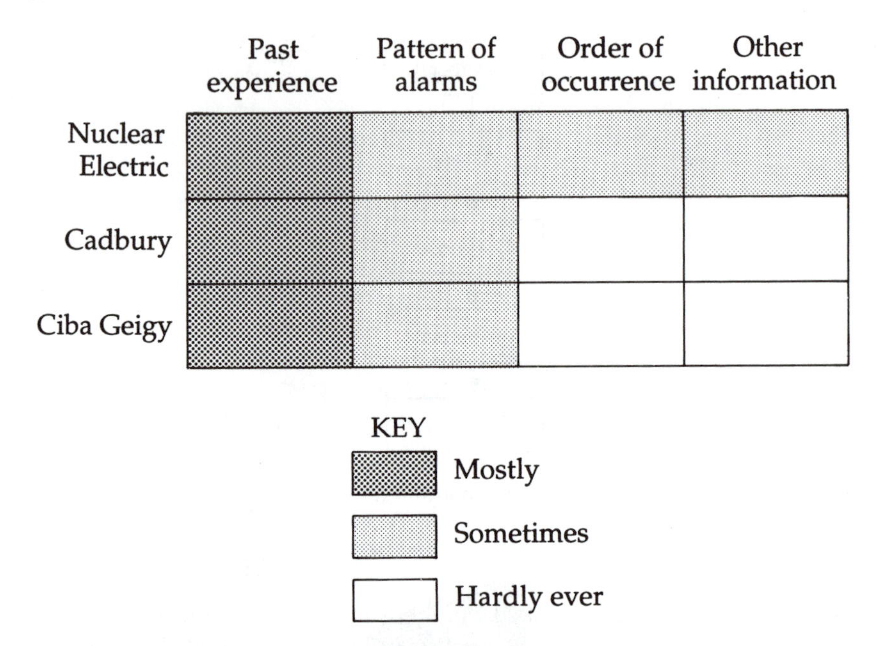

Figure 5.13 Diagnosis from alarm information.

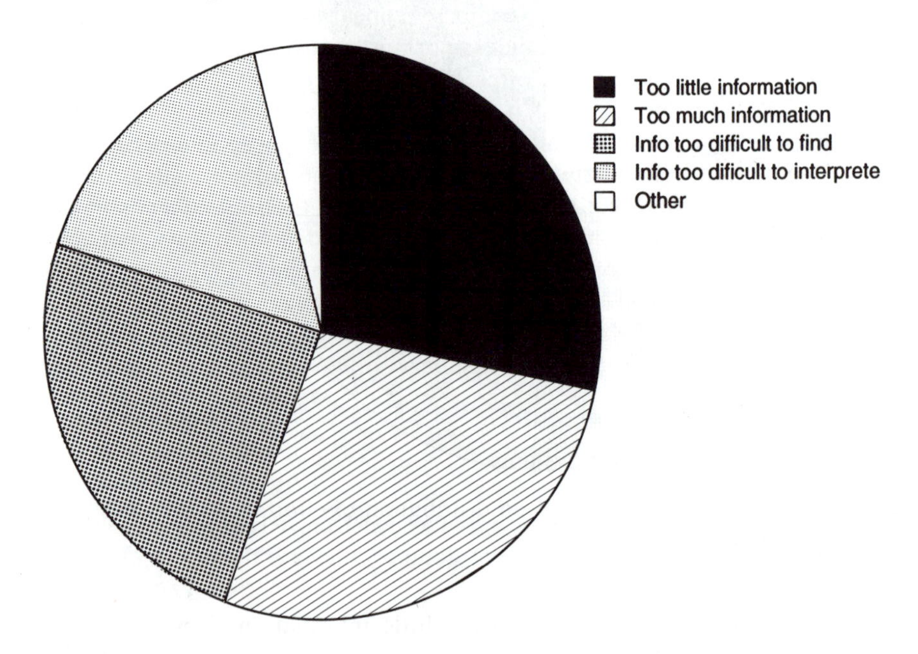

Figure 5.14 Problems in diagnosis.

determining causality difficult. This demonstrates that the alarm system does not appear to be supporting the investigative processes of the plant operators during the diagnosis stage of alarm handling.

The responses from the confectionery company suggest that there is too little information to diagnose from, and what information is made available is difficult to interpret. This is an often cited problem of using essentially 'raw' plant data (Goodstein, 1985). The responses from the chemical manufacturer suggest that there is both too little and too much information to diagnose from. This suggests that the information is not of the right type to aid diagnosis.

Item 13 asked plant operators to estimate the percentage of decisions that they considered to be rushed. Just over half of the RDEs at the nuclear utility thought this accounted for about 20% of their decisions. The confectionery company CDEs thought this accounted for most of their decisions. The chemical manufacturer CDEs thought this accounted for about 20% of their decisions. This adds to the plant operators' problems; not only do they have difficulties in finding the relevant information, but they are occasionally expected to find it too quickly.

Related to item 13, item 14 asked what the nature of these rushed decisions related to. The responses from the nuclear utility appear to be related to (in order of frequency) product, safety, time schedules and work practices. The responses from the confectionery company appear to be related to (in order of frequency) product, time schedules, work practices and safety. The CDEs from the chemical manufacturer offered no opinion on this item.

Item 17 asked plant operators if they ever felt under pressure to clear an alarm. The responses from the nuclear utility and the confectionery company were roughly split in half between 'yes' and 'no'. Four respondents from the nuclear utility reported that they felt pressure from the shift supervisor, but other sources were also mentioned. These included their own personal working practice. Two of the confectionery company CDEs' responses reported that they felt pressure from the shift supervisor, but other sources were also mentioned. Two teams from the chemical manufacturer reported that they felt pressure from the shift supervisor.

5.2.3 Problems with Alarm Systems

The final aim of the questionnaire was to elicit from RDEs what problems they encountered, and what might be done to alleviate them.

5.2.3.1 *Missed Alarms*

The responses to item 9 made it reasonable to suppose that alarms are missed, and 5 examples occurred recently. Reasons encountered by RDEs for missed alarms are illustrated in Figure 5.15.

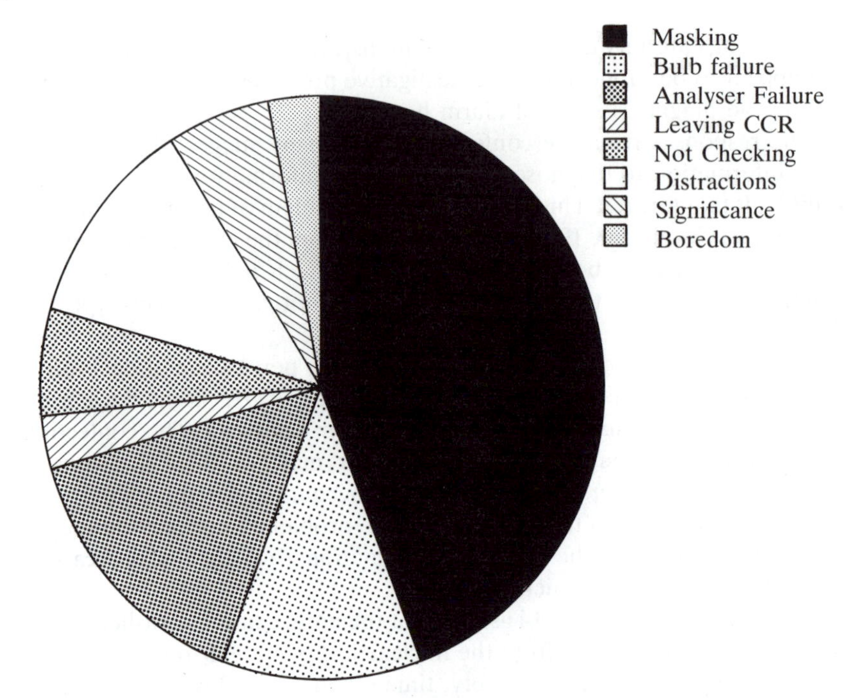

Figure 5.15 Frequencies with which causes for missed alarms were mentioned at the nuclear utility.

The major reported reason for missing alarms given was the masking phenomenon, where the alarm was masked by the presence of other alarms. This perhaps gives some insight into why the assessment stage of alarm handling takes up so much of the RDEs' time. Nearly half of the reasons given for masking were related to the number of alarms present; the rest of the reasons were as follows: non-urgent, non-related, repeating, standing, and shelved alarms.

Failures of the alarm system (bulb failures and alarm analyser failure and reset) contributed to approximately a quarter of reported reasons for missing alarms. It seems that an improvement in the robustness of the system could significantly reduce the number of alarms missed this way but, ironically, might contribute to the masking phenomenon.

The rest of the reasons given were: leaving the CCR, not checking at start of shift, distraction in the environment (people and tasks), the significance not appreciated and boredom with the tasks.

The responses to item 9 from the confectionery company made it reasonable to suppose that alarms are missed, and four examples occurred recently. The main reasons for alarms being missed were related to inadequate training, not being present in the control room, jamming the audible alarm, system failure and the distraction of attending to other tasks. These responses are illustrated in Figure 5.16.

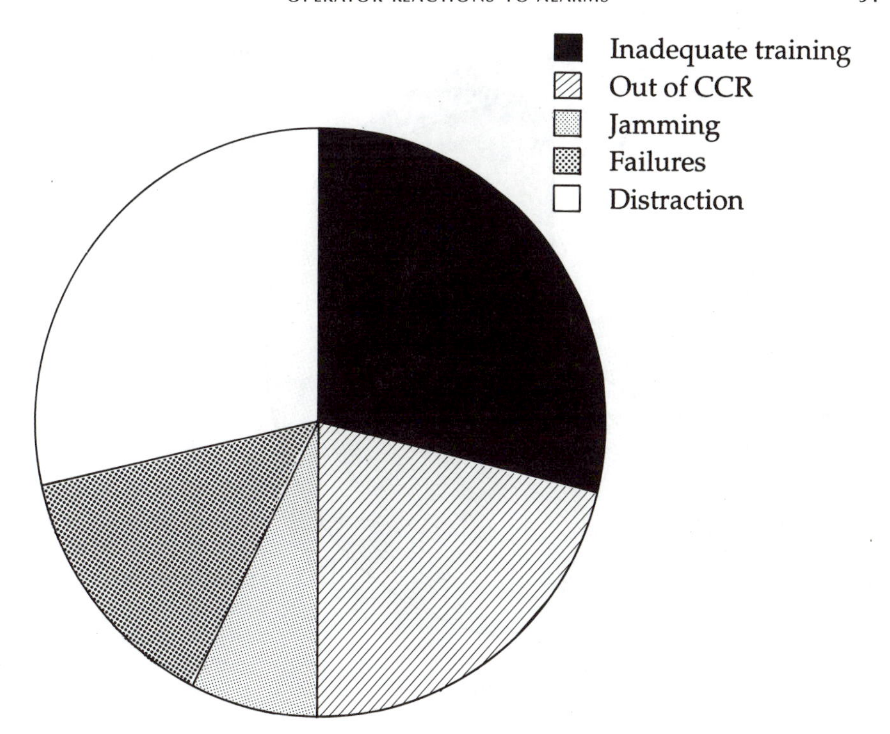

Figure 5.16 Frequencies with which causes for missed alarms were mentioned at the confectionery company.

If alarms are being missed, then it follows that they have not been detected. This may provide a starting-point for consideration of how to support the detection stage of alarm handling. If the alarm has not been detected, the information cannot be passed to the subsequent alarm-handling stages.

No missed alarms were reported by the chemical manufacturer teams, but this is difficult to believe. The CDEs' awareness that the questionnaire was going to be seen by members of the management team before leaving the site may well have led them to be cautious about what they were prepared to reveal.

5.2.3.2 *Improvements in Design*

Nine of the eleven RDEs from the nuclear utility who responded to the questionnaire thought that their alarm system could be improved. Responses to item 19 are illustrated in Figure 5.17. These ideas were largely reiterated in response to item 22. Starting at the top right of Figure 5.17, the suggestions were: highlight initial cause (to aid *diagnosis*), suppress irrelevant information, improve clarity of presentation, increase time to read information (to

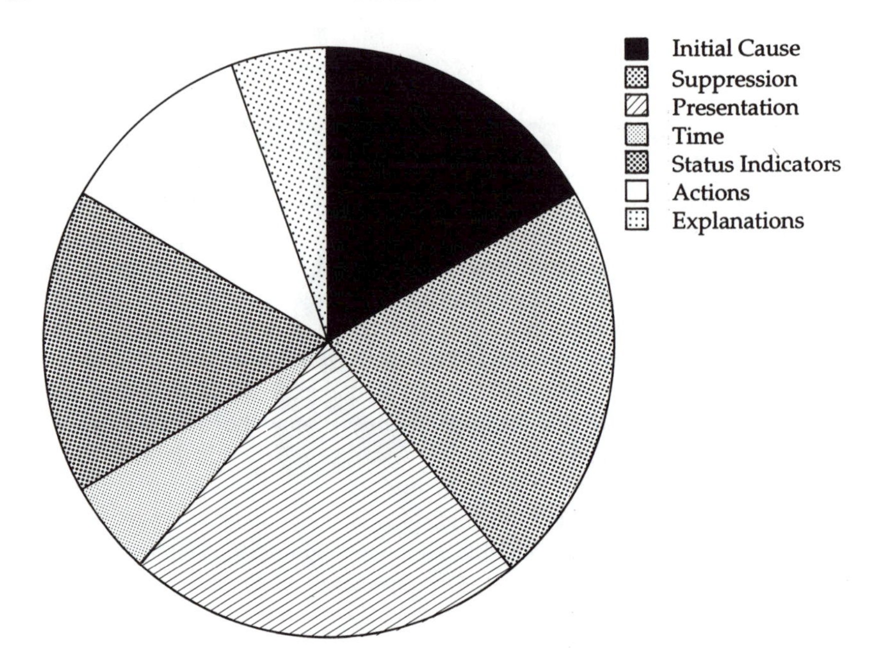

Figure 5.17 Improvements in design.

aid *detection*, *assessment* and *evaluation*), not using alarms as 'normal' (within given context) status indicators, provide clear reference of what action is required on receipt of alarm (to aid *compensation*) and an improved means of communicating the situation when handing over the control desk.

This may figure as the RDEs' 'wish list' for their alarm information system, but none of these requests seems unreasonable. However, their implementation may not be easy, and might require a redesign of the whole system. The issues presented here are supported by Andow and Roggenbauger (1983) who suggest that the alarm system is often a source of problems during operation. As Sorkin (1989) notes, 'liberal response criteria' (the notion that a missed alarm costs more than a false alarm) for alarm systems can produce high alarm rates and subsequent reductions in performance. In addition, the manner in which information is presented can make it difficult to determine the underlying condition of the plant.

Four of the eight of the confectionery company CDEs who responded to the questionnaire thought that their alarm system could be improved. Suggestions for change include: aiding the fault-finding process, keeping a permanent record of alarms as a 'look-back' facility, and the improved use of coding (i.e. tones) to speed up the identification and classification process.

Ironically none of the teams from the chemical manufacturer thought that their alarm system could be improved. This is a little incongruous considering the points they raised in previous sections.

5.2.3.3 *Resistance to Change*

In response to item 20 only one RDE said that he would resist change: 'Because I know this system well'. However, another RDE suggested that although he would not resist change he said 'Not sure what new system could offer' and 'Good systems are difficult to design'. Six CDEs had experienced other systems (Item 21) and only two preferred the present system. It may be inferred from the comments that early CRT-based displays were poor (slow, poor quality and low on detail), but annunciator systems were good (spatial information, references, clear and concise). In response to item 20, only one confectionery company CDE said that he would resist change: 'Because I know this system well'. One CDE had experienced another system (Item 21) but preferred the present system. It may be inferred from the comment that the early alarm displays were poor (low on detail), but that the current system gives more accurate information. None of the teams from the chemical manufacturer said that they would resist change, but two qualifying statements were added. The first comment was: 'resistance is futile', which might suggest that the relationship between management and the CDEs leaves something to be desired. The second comment was: 'if it was to improve', which suggests that the CDEs are aware that not all changes are necessarily improvements.

5.3 Summary

The responses in the questionnaires returned from the three industrial plants are summarised below. In all cases the alarm systems were VDU based. The 'reactions' of CDEs were the reports of:

- alarms as aids in the central control room (CCR);
- priorities in the CCR;
- problems related to alarms;
- problems related to fault diagnosis;
- suggested improvements to the alarm system.

Table 5.1 illustrates that the use of alarms as aids in the CCR varies in the emphasis on activities they support. The order of priorities was found to be similar across industries: ensuring safety being rated as the most important and reducing alarms typically rated as the least important. Problems associated with the alarm systems appear to be largely context specific, whereas problems related to diagnosis appear to be similar. Finally, suggested improvements appear to be related to the particular implementation of the alarm system. These studies provide practical illustrations of shortcomings of alarm systems that may be industry-wide, and not just related to one particular implementation.

Table 5.1 Summary of questionnaire responses

Alarms	Chemical	Nuclear	Confectionery
Presentation	VDU and audible tone	VDU and audible tone	VDU
Aids	Support, prediction and control	Support, monitoring and control	Fault finding
Priorities	Safety SOPs Efficiency Targets Alarms	Safety SOPs Targets Efficiency Alarms	Safety SOPs Efficiency Targets Alarms
Problems	None	Masking Technical failure Leaving CCR Other	Inadequate training Not in CCR Jamming System failure Distractions
Diagnosis	Too much info Too little into	Too much info Too little into Info difficult to find Info difficult to interpret	Too little info Info difficult to interpret
Improvements	None	Highlight cause Better presentation Suppression	Aid fault finding Record alarms Aid analysis

However, despite differences in the implementations of alarm systems (as illustrated by the different reactions to alarm systems illustrated in Table 5.1) there do appear to be distinct similarities in the way in which CDEs describe how they deal with alarms. The model derived from the content analysis of the question asking what operators typically do in response to alarms and what they do in critical situations is more complex than the model originally conceived prior to the survey. Before these studies were undertaken, a simple five-stage model of alarm handling was conceived, comprising: detection, assessment, diagnosis, compensation and evaluation.

Little advancement has been made upon the model of alarm handling proposed by Lees (1974) over two decades ago. This model comprised three stages: detection (detecting the fault), diagnosis (identifying the cause of the fault) and correction (dealing with the fault). This model appears to be very similar to the process model put forward by Rouse (1983) comprising detection (the process of deciding that an event has occurred), diagnosis (the process of identifying the cause of an event) and compensation (the process of sustaining system operation). A similar model, comprising detection (detection of the onset of a plant disturbance), diagnosis (diagnosing the particular disturbance from presented symptoms) and remedial actions (selecting and implementing the appropriate actions to mitigate the disturbance), was proposed by Marshall and Baker (1994) as an idealised three-stage decision model to describe how operators deal with faults. There appears to be little to distinguish these three models, apart from the idiosyncratic labelling of the last stage in all cases! Rouse (1983) offers an expanded version of the process model, comprising three levels: recognition and classification (in which the problem is detected and assigned to a category), planning (whereby the problem-solving approach is determined), execution and monitoring (the actual process of solving the problem). Arguably this is reducible to the original three-stage model, but rather more emphasis has been placed upon the interpretation of the problem. What is not clear from any of the analyses presented with these models, is whether they accurately reflect processes undertaken by the operator within the alarm-handling task.

5.4 A Model of Alarm Handling

An alarm-handling sequence can be described as consisting of a number of generic activity stages. These generic activities have been assembled into an analysis of alarm handling (Stanton *et al.*, 1992; Stanton, 1994) as shown in Figure 5.18. The analysis distinguishes between routine events involving alarms and critical events involving alarms. Although the two types of events have most activities in common, critical events are distinctive by virtue of an investigative phase. It is proposed that the notion of AIAs is used to describe the collective of the stages in alarm event handling. The term 'activities' is used to refer to the ensuing behaviours triggered by the presence of alarms. It is posited that these activities would not have been triggered without the alarm being present, thus they are alarm-initiated activities. The AIAs are linked to other supervisory control activities which can be typified as continuous tasks (such as visual scanning of instruments and fine tuning of plant variables in response to minor variations in plant) and discrete tasks (such as putting plant into service and taking plant out of service). In such tasks, alarm information may be used instead of, or in conjunction with, other information (such as data on plant variables, some internal reference to plant state, comments from other operators or reports from engineers who are in direct contact with plant). Whilst it may be difficult to distinguish between some activities, be they triggered by alarms or otherwise, examination of alarm-handling activities can be justified in terms of providing useful information regarding the design of alarm systems. This will, of necessity, involve the consideration of activities where alarm information is of primary importance (such as in a critical event) and activities where the alarm information is of secondary and/or supplementary importance to the task (such as in a routine event).

In alarm handling, operators report that they will observe the onset of an

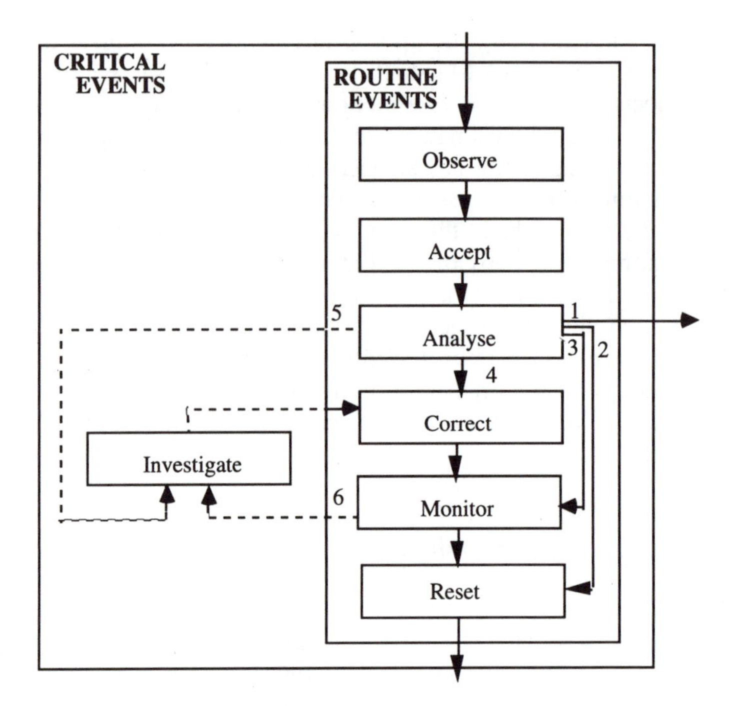

Figure 5.18 Alarm-initiated activities.

alarm, accept it and make a fairly rapid analysis of whether it should be ignored (route 1), reset (route 2), monitored (route 3), dealt with superficially (route 4) or require further investigation (route 5). If it cannot be cleared (by superficial intervention), then they may also go into an investigative mode (route 6). In the penultimate mode, the operators will monitor the status of the plant brought about by their corrective actions and ultimately reset the alarm. Routine behaviour has a 'ready-made' response, whereas critical behaviour needs knowledge-based, deductive reasoning. The taxonomy (observe, accept, analyse, investigate, correct, monitor and reset) is proffered as a working description of alarm-handling behaviour, rather than intending to represent a fully validated psychological model.

Activity in the control room may be divided broadly into two types: routine and critical. Incident-handling activities take only a small part of the operator's time, approximately 10% (Reinartz and Reinartz, 1989; Baber, 1991) and yet they are arguably the most important part of the task. This is particularly true when one considers that the original conception of the operator's task was one of operation by exception (Zwaga and Hoonhout, 1994). In order to develop a clearer understanding of the alarm handling, a toxonomy of alarm handling was developed (Stanton and Baber, 1995) on the basis of a literature review, direct observation and questionnaire data. This taxonomy reveals 24 alarm-related behaviours subsumed under seven

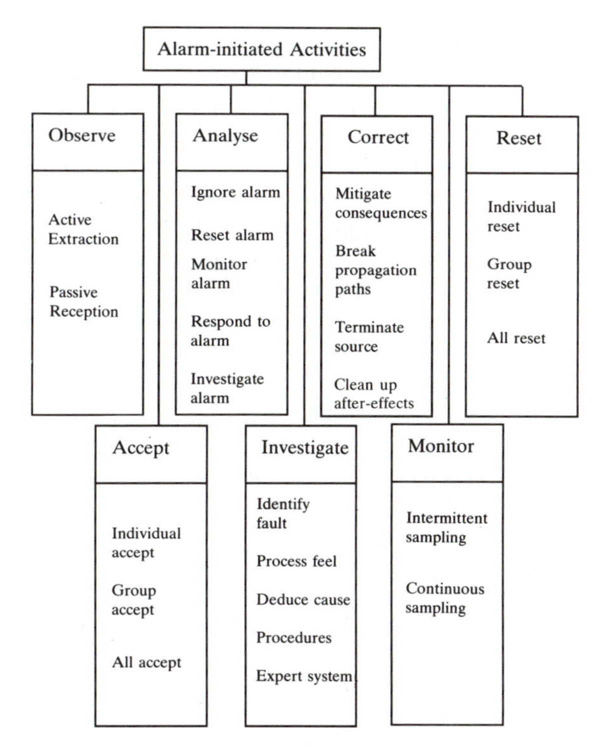

Figure 5.19 Taxonomy of alarm-handling activities.

categories: observe, accept, analyse, investigate, correct, monitor and reset (see Figure 5.19).

The development of a taxonomy and model of alarm handling provides a focus on the design formats required to support alarm-initiated activities. The human requirements from the alarm system may be different, and in some cases conflicting, in each of the seven categories of the taxonomy. The difficulty arises from the conflicting nature of the stages in the model, and the true nature of alarms in control rooms, i.e. they are not single events occurring independently of each other but they are related, context-dependent and part of a larger information system. It is not easy to separate alarm-initiated activity from general control room tasks, but it is useful to do so. Whilst many of the activities may also be present in more general aspects of supervisory control tasks their consideration is justified by relating them to the alarm system in particular. This leads to the following observations. First, designers of alarm systems need to consider all manner of alarm-initiated activities. Often there is little consideration of alarm handling beyond the initial 'observe' stage. Second, one could question the need to consider the alarm system as a separate entity from the other displays in the control system. It seems apparent that 'active extraction' of information would be more suitably supported by a permanent visual display. This would suggest three main changes: the integration of alarm information, the move away from the trend for sole reliance upon scrolling text-based alarm displays and the provision of parallel displays rather than sequential displays.

These changes could support operator heuristics by making the information available to the operator at all times, without requiring extensive searching activities. With the continued drive to replace traditional annunciator panels and wall-based mimics with VDUs in control rooms, there is good reason to question the use of text-based alarm displays.

References

ANDOW, P. K. & ROGGENBAUGER, H. (1983) Process plant systems: general considerations. In Buschmann, C. H. (ed.), *Loss Prevention and Safety Promotion in the Process Industries*, pp. 299–306. Amsterdam: Elsevier.

BABER, C. (1991) *Speech Technology in Control Room Systems*. Chichester: Ellis Horwood.

BAINBRIDGE, L. (1984) Diagnostic skill in process operation, *International Conference on Occupational Ergonomics*, 7–9 May, Toronto.

GOODSTEIN, L. P. (1985) *Functional Alarming and Information Retrieval*. Denmark: Risø National Laboratory, August, Risø-M-2511.18.

LEES, F. P. (1974) Research on the process operator. In Edwards, E. and Lees, F. P. (eds), *The Human Operator in Process Control*, pp. 386–425. London: Taylor and Francis.

MARSHALL, E. & BAKER, S. (1994) Alarms in nuclear power control rooms: current approaches and future design. In Stanton, N. A. (ed.), *Human Factors in Alarm Design*, pp. 183–91. London: Taylor and Francis.

OPPENHEIM, A. N. (1992) *Questionnaire design, Interviewing and Attitude Measurement*. London: Pinter Publishers.

REINARTZ, S. J. & REINARTZ, G. (1989) Analysis of team behaviour during simulated nuclear power plant incidents. In Megaw, E. D. (ed.), *Contemporary Ergonomics*, pp. 188–93. London: Taylor and Francis.

ROUSE, W. B. (1983) Models of human problem solving. *Automatica*, **19**, 613–25.

SORKIN, R. D. (1989) Why are people turning off our alarms? *Human Factors Bulletin*, **32**, 3–4.

STANTON, N. A. (1994) Alarm initiated activities. In Stanton, N. A. (ed.), *Human Factors in Alarm Design*, pp. 93–117. London: Taylor and Francis.

STANTON, N. A. & BABER, C. (1995) Alarm initiated activities: an analysis of alarm handling by operators using text-based alarm systems in supervisory control systems. *Ergonomics*, **38**(11), 2414–31.

STANTON, N. A., BOOTH, R. T. & STAMMERS, R. B. (1992) Alarms in human supervisory control: a human factors perspective. *International Journal of Computer Integrated Manufacturing*, **5**(2), 81–93.

YOUNGMAN, M. B. (1982) *Designing and Analysing Questionnaires*. Maidenhead: TRC.

ZWAGA, H. J. Z. & HOONHOUT, H. C. M. (1994) Supervisory control behaviour and the implementation of alarms in process control. In Stanton, N. A. (ed.), *Human Factors in Alarm Design*, pp. 119–34. London: Taylor and Francis.

CHAPTER SIX

Procedures in the nuclear industry

PHILIP MARSDEN

Human Reliability Associates, Wigan, UK

6.1 Introduction

It is generally recognised that procedures play an important role in the day-to-day management of nuclear power plant operations (Jenkinson, 1992; Livingston, 1989; NUREG, 1980; Swezey, 1987; Trump and Stave, 1988). Consequently, a significant amount of time and effort is devoted within the nuclear industry towards the production of high-quality work instructions and job aids designed specifically to optimise the work activities of employees. Yet despite the massive investment which typically goes into the preparation of procedures it is becoming increasingly evident that procedure-related system failures constitute a major source of vulnerability to the integrity of NPP systems. For example, in an analysis of the abstracts of 180 significant events reported as having occurred in the United States during 1985, it was found that of the 48% of incidents originally attributed to failures of the human factor, almost 65% could be further classified as involving a procedure deficiency (INPO, 1986). A similar finding was obtained in a study reported by Goodman and DiPalo (1991). These investigators surveyed the reports of almost 700 NPP incidents and concluded that failures with procedures were implicated in more than 69% of cases. Other significant contributory factors identified in the Goodman and DiPalo study were poor training and communication failures.

These studies raise serious doubts about the way procedure systems are designed and operated at many nuclear facilities. The purpose in this chapter is to review the common problems associated with procedure systems to determine what, if anything, can be done to improve the quality of procedure provision in a nuclear setting. It is suggested that procedure-related failures largely have their origins in weaknesses inherent within the infrastructure of the organisation. Consequently, it seems reasonable to suppose that it is only by making interventions at this level that we can possibly hope to reduce the overall numbers of incidents that involve a procedure component.

6.2 Procedure Failures: Human Error or Organisational Weakness?

Several attempts have been made over the years to classify the common failures associated with procedure usage in the nuclear domain. For example, Henry Morgenstern and his colleagues (Morgenstern *et al.*, 1987) developed a classification of operator errors in which the majority of failures with procedures were characterised in terms of four main types of human action: use of an incorrect procedure, use of the correct procedure but at the wrong time, implementing a procedure too late, and the failure to carry out a procedure in accordance with a written instruction. Each of these categories of operator errors relate strongly to the concept of errors of commission that have figured prominently in the work of Alan Swain and his colleagues (e.g. Swain, 1963; Swain and Guttman, 1982). Consequently the Morgenstern *et al.* classification has found favour among HRA practitioners with an interest in procedure systems.

A more comprehensive classification of procedure failures was proposed by Green and Livingston (1992). These authors devised a taxonomy in which the main aim was to specify the fundamental root cause for the operator error. They argued that consideration of root causes requires an eightfold classification of procedure deficiencies covering items ranging from areas where the technical accuracy of a procedure was at fault at one end of the spectrum to breakdowns of the interrelationship between training and procedures at the other. The full list of items included in the Green and Livingston taxonomy is as follows:

1 technical accuracy/completeness of procedure at fault;
2 document poorly formatted;
3 language/syntax problems;
4 procedure poorly located/cross-referenced;
5 poor procedure development process;
6 insufficient verification/validation of documentation;
7 failure to revise procedure;
8 training and procedures poorly interrelated.

Classification schemes such as the ones described above clearly can provide a useful conceptual framework to help identify some of the ways that procedure systems go wrong in the nuclear domain. Moreover, where necessary they can also provide an invaluable aid to the estimation of the frequency with which each failure type is likely to occur in a specific location (see for example Macaw and Mosleh, 1994). Where they are weak, however, is with regard to their ability to suggest possible organisational solutions relative to the known problem areas and to provide concrete guidance to assist managers to implement effective policy aimed at reducing the occurrence of incidents involving a procedure component. In order to achieve this objective what is needed is a more functionally orientated classification scheme in which the external manifestations of each failure are better related to the factors that precipitate their occurrence. In our own work (e.g. Marsden and Whittington, 1993; Marsden and Green, 1995) we have tended to assume that these factors are to be found more within infrastructure of the organisation than the fallible nature of the human operator. Consequently we have attempted to devise models of procedure provision in which procedure failures are directly related to underlying features of the organisation. The current revision of the proposed taxonomy is presented in Table 6.1. The table also provides some examples of failure types that are typically associated with each area of the organisation.

Described in overview, the classification presumes that there are three common types of organisational weakness that are strongly implicated in the malfunction of a procedure system. One type of problem arises because the company fails to specify a formal procedure development process. The most usual outcome of a failure of this type is either that a procedure is prepared that is incomplete relative to the task it describes, or alternatively a procedure is prepared that contains a technical fault. The case where no procedure exists can also be attributed to failures of the procedure preparation process insofar as they suggest that the analysis of the work role has failed to indicate the need for a procedure in relation to a particular aspect of work activity. A second type of problem identified in the taxonomy arises because of the finding that people frequently fail to follow the precise instructions which are provided in the procedure. Although there may be many reasons why people fail to comply with a written instruction one possible explanation is that the organisation fails to encourage the development of a 'procedures culture' in which working with a procedure is viewed as a positive attribute of the work role. Thus, the user often feels alienated from the procedure system and there is no feeling of ownership which could help encourage the development of work practices which comply with an established safe system of work. The final problem type identified within the taxonomy relates to breakdowns in the interrelationship between the procedure system and other aspects of the organisation which should be providing a degree of operator support. The two most important examples of such a breakdown are when training and procedures are poorly related, or where a procedure is not revised following some type of operational change.

6.3 Estimate of the Scale of the Problem

The extent to which these three problem types pose a threat to the integrity of NPP operations can be estimated by using the taxonomy to classify

Table 6.1 Summary of organisational weaknesses implicated in procedure failures

Organisational weakness		
Procedure preparation process	User compliance	Procedure system infrastructure
Incorrect procedure prepared Inaccurate procedure prepared No procedure prepared	Intentional non-compliance ■ bad habits ■ routine violations Unintentional non-compliance ■ slips ■ lapses	Procedure/training failure Failure to revise procedure ■ after changes to work practice ■ after plant modifications

accident and near-miss data. When this is done a graph such as that shown in Figure 6.1 is the result. This figure is based on the analysis of the 180 incidents reported to INPO for 1985 which have been reclassified for the purposes of the present chapter according to the proposed classification scheme.

As can be seen from Figure 6.1, of the 48% of incidents attributed to human performance failures, approximately 89% could be further reclassified as involving a failure of the organisation to provide an acceptable degree of operator support, procedures or otherwise. In 43% of cases it was assessed that the specific failure was the preparation of a procedure which contained some kind of deficiency. The predominant deficiency type was the preparation of a procedure which was incomplete relative to the task accounting for 69.77% of the category, while development of an incorrect procedure occurred for approximately 12% of instances. Weakness in the training and revision functions together accounted for 30% of the reported incidents. In this case, it was found that breakdowns in the relationship between training and procedure systems was a factor in the predominant failure mode (60%), while failure to monitor or revise a procedure arose in 31% of cases. Finally, non-compliance was identified as a causal factor in 16% of cases. The most interesting aspect of this result was the finding that the majority of procedure deviations reflected unintentional deviations from procedures (67%) in which operator errors resulted primarily from a breakdown in the natural course of information-processing (e.g. Norman, 1981; Reason and Mycielska, 1982). In contrast to this, intentional deviations (bad habits, violations of regulations) accounted for 31% of incidents involving non-compliance.

If the above figures are accepted, then they are strongly suggestive of a

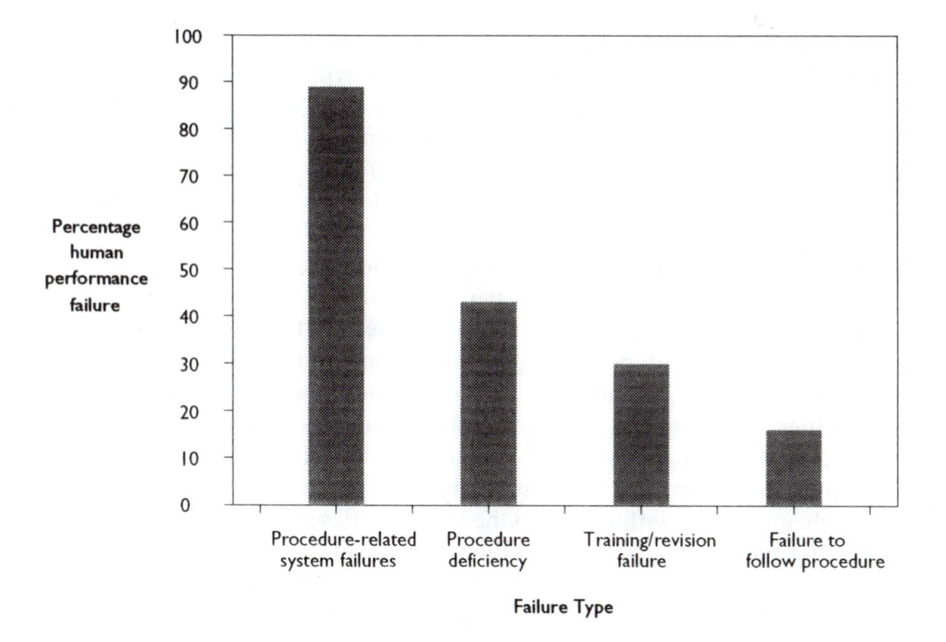

Figure 6.1 Procedure-related failures by organisational factors.

situation in which the majority of problems with procedures result from three basic failure types: (a) weaknesses inherent within the procedure preparation process; (b) problems resulting from the failure of employees to follow an established procedure; (c) breakdowns inherent within the infrastructure of the procedure system. In the remaining sections each of these problem areas is discussed in more detail and a potential human factors solution is outlined.

6.4 Procedure Preparation

On 10 March 1983, an engineer working at Quad Cities in the United States prepared a note explaining the procedure for performing a shutdown scheduled to be undertaken later that night by the evening shift. The shift supervisor incorrectly interpreted the procedure as instructing him to insert control rods in an order that was the reverse of that required under normal circumstances. Prior to the arrival of the night shift supervisor, the evening shift operators incorrectly installed 33 rods. In order to do this they needed to bypass a safety critical system (the rod worth minimiser, or RWM) designed specifically to prevent improper rod insertion. Following shift handover, the night shift supervisor initially restored the integrity of this system. However, because the system prevented further rods from being inserted into the core in the manner specified, the decision was taken that the RWM was faulty and the system was again bypassed. A further 10 rods were wrongly inserted into the core and the reactor was shut down without incident. An NRC investigation concluded that this near-miss incident had created the basic conditions for a rod drop accident and imposed a $150 000 fine on the owning company, Commonwealth Edison (Annual Report of the Nuclear Regulatory Commission, 1984).

Many reported incidents involving procedures occur because of weaknesses inherent within the preparation process. The near-miss accident cited above, for example, serves to illustrate what can go wrong when procedures are developed using informal methods which rely on the knowledge of a single individual who is familiar with the functions and capabilities of the system, but not with the actual behaviours required to operate it. Procedures prepared by such 'technically qualified' personnel working in isolation are frequently found to be incomplete, incorrect, or generally unrealistic with regard to the task they describe, while unfamiliarity with information-presentation methods can lead to a situation in which critical information is effectively masked in documentation which is poorly formatted. Each of these problems can render a procedure unusable as an operator aid.

One way to reduce the likelihood of procedural deficiencies causing failures of this sort involves the specification of a formal procedures development process such as the one proposed by Embrey (1986). He argued that optimal procedures development involves at least 6 stages of preparation such as those shown in Figure 6.2. Each of these stages is discussed in turn.

6.4.1 Task and Error Analysis

In this scheme the first step in procedure preparation involves carrying out a systematic analysis of the task using a formal method such as hierarchical task analysis (HTA). In essence, HTA is a methodology for breaking tasks down into sub-goals and specifies the way these goals are to be combined in order to achieve the overall objective of the task. At the highest level of analysis tasks are defined in terms of work plans which take the form of sub-goal sequences (e.g. 1. Set up the gland water system. 2. Establish ACW supply, etc.). At the lowest level, each sub-goal is described in terms of sequences of object–action pairings such as '1. Open valve PV101', '2. Close valve PSV 54'. There are two major advantages to be gained by using a method such as HTA for procedures development. First, it forces the expert to make what he or she knows verbally explicit and this permits the accuracy and completeness of operational knowledge to be tested objectively prior to implementation as a procedure. Second, it provides an opportunity to involve end users in procedure preparation and this can help encourage feelings of ownership.

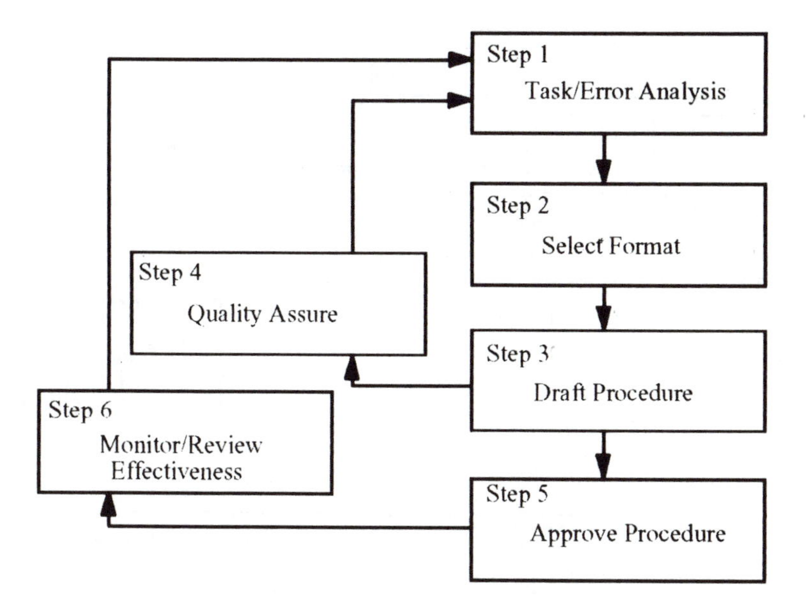

Figure 6.2 A flow-chart representation of a procedure development process.

The results of an HTA analysis are conventionally expressed in terms of flow charts such as that shown in Figure 6.3. This diagram shows four hypothetical steps which are required to warm up a furnace prior to switching the system to automatic. The plan specifies that these steps should be performed in strict sequence. According to the HTA, the first sub-task to be performed is the preparation of plant and ancillary services. This task can be further decomposed into three discrete activities which may be performed in any order. One activity is to ensure that the plant itself is ready. The line drawn underneath the box 0.1.1 indicates that operators are trained to perform this step and thus no further decomposition of sub-task is required. A second activity is to ensure that gas and oil are available. Again operators are trained to make this check and thus no further guidance is required. Finally, a check should be performed on the oxygen analysing equipment as this measuring device will provide data to be used during the later stages of the task. Once the plant and services have been prepared for operation then the next requirement is for the operator to start the air blower and oil pump. Finally the operation is concluded when the furnace becomes heated to 800°C when the system can be switched to automatic. The HTA indicates that four particular steps are required to achieve this objective and the work plan specifies how each of the heating sub-steps is required to mesh together.

Once an HTA for the job has been prepared then the next step involves

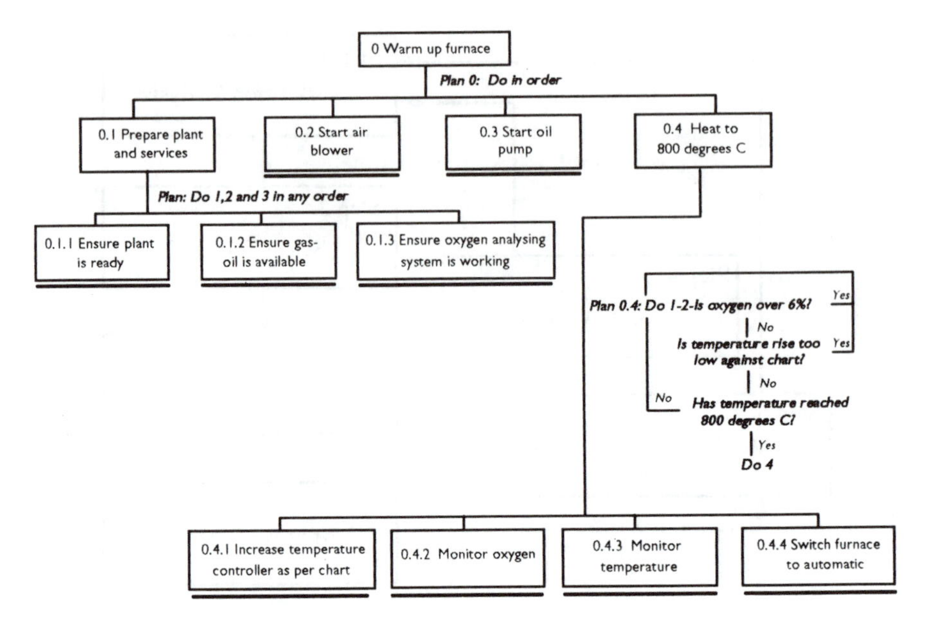

Figure 6.3 Prototypical HTA analysis.

the construction of an error analysis matrix in which plausible human error scenarios are considered in relation to each task step. This analysis indicates areas where the job is particularly vulnerable to the actions of the human operator. The error analysis plays a vital role in safety management because it enables the analyst to consider procedures, training and equipment design implications of the task. Specification of remedial measures can be used to reduce the likelihood of the incidence of the errors identified. An example of the results of an error analysis for the furnace warming task is provided in Table 6.2. In this table the error modes are specified along with consideration of the consequences of error for the system in question and the possibility that the specific error mode can be recovered before the consequences obtain. Analyses of remedial strategies for each error are shown in columns 4 to 6. Column 4 shows the steps that can be taken within the procedure to reduce the likelihood of the error occurring. Columns 5 and 6 show the corresponding information for training considerations and the ergonomic design of the equipment.

6.4.2 Select Format

Once the task has been formally analysed the next step in the process is the selection of an appropriate format. Several factors may need to be taken into consideration when choosing a format for the procedure but two of the most important considerations are task complexity, and user familiarity. As a rule, as the operator's familiarity with a task increases and task complexity decreases, the more likely it is that operators will need a job performance aid (JP) which uses a checklist format. Conversely, as task complexity increases and familiarity diminishes, then a much greater level of operator support will be required and this would best be provided by using a step-by-step procedures format. Alternative procedural formats may be required for jobs involving more than one person (e.g. play script formats etc.; Berry, 1981). A simple matrix for determining the selection of appropriate procedures format is provided in Table 6.3.

6.4.3 Preparation of Draft Procedure

The third stage of development involves preparing a draft document based upon task elements determined in step 1, presented in the format determined in step 2. In preparing the draft document several decisions may need to be made regarding the design of the layout of the procedure. Once a layout has been devised Embrey suggests that it should be optimised by checking for consistency with human factors guidelines which specify best practice for information presentation.

6.4.4 Quality Assurance

Prior to approving the procedure for general use, several tests of the draft procedure will need to be undertaken and these should be performed by suitably qualified inspectors. Wherever possible testing should take place in the actual operating environment. Embrey suggests that it is desirable that a series of end-user consultations is also undertaken during this phase of development to ensure that the proposed method of work is consistent with current practices. All results should be fully documented and this information can be used to make modifications to the draft procedure. Incremental refinement of the procedure may necessitate several design iterations between steps 1 to 4.

6.4.5 Approving the Procedure for General Use

In the fifth stage the procedure is approved for general use. New procedures will often need to have special controls and/or limitations placed on their use such that only certain qualified personnel would be allowed to perform the procedure on the first few occasions. This qualification would be especially relevant to tasks involving a significant hazard component.

Table 6.2 Example of an error analysis for hypothetical furnace heating task

	Task step	Error mode	System consequence	Error recovery	Procedure implication	Training implication	Design implication
0.4.1	Increase temperature controller	Controller adjusted to produce too large an increase	When temperature increases too rapidly it can place strain on system	Operator notices that temperature is rising too fast and decreases temperature controller	Ensure that procedure contains bandwidth range values in relation to controller adjustment step and check that chart is up to date	Operators must demonstrate knowledge of setting requirement, and competence to adjust controller accordingly	Place a label which states the required settings on or near the temperature controller
0.4.2	Monitor oxygen	Right action performed on wrong object (operator reads a wrong, but similar, dial)	If the dial is misread (or the wrong dial read) the oxygen levels will become too high and may result in the temperature in the furnace rising too quickly	Supervisor may check oxygen levels at local control panel to confirm actions of operations staff	Insert warning information prior to step 0.4.2 in procedure	Emphasise in training programme the importance of obtaining the correct level of oxygen. Trainees should demonstrate ability to read and interpret the instrumentation properly	Relocate the oxygen dial because of its proximity to other dials which are similar and may result in the wrong dial being read when workload is high

0.4.3	Monitor temperature	Wrong action	If the temperature is below 800°C and the system is switched to automatic the system must fail	No recovery scenario once the system has been erroneously switched to automatic	Introduce a new step in the procedure prior to action 0.4.3 which requires an independent check to be made of the status of the system prior to the switch to automatic	Emphasise the importance in training of attaining the correct temperature prior to switching the system to automatic. Trainees should demonstrate ability to read and interpret the instrumentation properly	1 Relocate temperature dial if close to other similar dials 2 Colour code dial to reflect critical areas where switch to automatic can be made
0.4.4	Switch system to automatic	Right action on wrong object (wrong switch activated)	Unspecified system state	Recovery from error dependent upon the ability of operators to diagnose fault	Procedure should indicate the type of feedback expected from control panel when system is switched to automatic	1 Demonstrate to trainees automatic switching process 2 Drill trainees in switching task 3 Trainees should demonstrate that they recognise appropriate plant feedback (automatic lights, etc.)	Install light which shows operator that plant switch to automatic has been successful

Source: Embrey, 1986.

Table 6.3 Simple matrix for helping determine correct format for procedure

		Task complexity	
		High	Low
Familiarity of user with task	High	Checklist format plus supplementary aids (e.g. decision trees, flow charts, logic diagrams, visual aids, etc.)	Predominantly checklist procedures
	Low	Step-by-step procedure format	Narrative procedure format

6.4.6 Long-term Maintenance

The final aspect of Embrey's (1986) proposed process involves the long-term maintenance and updating of the procedure over the lifetime of the system. He suggests that routine checks of the effectiveness of the procedure be made periodically to assure the long-term applicability and usability of the procedure over time.

6.5 Non-compliance

On 18 September 1988, at Stade PWR in the Lower Saxony region of West Germany, an electrical failure caused a valve to close in one of the four main steam lines. As a direct result of the closure, other valves in the remaining three lines automatically closed, blocking all four lines. The automatic reactor protection system should at this point have scrammed the reactor and turbine. Contrary to procedures, however, operators overrode the system and reopened the valves in an attempt to keep the reactor running. The valves quickly closed for a second time and the system scrammed. The opening and closing of the steam line valves in rapid succession led to a violent pressure wave in the main steam lines causing a maximum displacement of 20 cm in the pipework. Displacement of 20 cm was more than double the acceptable limit and an investigation concluded that the actions of the operators brought the steam lines close to failure (cf. May, 1989).

Non-compliance with procedures features prominently in statistics covering accident and near-miss reports in the nuclear domain. In a survey of 180 significant nuclear events cited by Green (1991), for example, more than 16% of all incidents were attributed in some way to the failure of personnel to follow procedures. Similarly, Brune and Weinstein (1982) found failure to comply with the prescriptions contained within procedures contributed significantly to procedure-related performance errors in activities involving the maintenance, test and calibration of equipment. From the point of view of the user, it is possible to distinguish between two broad classes of non-compliant behaviour: intentional non-compliance (e.g. routine violations, unsafe acts, etc.) and unintentional non-compliance (e.g. a procedure was used but an action was omitted, a procedure was used but an action was performed incorrectly, etc.) Of these two types, intentional non-compliance is potentially the most serious and it is with this aspect of behaviour that we are primarily concerned here.

In attempting to provide an answer to the question 'Why don't people follow procedures?', it is useful to begin by considering what users typically think about procedures. As part of a study of employee attitudes reported by Zach (1980), a large group of operators were asked to say why they thought procedures were needed. Interestingly, the most conspicuous answer was that procedures were required primarily for the purpose of regulatory inspections and little mention was made of the fact that they serve as operator aids. In fact, procedures were commonly viewed as the bane of the operators' life and were perceived to devalue the work role. In particular, operators reported that working with procedures made work much less rewarding and the job more difficult than it would otherwise be.

These few comments provide an important insight into why people sometimes fail to comply with a procedure. Procedural systems are typically viewed by operators as a system of work control designed essentially to protect the company in the event of an accident. Consequently, operators report that they have little confidence in the fidelity of procedures as usable work aids. Unfortunately, this view is not without some foundation. When pressed most operators are able to recall occasions where non-compliance with a procedure was sanctioned (either implicitly or explicitly) by the operating company and this occurred particularly when commercial pressures to produce were high. Inconsistencies such as these tend to undermine a compliance policy and leave areas of ambiguity about when procedures are to be followed.

What changes then would users like to see made to procedures? When this question was put to operators in two studies carried out by Zach (1980) and Seminara *et al.* (1976) answers tended to be concerned largely with the ongoing debate between the relative merits of 'lock-step' (i.e. high compliance) vs guidance (low compliance) procedures. The greater proportion of answers indicated a preference for guidance documentation which places much more reliance on the personal initiative and experience of the user. Consequently, operators expressed a desire to see more concise forms of procedures which described tasks only in terms of important details. Along similar lines users indicated a preference for flow-diagram (as opposed to text-based) procedures. They suggested that flow charts make it easier to identify key elements of the task quickly and to relate specific actions to overall objectives. It was acknowledged, however, that on the negative side there is some loss of information involved in using flow diagrams for complex tasks. Finally, operators indicated that they would like more formal training relating to procedures of all types. Training with procedures in mind is an important issue in its own right and will be considered more fully in the following section.

The major point which is being made here is that none of the above changes are unreasonable provided that adequate user support (such as training resources) are made available. Furthermore, attempting to address these problems may help ameliorate the potential threat of serious incident arising from the failure of users to comply with a written instruction.

6.6 Procedure Infrastructure

On 11 February 1981, at the Tennessee Valley Sequoyah I plant, a unit operator instructed a recently qualified employee to check the containment spray valve and ensure that it was closed. Instead of closing the valve the operator opened it, and as a direct result 40 000 gallons of primary cooling water was released into the containment area, along with a further 65 000 gallons of water held in a storage tank. Eight workers were contaminated as a result of the accident which involved low-grade radioactive materials. The reactor was fortunately in a cold shutdown condition at the time of the incident, thus preventing the occurrence of a more serious incident (*New Scientist*, 10 December 1981).

The final source of vulnerability considered here covers those cases where procedures are not directly at fault, but where weaknesses in supporting systems could be said to have increased the probability of operator error. Of particular concern are those situations where incidents occur because operators are inadequately trained in the use of particular procedures, as in the case above, or where problems arise because of a failure to monitor the effectiveness of procedures. The most common scenario for a failure of this latter kind is the situation where a procedure is not revised following an operational failure of some type.

The basic design philosophy proposed to counter problems of this type advocates the development of a procedural system comprising three interrelated elements: a procedures database or inventory, a training function and a system for monitoring and updating. These three elements and their interrelationships are shown in overview in Figure 6.4.

In this type of system, procedures are explicitly related to the training, insofar as they are developed specifically with training needs in mind. For example, analysis of the structure of the procedure will imply particular training requirements and this information can be used to formulate training strategies. Similarly, the procedures themselves can be used as a primary source of training materials and these are presented to trainees as part of a

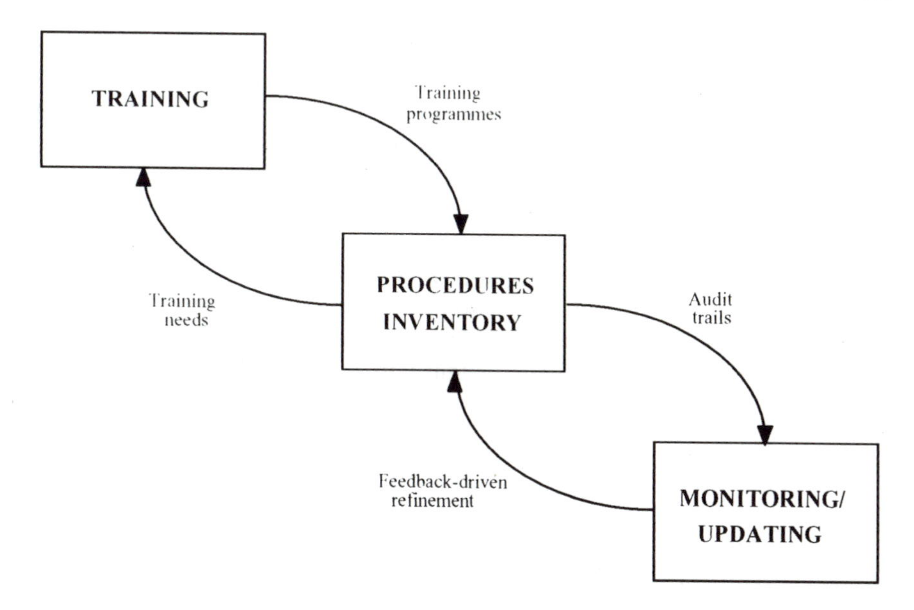

Figure 6.4 Minimum structural requirements for an effective procedures infrastructure.

structured training programme. Employee qualification is based upon methods of testing in which the trainee must demonstrate competence relative to particular sets of procedures.

Procedures are also formally linked in the system to a monitoring function. In this case it is proposed that quality points be built into procedures that lay down a clear audit trail each time the procedure is used. The audit trail can be periodically examined and checked for compliance with safety standards and production targets. The information generated by the periodic reviews of procedure quality should be used to modify the procedures in a process of feedback-driven refinement. There are several important points which follow from the implementation of a procedures infrastructure of the type proposed:

- First, it is a mistake to think of procedures as complete. Rather, they should be thought of as 'living documents' which always exist in a state of latest revision. People should always be encouraged to challenge the prescriptions contained within procedures with a view to upgrading safety or production targets.
- Second, employee training must work for the benefit of both the company and employee alike. Such objectives are best served by the use of proactive training methods involving a blend of on-the-job vs formal training methods which deliver both technical (relating to the process)

and behavioural (relating to task skills) knowledge. With regard to employee training it is not sufficient to assume competence based upon the principle of time served. Rather, formal methods of competence testing should be employed which require the trainee to demonstrate his or her skills.

- Finally, the process of rapid prototyping and feedback-driven refinement is more productive when the monitoring process is implemented in a blame-free environment. People should always be encouraged to report their mistakes and near-misses so that procedural implications can be evaluated and error-tolerant solutions designed in an attempt to minimise the likelihood of catastrophic system failure.

6.7 Conclusions

In this chapter it has been suggested that there are three organisational problem areas associated with the use of procedures in the nuclear industry. In particular, procedural systems commonly fail (a) due to weaknesses inherent within the preparation process, (b) because people fail to follow procedures, and (c) because systems which should provide support for procedures are poorly designed and implemented. It was suggested that improvements in each of these areas can be made by (i) adopting a formal development process, (ii) listening to the views of users and involving them in methods of work control, and (iii) developing a strong procedures infrastructure which explicitly relates procedures to a training and monitoring function. It is proposed that the changes recommended here will not guarantee that procedures will in the future be perfect. Experience suggests, however, that significant improvements can be made in each of the above areas if the ideas which have been proposed are properly implemented.

References

BERRY, E. (1981) How to get users to follow procedures. *IEEE Transactions on Professional Communication*, **25**, 21–6.

BRUNE, R. L. & WEINSTEIN, M. (1982) *Procedures Evaluation Checklist for Maintenance, Test and Calibration Procedures Used in the Nuclear Industry*, NUREG CR-1369. Albuquerque, NM: Sandia National Laboratories.

EMBREY, D. (1986) *Guidelines for the Preparation of Procedures for the High Reliability of Ultrasonic Inspection Tests*. Wigan: Human Reliability Associates.

GOODMAN, P. C. & DIPALO, C. A. (1991) Human factors information system: A tool to assess error related to human performance in US Nuclear Power Plants, *Proceedings of the Human Factors Society*, 35th Annual Meeting, San Francisco.

GREEN, M. (1991) *Root Cause Analysis*, Parbold. Wigan: Human Reliability Associates.

GREEN, M. & LIVINGSTON, A. D. (1992) Procedures development: Common deficiencies and possible solutions, *International Atomic Energy Authority International Working Group on Nuclear Power Plant Control and Instrumentation, Specialist Meeting on Operating Procedures for Nuclear Power Plants and their Presentation*, Vienna, Austria.

INPO (1986) *A Maintenance Analysis of Safety Significant Events*. Atlanta, GA: Institute of Nuclear Power Operations (INPO).

JENKINSON, J. (1992) The application of symptom based operating procedures to UK gas cooled reactors, *International Atomic Energy Authority International Working Group on Nuclear Power Plant Control and Instrumentation, Specialist Meeting on Operating Procedures for Nuclear Power Plants and their Presentation*, Vienna, Austria.

LIVINGSTON, A. D. (1989) The use of procedures for decision support in nuclear power plant incidents. In *User Requirements for Decision Support Systems Used for Nuclear Power Plant Accident Prevention and Mitigation*, IAEA-TECDOC-529.

MACAW, A. & MOSLEH, A. (1994) A methodology for modelling operator errors of commission in probabilistic risk assessment. *Reliability Engineering and Safety Systems*, **45**, 139–57.

MARSDEN, P. & GREEN, M. (1996) Optimising procedures in a manufacturing system. *International Journal of Industrial Ergonomics*, **17**(1), 43–51.

MARSDEN, P. & WHITTINGTON, C. (1993) Work instruction provision: Some common problems and human factors solutions, *Human Factors in Nuclear Power Conference*. London: IBC.

MAY, J. (1989) *The Greenpeace Book of the Nuclear Age*. London: Victor Gollancz.

MORGENSTERN, M. H., BARNES, V. E., MCGUIRE, M. V., RADFORD, L. R. & WHEELER, W. A. (1987) *Study of Operating Procedures in Nuclear Power Plants*, NUREG CR-3968. Seattle, WA: Battelle Human Affairs Research Center.

NORMAN, D. A. (1981) Categorisation of action slips. *Physical Review*, **88**, 1–15.

NUREG (1980) *NRC Action Plan Developed as a Result of the TMI-2 Accident*, NUREG-0660. Washington, DC: USNRC.

REASON, J. T. & MYCIELSKA, K. (1982) *Absent-mindedness: The Psychology of Mental Lapses and Everyday Errors*. Englewood Cliffs, NJ: Prentice Hall.

SEMINARA, J. L., GONZALEZ, W. R. & PARSONS, S. G. (1976) *Human Factors Review of Nuclear Power Plant Control Room Design*, Electric Power Research Institute Report NP-309, Chapter 17, November.

SWAIN, A. D. (1963) *A Method for Performing a Human Factors Reliability Analysis*, Monograph SCR-685, Albuquerque, NM.

SWAIN, A. D. & GUTTMAN, H. E. (1982) *Handbook of Human Reliability Analysis with Emphasis on Nuclear Power Plant Applications*, NUREG CR-1278. Washington, DC: NRC.

SWEZEY, R. W. (1987) Design of job aids and procedure writing. In Salvendy, G. (ed.), *Handbook of Human Factors*. New York: Wiley.

TRUMP, T. R. & STAVE, A. M. (1988) Why what isn't how: A new approach to procedure writing, *IEEE 4th Conference on Human Factors and Power Plants*, Monterey, Canada.

Zach, S. E. (1980) Control room operating procedures: content and format, *Proceedings of the Human Factors Society Annual Meeting*.

Additional Reference

Green, M. & Marsden, P. (1992) The design of effective procedures for industry, *1992 Ergonomics Society Annual Conference*. Birmingham: University of Aston.

CHAPTER SEVEN

Simulators: a review of research and practice

NEVILLE STANTON

Department of Psychology, University of Southampton

7.1 Introduction

Simulators are widely used in the training and licensing of nuclear power operators, as well as for research into the behaviour of operators (see Chapter 8). Typically these simulators have very high levels of fidelity, i.e. they look and behave just like the technology that they are mimicking. This high level of fidelity presumably offers reassurance to the training community that the reactions of the operator in this simulated condition will parallel those in the operational environment. Simulators have been employed extensively in the nuclear industry since the early days of nuclear power (Jervis, 1986). They have served many purposes, including the following:

- to aid the design of the plant;
- for ergonomic development;
- as mock-ups;
- as training aids;
- to predict emergency response;
- for alarm analysis;
- to test control systems;
- to test transient fault performance systems;
- for reliability evaluation;
- to calculate and co-ordinate protection systems;
- to monitor and test the integrity of essential systems.

This chapter will address some of the issues pertinent to simulator design and use for the nuclear industry.

7.2 What are Simulators?

Simulation is a representation of reality, an imitation of a real process. This may be achieved through an operating representation of a real process obtained by means of a device which represents, in part or in full, the physical, dynamic, operational and decision-making elements of a system which is being provided. The representation may mimic every detail or only a subset of features. Typically there are three major facets to simulation: the model, the equipment and the application.

7.2.1 The Model

The model is created to allow a system to be manipulated for a purpose, such as to instruct, to test and organise empirical data, to prompt investigation, to test hypotheses and to measure performance. The model can take many forms from a complex mathematical computation to a simple representation for specific situations.

7.2.2 The Equipment

The sophistication of the equipment employed may range from paper-based materials (in the case of a static mock-up) up to full-scope, interactive, computer-based systems (in the case of a dynamic simulation). The degree of sophistication largely depends upon the resources available, the utility offered and the purpose of the simulation.

7.2.3 The Applications

Simulators have been employed for many applications areas, for example:

- *industry*, e.g. nuclear reactor simulators, power distribution simulators, oil-rig control-room simulators, etc.;
- *aerospace*, e.g. flight simulators, navigation simulators, moon-landing and moon-walking simulators, space vehicle rendezvous docking simulators, etc.;
- *surface transport*, e.g. train simulators, car-driving simulators, coastal navigation simulators, etc.;
- *armed services*, e.g. tank-driving simulators, aiming simulator, drill simulator, etc.;

- *sport*, e.g. simulators for throwing the ball (tennis, cricket), sailing simulator, etc.;
- *medicine*, e.g. phono-cardiac simulator (heartbeats), phantom head (dentistry), etc.

7.2.4 The Use of Simulation

The use of simulation falls into four broad categories:

1 *research*, e.g. computer simulations within the research laboratory to investigate human performance, e.g. as a source of data on human errors relevant to risk and reliability assessment;
2 *evaluation*, e.g. as a dynamic mock-up for design evaluation, controls and as a test-bed for the checking of operating instructions and procedures;
3 *investigation*, e.g. of performance limitations of the system and to provide an environment in which task analyses can be conducted, e.g. on diagnostic strategies;
4 *training*, e.g. in procedures, dynamic plant control, diagnostic skills, team training and as a means of testing and assessing of operators.

7.2.5 Types of Simulator

Clymer (1981) developed a classification of simulators based upon the type of representation. This classification indicates the different purposes to which simulators are put. Any one simulator may fall into one or more of the following classifications:

- *replica simulator* complete, exact duplication of the human–machine interface and realistic set-up of the system environment;
- *generic simulator* representative class of system but not a replica of one;
- *eclectic simulator* deliberately includes a mixture of features (such as several malfunctions) to provide a broader experience;
- *part task simulator* concerned only with part of the task or operational system;
- *basic principles simulator* a generic and/or part task simulator that omits many details in the interest of economy and simplicity.

Classification of simulators by the type of skill they are concerned with would place them in one or more of the following groups: perception, decision making, motor skills and information processing. The nature of the representation raises some important issues that are pertinent to the use of simulators. These issues may be summarised as the following points:

- Why are simulators used?
- What aspect of the situation should be represented?
- How should it be represented?
- How can we be sure that simulators are useful?

These questions will be answered in the following sections.

7.3 Why are Simulators Used?

There are a number of reasons why simulators are used in preference to the real operational environment. Simulators are safer than the real equipment. For example it may be too dangerous to practise in the real situation and some machines may require a minimum level of skill for safe operation. In pilotage, for instance, if simulators were not used it is highly likely that many more accidents would occur as pilots would be less able to train for emergency events. Nuclear power plants are another area where simulators provide an essential input into the training and retraining of human operators.

Infrequency of use or lack of availability to real environment provides another justification for the use of simulators. For example, space programmes rely almost exclusively on the use of simulation to train, instruct and practise operational procedures that the astronauts are likely to undertake.

Simulators provide a means of collecting data on the operator's performance, whether they are used in training, retraining or as a means of checking proficiency. They also offer a mechanism for compressing experience, to see how operators behave in a variety of scenarios and analyse their strengths and weaknesses. Data on performance may be collected by software logging, video-tape and strategy analysis by experts.

Simulators can save money in a variety of ways. Purchasing and operating costs of simulators can be substantially lower in comparison with actual equipment. For instance, if a pilot crashes an aircraft whilst practising an emergency procedure there would be loss of money in terms of cost of damage to the aircraft, fuel cost, cost of insurance and compensation. Added to these costs are the human costs, i.e. the death of the pilot. Evidence from aviation research supports the claim that simulator operating costs are approximately 10% of those for aircraft.

Complex environments may benefit from simulation by reducing task difficulty. This could be achieved through a variety of means, for example: supersensors (which present information normally unavailable), replay facilities (which enable the situation to be replayed so that the operators can gain insight into the effect of their actions), micro-simulation (simulating only a part of the task, which may reduce the information overload present in the real situation) and altering temporal aspects of the environment (e.g. speeding up or slowing down).

7.4 What Should be Represented?

Gagné (1962) suggested that three characteristics are common to simulators: they attempt to represent a real situation, they provide certain controls over the situation and they deliberately omit certain aspects of the real operational situation. Before designing a simulator it is first necessary to determine what needs to be represented. There are at least five phases to simulator design, which are: defining the design, organising information on the operational system, analysis of the tasks involving human participation, gross device definition and deciding upon characteristics of the operational environment to simulate (Andrews, 1983). Each of these phases have several component parts that form an integrated simulator design process. The phases are outlined below:

- *Phase 1* Defining the design
  - — Define gross characteristics of the environment
  - — Determine provisions for the management of the environment
- *Phase 2* Organising information on the operational system
  - — System description
  - — Human–machine interactions
  - — System documentation
- *Phase 3* Analysis of tasks involving human participation
  - — Operational task structure
  - — Develop simulation objectives
  - — Definition of simulation elements
- *Phase 4* Gross device definition
  - — Overall device layout
  - — Equipment required
  - — Complexity of simulation anticipated
- *Phase 5* Characteristics of the operational environment to simulate
  - — Representation of the operational environment
  - — Management of simulation and data collection

These phases in simulator design highlight the need to analyse the real operational environment in detail before the specification of what to simulate can be presented. Of particular importance is the need to define clearly the classification of human performance measures that are critical to the objectives of the application. Particular attention should be given to the dynamic versus the static aspects of the environment, the complexity of the behaviours and the human–machine interface, the complexity and the nature of displays and the key events to be simulated.

7.5 How Should the Task be Represented?

There is a lively debate about what aspects of the task should be represented. The debate ranges from full representations of everything to a microcosm of an environment and to an abstraction of the environment. The following sections provide an overview of the main debates.

7.5.1 Fidelity

The main issue is the degree of fidelity that is required. Fidelity is the degree of similarity between the simulator and the equipment that is being simulated. A two-dimensional definition of fidelity refers to the degree to which the simulator looks like (physical fidelity) and acts like (functional fidelity) the real operational equipment. The lack of systematic research into these issues makes comparison of the studies difficult (Hays, 1981), but the main opinions are either that the simulator should resemble the real equipment as closely as possible (Baum *et al.* 1982; Gonsalves and Proctor, 1987), or that reduced physical fidelity, whilst maintaining functional fidelity, will improve training (Stammers, 1979, 1981; Montague, 1982; Boreham, 1985). One study supporting the latter approach postulates that often 80% of the benefit can be achieved at 20% of the cost of full fidelity (Welham, 1986).

A classic example of poor simulator design is the former Esso simulator used to train manoeuvres in supertankers. Trainees manoeuvred a 1:25 scale model of the tanker, that was perfect in every detail, except that all commands were carried out five times as fast as in reality (in a square root relation to the scale of the model). This led to some significant errors when the trainees were allowed to pilot a real supertanker (Sanders, 1991).

The degree of fidelity required also appears to be dependent upon task type. For example, perceptual-motor tasks require greater physical fidelity than cognitive tasks (Baum *et al.*, 1982). Experimental studies have indicated a relationship between transfer and stimulus–response characteristics of the simulated and real operational equipment. This is shown in Table 7.1.

Positive transfer occurs when the stimulus and response elements are the same in the simulated and real environments, whereas negative transfer occurs if response elements are different (whilst retaining the same stimulus elements). When the stimulus elements are the same and the response elements are incompatible, high negative transfer occurs. No transfer occurs if both the stimulus and response elements are different. This means that for the simulator to be effective it should behave in essentially the same way as

Table 7.1 Relationship between simulator and task (adapted from Wickens, 1992)

Stimulus elements	Response elements	Transfer
Same	Same	High positive
Same	Different	Negative
Same	Incompatible	High negative
Different	Same	Positive
Different	Different	None

the real equipment, i.e. an operator input into the system should have the same effect as it would in the real system. This does not mean that all of the elements of the real operational environments need to be provided, but that those that are provided need to act in a manner that is functionally equivalent to the way in which they would behave in the real equipment. Comparisons can be drawn between studies of transfer in aircraft for example; there will be a high positive transfer between two aircraft with identical control layouts and movements, even if they have very different cockpit displays. Thus, it is the functional fidelity of the simulator that is of major importance to the transfer of operator behaviour.

The motivation behind high physical fidelity is, perhaps, in the absence of unequivocal evidence to demonstrate that simulators that are low in physical fidelity, whilst retaining high functional (psychological) fidelity produce a high degree of transfer: i.e. conclusive proof that there is a good (positive) correlation with performance on the low physical fidelity simulator and performance on the real operational equipment. The layperson may favour high physical fidelity because it has high face validity, i.e. if it looks like the real equipment then it must lead to better training, etc. Despite this gap in our knowledge, the argument in favour of the low-fidelity approach is gaining momentum.

Rolfe (1985) considered six components of fidelity: physical, dynamic, engineering, operational, instructional and psychological. Each of these will be considered in turn.

1 *Physical fidelity* This is the degree to which the physical characteristics of the simulator match the real task situation.
2 *Dynamic fidelity* This is the degree to which the simulator works in the same way as the real equipment.
3 *Engineering fidelity* This is the degree to which the simulator is built in the same way as the real operational equipment.
4 *Operational fidelity* This is the degree to which the simulator, over a period of time, can match the operation of the real task situation.

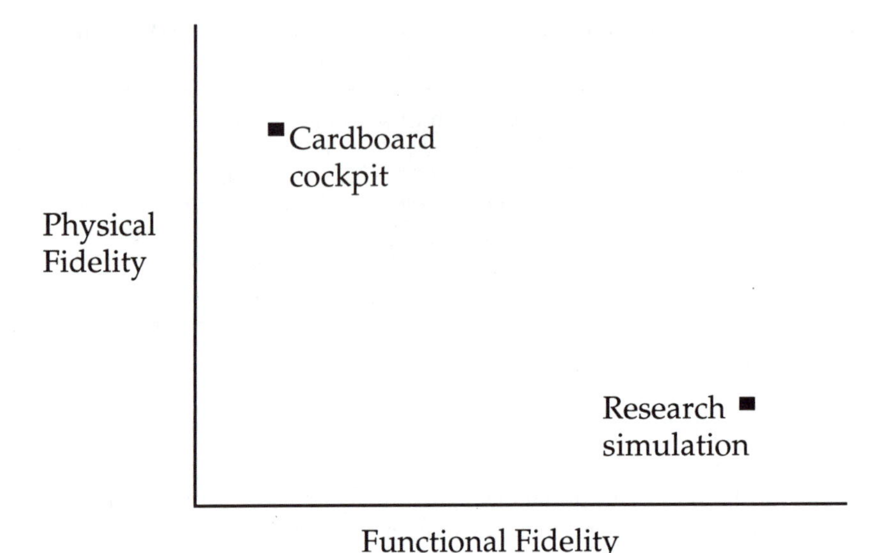

Figure 7.1 An illustration of two dimensions of fidelity.

5 *Task fidelity* This is the degree to which the task domain on the simulator reflects that in the real operational situation.
6 *Psychological fidelity* This is the degree to which transfer occurs, despite the lack of other aspects of fidelity.

The fidelity may be compared to a spectrum, and any one simulator may be at various points on the spectrum for various aspects of fidelity. For example Figure 7.1 indicates where two simulators (a cardboard cockpit and a research simulator) lie on two axes (physical and dynamic fidelity).

Figure 7.1 illustrates that research simulators fit into a category of high dynamic fidelity but low physical fidelity, whereas the cardboard cockpit has high physical fidelity but low dynamic fidelity.

7.5.2 Dimensions of Simulation

Stammers (1983, 1986) proposed nine dimensions of simulation, which will be considered here:

1 *Stimulus/displays* This dimension concerns the physical resemblance of the simulator in relation to the plant. How realistic are the displays that are presented and is the control room completely represented or only partially?
2 *Responses/controls* In a similar vein, to what extent are the physical characteristics of the control devices represented in the simulator? Again the emphasis is on the physical fidelity of the situation.

3 *Display–control relationships* Whilst it is possible to have a realistic control panel, just how do the displays and controls interact and represent the dynamics of the plant? This is both a hardware and a software problem with, for example, a considerable amount of effort in software modelling of plant dynamics.

4 *Task complexity* The interaction of hardware, software and the extent to which all aspects of the plant operation are represented in the system. It may be that only some parts are represented, the others being static, or it may be an attempt to produce a comprehensive simulation of all the possible states of the plant. An alternative may be to present examples of particular fault conditions or to allow only the practice of start-up and shut-downs.

5 *Temporal aspects* This dimension concerns the extent to which the simulator reacts in real time or is deliberately slowed down or speeded up. This is a feature that is often built into simulators. It is a deliberate attempt to make the simulator unlike the real plant. Altering the temporal aspects of the simulator makes it possible to create faster system responses to control actions or give the operator time to consider alternative courses of action.

6 *Environmental stressors* One of the reasons that simulators are used is to remove the operator from environmental stress. This would be in terms of such things as noise, bad lighting conditions, dangerous operating states, etc. Whilst this may be laudable for initial training of operators, it may not be as useful for judging how they will react under real operating conditions when the stressors will be present.

7 *Situational payoffs* An additional source of potential stress to operators is the consequence of error that may lead to a high negative payoff. This could be in the form of a plant shut-down, turbine or reactor trip which, when preventable, could lead to some sanctions applied to the operator. Removing these consequences could lead to some changes in behaviour.

8 *Social environment* The social environment of the workplace is also a factor that can be more or less simulated. This is particularly relevant to team task situations where social skills may be particularly important for successful performance.

9 *Additional control* This refers to the extent to which control and feedback loops are built into the process simulator. Measurements of performance can be taken from a variety of variables, which may not be possible with the real operational equipment. Extra features, for example system freeze and playback, may be built into the simulator. These features are extra characteristics that the simulator has, and it is possible that many of these features are far more complex than those available on the real plant, for example measurement of operator performance.

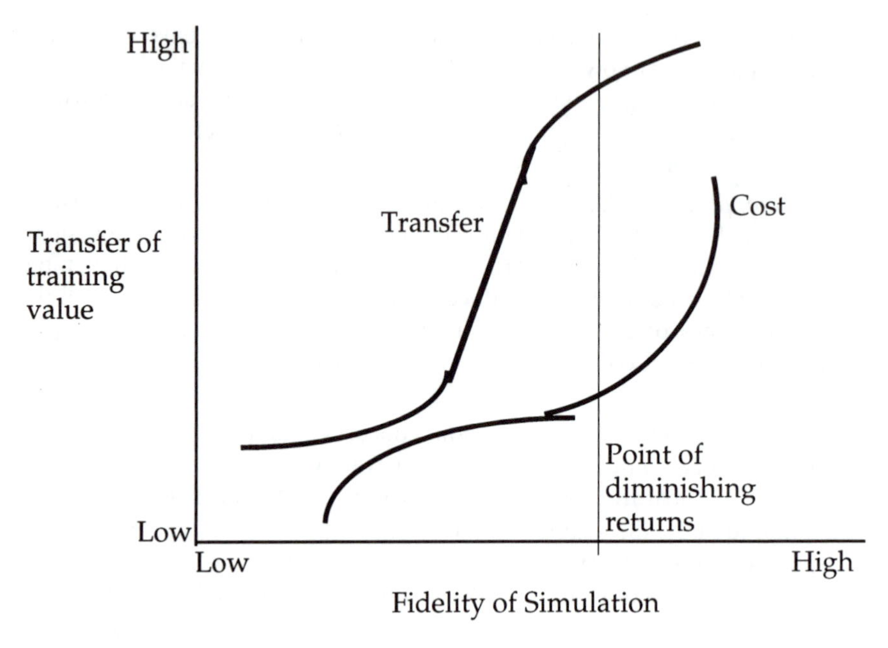

Figure 7.2 The relationship between cost, fidelity and transfer of training.

However, these features are built in to enhance the evaluation process rather than to represent the real equipment.

7.6 Training Simulators

It has been suggested that the transfer value diminishes as the cost of implementing the simulation increases beyond a certain point (Miller, 1954). This point is illustrated in Figure 7.2.

In training environments, maximum benefit appears to be derived from initial low-fidelity simulation followed by moderate-fidelity simulation for troubleshooting tasks. This mixed-fidelity approach has found favour in more general aspects of technical training (Rouse, 1982). The weight of evidence, admittedly mainly from the training literature, does support the notion that high functional fidelity does not have to be supported by high physical fidelity (and in many instances high physical fidelity would make performance worse) to have a beneficial effect on transfer performance (Stammers, 1979; Freda and Ozkaptan, 1980; Hays, 1981; Baum *et al.*, 1982; Yuan-Ling, 1984; Stammers, 1985, 1986; Welham, 1986). It is recognised that under certain circumstances it may be difficult to persuade organisations to adopt this viewpoint, for example licensing of operators (Gonsalves and Proctor, 1987).

The use of simulators may be justified in terms of cost effectiveness, for example Roscoe (1971) determined a transfer effectiveness ratio (TER) with the following equation:

$$\text{TER} = [(\text{TE} - \text{SIM}) - (\text{TE} + \text{SIM})]/(\text{TE in SIM})$$

where:

(TE − SIM) = training effort required to learn the job on the operational equipment without the aid of a simulator;

(TE + SIM) = the training effort needed to learn the job when some of the training is undertaken using a simulator.

The difference between (TE − SIM) and (TE + SIM) is a measure of the training resources saved by the use of the simulator. However, this needs to be considered in the context of the amount of effort expended in learning the tasks in the simulator: (TE in SIM). This measure of transfer is sensitive to both positive and negative effects, and thus is particularly useful. Highly realistic simulations tend to be very expensive, but their added realism may add little to their TER.

An analysis of transfer of training was reported for a tank training simulator by Moraal and Poll (1979) cited by Sanders (1991). They observed a strong positive effect from the simulator to the tank with respect to four tasks: shifting gear (A), steering (B), a combination of shifting gear and steering (C) and avoiding obstacles (D). The results are indicated in the following figure.

Figure 7.3 illustrates the utility of the training simulator, where the white bars represent the group without the simulator training and the black bars represent the group with both simulator and time in tank training. The overlapping black line shows the time needed in the task after time spent in the simulator substantially reduces training time on the operational equipment for all tasks A to D. This has important implications, not only for training, because it suggests that it is possible to create a representation that leads to a high degree of concordance between activities on the simulator and activities in the operational environment. This may occur at both a physical and cognitive level independently.

Simulators appear to be most appropriate when a dynamic environment is needed, i.e. when the operator needs to interact with a model of the system that approximates to the real operational system in some way and feedback from their actions is necessary. Other media, such as text, audio, slides and video, may not provide such a rich environment in terms of interaction, nor have the means to capture detailed inputs from the operator. There is an increasing realisation that simulation does not have to be the most costly option, as Caro (1972) succinctly stated:

> Even now, there is substantial applied research evidence that much of the training conducted in expensive simulator could be accomplished in less expensive devices.

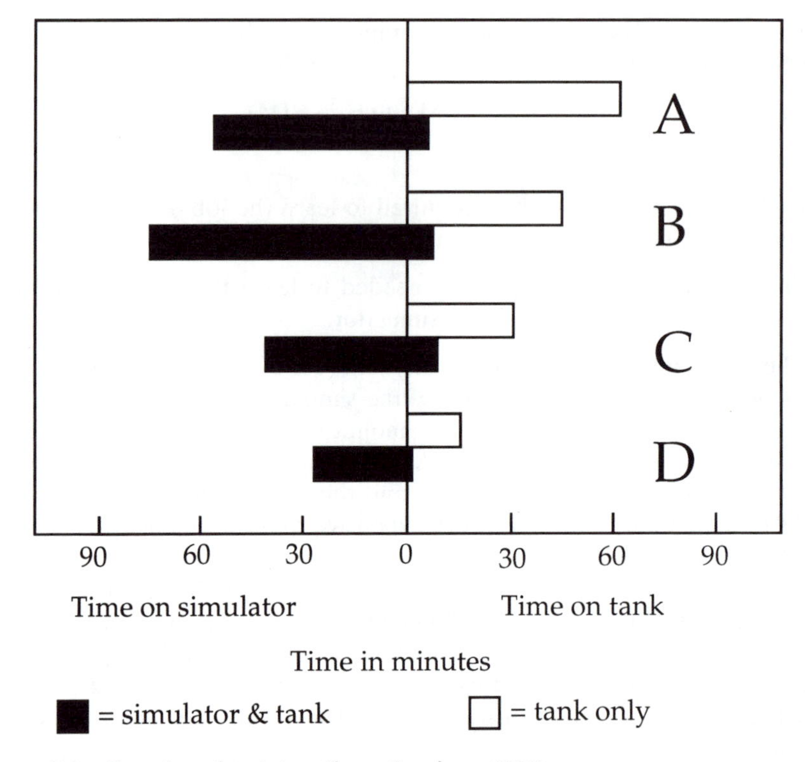

Figure 7.3 Transfer of training (from Sanders, 1991).

Microcomputers offer a relatively inexpensive means of achieving simulation at a fraction of the cost of traditional, full-scope, simulators (Sklaver, 1986). Adams (1971) made the point clearly when he stated:

> I would not consider the money being spent on flight simulators staggering if we knew much more about their training value, which we do not. We build flight simulators as realistically as possible, which is consistent with the identical elements theory of transfer of Thorndike, but the approach is also a cover-up for our ignorance about transfer because in our doubts we have made a costly device as realistic as we can in the hopes of gaining as much transfer as we can . . . the result has been an avoidance of the more challenging questions of how the transfer might be accomplished in other ways, or whether all that complexity is really necessary.

Some research suggests that as the competence of the operator increases, so does the demand for higher fidelity to optimise performance and transfer (Johnson and Rouse, 1982). This suggests that in the initial stages of training a very simple simulation is required, whereas when training is advanced greater fidelity is needed. One researcher recommends that it is more appropriate to have 20 low-fidelity simulators and one actual equipment trainer than two actual equipment trainers (Dallman, 1982).

7.7 How Can We Be Sure That They Are Useful?

A variety of justifications has been offered for the use of simulators, for example:

- high utilisation;
- lower purchase and operating costs;
- increased safety and controlled level of stress;
- conditions not readily available in the real world;
- less operational and environmental disturbance than using real equipment.

However, there are some deceptive justifications of simulator effectiveness (Rolfe and Caro, 1982), using such criteria as:

- user opinion;
- the realism of the simulator;
- the utilisation of the simulator;

User opinion is suspect because of three main reasons. First it assumes that the user is a good judge of what makes an effective simulator. Second, user opinion is based on previous experience, and unless there is a high degree of homogeneity of experience, opinions may vary widely. Finally judgements may be obtained from an unrepresentative sample.

The concept of realism (cf. fidelity) has problems because it is multi-dimensional, e.g. physical, dynamic, operational, temporal, etc. The amount of each dimension of realism required will vary with the appropriateness to the particular task type, e.g. the difference between cognitive and perceptual motor tasks indicated in Section 7.4. In order to explore generic cognitive functioning and reliability it may be necessary to deliberately depart from the aim of reproducing the operational conditions completely.

The utilisation of a device is often favoured as a measure of estimating effectiveness. However, it is a poor indication. There are always pressures to make sure that a device is shown to be in use for as great a part of its serviceable life as possible. This demand fosters the phenomenon of expanding the use to fill the time available. Although the device is in constant use, it does not mean that this time is being used effectively and efficiently.

More credible measures of simulator effectiveness are the degree to which performance on the simulator correlates with activities in the real operation environment. This may be measured in a variety of ways. Reliability and validity are the concepts that are appropriate for measuring effectiveness.

7.7.1 Reliability

The reliability of the data gathered from the simulator may be tested in a variety of forms. Essentially it concerns the degree to which the simulation can be repeated with the same reliability relative to the outcome. Three methods of determining reliability of the data are:

- Test–retest reliability
- Alternative form reliability
- Internal consistency reliability

Test–retest reliability is the correlation of performance on the simulator with the performance obtained by the same group of people retested after some period of time. Alternative form reliability is an estimate of the extent to which different forms of simulator measure the same performance. Internal consistency is a means of determining homogeneity of performance by using split-half coefficients resulting from comparative sessions of the simulation.

7.7.2 Validity

Similarly there are a number of methods for establishing validity of simulator, some more credible than others. Types of validity are:

- face validity;
- faith validity;
- content validity;
- concurrent validity;
- predictive validity;
- construct validity.

Face validity concerns whether the simulator appears to relate to the task it is supposed to represent. Face validity can help the acceptance of the simulator, but does not necessarily guarantee that transfer of performance will occur. Therefore this can be very misleading, as was discussed earlier. Views on simulators can be extreme. The blind acceptance or rejection of a simulator without recourse to evidence is not recommended. The simulator may prove to be a useful tool in testing the control system hardware, but this does not necessarily mean that it will be equally useful in training operators or collecting human reliability data. The acceptance of a simulator without objective validation is termed faith validity. Content validity of a simulator is the extent to which the simulator can be rationally related to the activities that make up the real operational task. Content validation normally rests upon task analysis exercises. Concurrent validity is the relationship between the performance obtained on the simulator and some criterion of actual task performance that can be obtained at the same time. Thus, if we were to correlate the activities of operators on the simulator with the results of ratings of actual work performance, we would have undertaken a concurrent validity study. Predictive validity is the extent to which performance on the simulator correlates with some future outcome. This is particularly important if the behaviour cannot be observed concurrently, for example due to its infrequency (such as real emergency events). There may be some considerable delay before validation is complete, but we may still wish to exploit the data. Therefore we would attempt to satisfy other relevant forms of validity, such as content validity, and concurrent validity on a full-scope simulator, before we are able to check predictive validity. It can sometimes be difficult to establish comparable measures of performance, and it is also important to have a cross-section of performance to avoid suffering from problems related to restricted sample range. Construct validity is more abstract than other forms of validity, and is the extent to which the simulation is based upon some theoretical construct. For example, we may have a theoretical notion of human information processing and error which we seek to examine. Building up a picture of construct validity is typically a long process and goes beyond the mere visual inspection of intercorrelations.

The greater the degree of correlation between behaviour on the simulator and behaviour in the real operational environment, the more confidence can be placed in the data derived from the simulator. If the same types of activities are being compared, i.e. the operators in the simulated environment are involved in the same type of generic activities as those in the real operational situation, we may use a non-parametric correlation coefficient to determine the degree of correspondence. Thus it is possible to infer whether the simulated environment involves the operator in the same types of mental activities as those inherent in the real situation. Validation is an essential part of the simulator development process; without it there is no way of being sure that the simulator actually results in representative activity of the operational environment. Once this has been established, it may be possible to use the data derived from the simulated environment to predict the success of behaviour in the real environment.

7.8 The Use of Simulators to Collect Data on Human Performance

A satisfactory simulation for one purpose cannot be assumed to be acceptable for a different purpose without further validation (see faith validity). Simulators have been used to provide data for human reliability assessment. Electricité de France and the Electric Power Research Institute have been using requalification training simulators to test the human cognitive reliability (HCR) technique (Kantowitz *et al.*, 1991).

Simulators have been used for the collection of data for a wide range of purposes under experimental settings, and are well reported in the technical literature. The following consists of a representative sample of the type and nature of such studies. A full-scope, real-time, simulator of an 1100 MWe BWR power plant was used to evaluate an operator decision aid. The aim of the experiment was to evaluate the effect of the decision aid on the operator's decision-making process under simulated malfunctions. Analysis of the data was classified through a model of the operator's decision-making process consisting of:

- detection of a deviation from normal state;
- interpretation of the situation;
- selection of a goal state;
- planning the task strategy;
- execution of the procedure based on the strategy.

The results suggested that the provision of the decision support system aided operators in diagnosing adverse plant conditions and in establishing counteractive strategies. Thus the use of a simulator provided a means of collecting data for use in evaluating the effectiveness of the decision support aid. Alternatively, it may be used to examine operators' behaviour. In this latter study the simulated trials were used to collect data on operator error. There were 200 errors observed in 31 tests, and 50% of the errors were regarded to be of significant enough concern to be worthy of further analysis.

A more recent investigation into human error through the use of simulators has classified errors into Rasmussen's (1986) model of human performance. The researchers assign the errors they observed into skill-based, rule-based and knowledge-based (Ujita *et al.*, 1989). The distinctions between these types of errors have been summarised eloquently by Reason (1990) and are presented in Table 7.2.

This type of classification system provides a useful diagnostic tool to prompt investigation into the cause of human error in system operation. Ujita *et al.* (1989) were able to classify generic types of error in this way by comparing actual performance with the correct procedure to generate error level statistics.

7.9 Simulators and Teams

Other studies have focused on team behaviour (Kawano *et al.*, 1991; Sasou *et al.*, 1991). From the analysis of operators' behaviour during emergency situations using simulators, Kawano *et al.* (1991) were able to draw the following conclusions:

- Errors were observed during the minutes following a 'scram'.
- Some errors were related to high workload.

Table 7.2 The distinctions between skill-based, rule-based and knowledge-based errors (from Reason, 1990)

Dimension	Skill-based errors	Rule-based errors	Knowledge-based errors
Type of activity	Routine actions	Problem-solving activities	
Focus of attention	On something other than the task at hand	Directed at problem-related issues	
Control mode	Mainly automatic processes (schemata)	(stored rules)	Limited conscious processes
Predictability of error types	Largely predictable (actions)	'strong but wrong' (rules)	Variable
Ratio of error to opportunity for error	Though absolute numbers may be high, these constitute a small proportion of the total number of opportunities for error		Absolute number small but error ratio high
Influence of situational factors	Low to moderate; intrinsic factors (frequency of prior use) likely to exert the dominant influence		Extrinsic factors likely to dominate
Ease of detection	Detection usually fairly rapid and effective	Difficult and only often achieved through external intervention	
Relationship to change	Knowledge of change not accessed at proper time	When and how anticipated change will occur unknown	Changes not prepared for or anticipated

- Some errors were related to poor man–machine interfaces.
- Differences in team performance were observed.

Sasou *et al.* (1991) have developed a method for describing operator communications, activities and observation points for analysing team activities. The 'team activity description method' (TADEM) consists of taxonomic classifications for each of the three activities (communications, activities and observation points). These are represented separately as three time-line diagrams. Sasou *et al.* (1991) propose that this form of analysis will be useful for describing team activities when analysing abnormal operating conditions. However, since it is behaviourally based, there does not appear to be any means of annotating the cognitive elements of the task, which will ultimately limit the usefulness of the method.

It is generally assumed that the stress experienced by operators under simulated conditions is not the same as that in the real operational environment. This may be largely due to the lack of risk in the simulated

1. TASK COMPLEXITY FACTORS

1.1. System demands
- High risk environment
- Short system time window
- Workload and multi-tasking
- Unfamiliar situations

1.2. Information uncertainty
- Difficulties and delays in obtaining information
- Ambiguous information
- Conflicting information
- Information overload

1.3. Planning difficulties
- Rigid safety priorities
- Conflicting goals
- Uncertainty about outcome
- Negative performance feedback
- Lack of means or facilities

2. PHYSICAL OR ENVIRONMENT FACTORS
- Noise
- Distractions
- Extreme thermal conditions
- Fatigue and sleep loss
- Hours of work been on duty

2. SOCIAL & ORGANISATIONAL FACTORS
- Role conflict
- Role ambiguity
- Poor management decisions
- Poor group communication
- Hierarchical team structure
- Rigid company culture

3. PSYCHOLOGICAL FACTORS
- Type A personality
- External locus of control
- Lack of experience with stress

Figure 7.4 List of stressors (from Kontogannis and Lucas, 1990).

environment. However, it is first worth considering which elements of the operational environment are likely to be the source of stress. This is indicated in Figure 7.4.

From the list of stressors above, however, it may, to some extent, be possible to re-create some of these in the simulated environment. For example, time pressure-related stressors are relatively easy to re-create, by demanding that the operators respond within a limited time window. At a more general level, some level of stress is likely to accompany the task as operators will be aware that their responses are being recorded for analysis. Task-related stress may also be experienced if the task being performed is of a critical nature, for example mitigating the effects of a LOCA. Thus it is possible to induce some forms of stress into the simulated environment.

7.10 Process Industry Applications

Simulators have also been used in more experimental settings, such as to investigate the transfer of fault-finding skills (Patrick *et al.*, 1989), perform research into methods of representing process plant information for fault diagnosis (Praetorius and Duncan, 1991) and compare annunciator systems (Kragt, 1983). In such paradigms, lower-fidelity simulation appears to be acceptable, and yet the data derived from these studies are used to make informed recommendations and decisions regarding the design of complex control room environments. For example: Patrick *et al.* (1989) found qualitative differences in transfer performance, which indicated the processes involved; Praetorious and Duncan (1991) supply evidence to suggest that functional plant representation supports the kind of reasoning used in fault diagnosis; and Kragt (1983) reports a study to show that reliance upon a single method of alarm presentation should be avoided.

The argument in favour of low-fidelity simulation is that it enables control over variables that would not be possible in more natural settings. This sacrifice of realism is in the interests of ensuring that we may be certain which variables are influencing behaviour. Recent research illustrates this point well. In a study to investigate the differences in performance for three types of visual alarm media (text, annunciator and mimic), Stanton (1992) presented subjects with a simulated process control task. It was proposed that subjects in a mimic alarm condition would be more able to optimise the process because the proximity of the alarms would be within the current field of view when controlling the process and they would have no spatial conversion to make. However, subjects in the text alarm condition are likely to have better fault-management prioritisation, because they should be able to compare their priorities from within the scrolling text display. Subjects in the annunciator condition are likely to respond quicker to presented alarms because the flashing annunciator is more likely to draw their attention quickly than the other two presentation methods.

To conclude, these studies have demonstrated the necessity for task decomposition and low-fidelity simulation for research into human performance. Without this degree of control it is difficult, if not impossible, to determine where the effects lie. Employing this approach does not, however, sacrifice the worth or applicability of the findings. Rather, it serves to illustrate that the results of the studies can be used to make informed decisions, rather than being hopelessly non-significant or dangerously wrong. In the first study, much of the fine-grained information elicited in the second study was either lost in the noise or impossible to measure.

The use of low-fidelity simulation has therefore demonstrated benefits in collecting data on human performance. In the last illustration, it was noted that it enables researchers to concentrate their efforts upon the fundamental aspects of the task.

7.11 Simulations and Human Reliability Data Collection

The use of simulators to collect data for use in human reliability assessments is not a new idea. Embrey and Ainsworth (1983) proposed this method of data collection over a decade ago and suggest that the process of simulation and data collection needs to be structured by the application of a taxonomy or classification system, so that then data can be generalised to all categories of that task for which the assessments are required. This approach is in contrast to the task decomposition methods, which would seek exhaustively to identify every conceivable scenario event. The two approaches may be characterised as either nomothetic (looking for commonality and deriving generality) or idiographic (scrutinising individual events and arriving at specifics). The latter approach suffers from lack of use beyond the particular event under investigation, whereas the former approach may be too general to apply to specific scenarios. However, since it is impossible to examine every single event in sufficient detail to make the idiographic approach universal (as this would require almost an infinite number of experiments to be conducted) the most promising means of collecting the data required is the nomothetic approach.

Spurgin (1990) report the use of simulators to collect human reliability data during disturbances at nuclear power plants. They collected data from a variety of simulators under various scenarios using high-fidelity, full-scope simulators. The principal aim of the studies was validation of the HCR model of human reliability. The authors of the report suggest that the data gained in the course of the studies provide a useful means of gaining insight into crew behaviour in handling incidents. The data have led to some revision of the HCR model, an improved data collection methodology and a database of crew responses from seven plants. However, the authors also question the validity of the data gathered. The simulators used were physically similar to the actual plants, but some differences remain. This means that there could, in principle at least, be some differences in the way that crews behaved in the simulated conditions, and how they would behave in the real conditions. One could question the rationale of physical similarity, as it is likely to produce situational-specific results: i.e. data that are unlikely to be useful beyond predicting behaviour in all but the very special circumstances in which they were collected. Spurgin *et al*. (1990) claim that the behaviour observed was not very different to recorded behaviour of actual incidents, but only a limited data set of actual incidents exists. Data collected from a higher-level, cognitive-task simulator, may be able to produce data with wider implications and a broader reference set. The use of simulators certainly offers a more objective means of data collection than the alternative methods of human reliability assessment. Thus it is an attractive proposition.

The ACSNI study group on human factors made two main pleas regarding data collection on simulators for use in human reliability assessment. First they argue that there is a limited amount of information regarding the cognitive functioning of humans, despite this being a major contributing factor in man–machine system failures. This is in part due to the limited number of occasions available to collect data. Second they state that whilst it may be desirable to collect data from the real world (as this holds 'ecological' validity) in practice it would be better to collect data from experiments specifically designed for the purpose of human reliability assessment, rather than relying upon data collected for other purposes. These recommendations suggest that a series of experimental investigations in human reliability through the use of simulated events could provide a useful, and much needed, input into human reliability databases. These data may be validated in a variety of ways. First by comparison with existing human reliability assessment techniques (see Kirwan, this volume), secondly by comparison with data obtained from full-scope simulation, and thirdly with data collected from reporting systems in the operational environment. Although statistical techniques allow for concurrent validation of low-fidelity simulator data with human reliability assessment techniques, this does not guarantee that the output is predictive. Data that are comparative with high-fidelity simulation may be treated with some confidence if the high-fidelity simulator is high on all dimensions of fidelity. It may be possible to compare outputs from the low-fidelity simulation with frequently performed operations in the real environment. This would provide some indication regarding the transferability of the findings from the simulator studies.

7.12 Conclusions

The criticism that low-fidelity simulators are inappropriate because they lack ecological validity (real-world validity) has lost some force in recent years. There are examples where data derived in such experiments have been usefully applied to 'real-world' settings. There is also a good deal of scepticism amongst the community of psychologists of a wholesale abandonment of experimental rigour in favour of a totally naturalistic approach. There are so many variables influencing behaviour in the 'real world' and it is so difficult to manipulate them systematically, that it can become impossible to assess the relative importance of each variable in determining behaviour. It is no easy matter to obtain the required combination of experimental rigour and ecological validity, but consideration of the technical literature suggests that low-fidelity simulation may be an appropriate medium for the collection of human performance data. Microcomputers offer a relatively inexpensive means of implementing low-fidelity simulation. However, caution is advised on the acceptance of any particular implementation of the simulation before it has been properly validated. Thus a methodological approach to simulator design and evaluation is recommended, before data collection begins. This should ensure that the simulator represents only the pertinent aspects of the real operational environment in a manner that is appropriate. Psychological and dynamic fidelity are considered to be of greater benefit than engineering and physical fidelity for research into human performance. Thus it is essential to employ a human-centred approach to simulator design. Initial simulator design might be evaluated and fine-tuned through the use of the 'Wizard of Oz' technique. This relies upon the experimental team controlling the script of the simulator, although the subjects are unaware of this. In this way the initial bugs in the experimental method and data collection systems may be ironed out early on in the research programme.

References

ADAMS, J. A. (1971) A closed loop theory of motor learning. *Journal of Motor Behaviour*, **3**, 111–49.

ANDREWS, D. H. (1983) The relationship between simulators, training devices and learning: a behavioural view, *Proceedings of the International Conference on Simulators*, pp. 70–5, Brighton.

BAUM, D. R., RIEDEL, S., HAYS, R. T. & MIRABELLA, A. (1982) *Training Effectiveness as a Function of Training Device Effectiveness*, ARI Technical Report 593. Alexandria, VA: US Army Research Institute.

BOREHAM, V. C. (1985) Transfer of training in the generation of diagnostic hypothesis: the effect of lowering fidelity of simulation. *British Journal of Educational Psychology*, **55**(3), 213–23.

CARO, P. W. (1972) Transfer of instrument training and synthetic flight training system, *Proceedings of the Fifth Naval Training Device Center and Industry Conference*, Naval Training Device Center, Orlando, FL.

CLYMER, A. B. (1981) Simulation for training and decision making in large scale control systems. In Cheremisinoff, P. N. and Perlis, H. J. (eds), *Automated Process Control*. London: Arbor.

DALLMAN, B. (1982) Graphics simulation in maintenance training – training effectiveness at cost savings. *Conference Proceedings of the Association for the Development of Computer Based Instructional Systems*, June, Vancouver, Canada.

EMBREY, D. E. & AINSWORTH, L. (1983) *Collecting Human Reliability Data via Simulators*, Wigan: Human Reliability Associate.

FREDA, J. S. & OZKAPTAN, H. (1980) *An Approach to Fidelity in Training Simulation*, Report No. ARI-RN-83-3. Alexandria, VA: Army Research Institute for the Behavioural Sciences.

GAGNÉ, R. M. (1962) Simulators. In Glaser, R. (ed.), *Training Research and Education*. New York: Wiley.

GONSALVES, T. B. & PROCTOR, J. E. (1987) Consensus and analysis of the population of continuous real-time simulators in the nuclear power plant industry, Simulators IV, *Proceedings of the SCS Simulators Conference*, 6–7 April.

HAYS, R. T. (1981) Research issues in the determination of simulator fidelity, *Proceedings of the ARI sponsored workshop*, 23–24 July 1981. Report No. ARI-TR-547. Alexandria, VA: Army Research Institute for the Behavioural Sciences.

HSC (1991) Advisory Committee on the Safety of Nuclear Installations: study group on human factors. Second Report: *Human Reliability Assessment – A Critical Overview*. London: HMSO.

JERVIS, M. W. (1986) Models and simulation in nuclear power: station design and operation. In Lewins, J. and Becker, M. (eds), *Advances in Nuclear Science and Technology*. New York: Plenum Press.

JOHNSON, W. B. & ROUSE, W. B. (1982) Training maintenance technicians for troubleshooting. Two experiments with computer simulations, *Human Factors*, **24**(3), 271–6.

KANTOWITZ, B. H., BITTNER, A. C., FUJITA, Y. & SCHRANK, E. (1991) Assessing human reliability in simulated nuclear power plant emergencies using cascaded Weibull functions. In Karwowski, W. and Yates, J. W. (eds), *Advances in Industrial Ergonomics and Safety III*. London: Taylor and Francis.

KAWANO, R., FUJIIE, M., UJITA, H., KUBOTA, R., YOSHIMURA, S. & OHTSUKA, T. (1991) Plant operator's behaviour in emergency situations by using training simulators. In Quéinnec, Y. and Daniellou, F. (eds), *Designing for Everyone*. London: Taylor and Francis.

KONTOGANNIS, T. & LUCAS, D. (1990) *Stress in Control Room Operations*. Wigan: Human Reliability Associates.

KRAGT, H. (1983) *Operator Tasks and Annunciator Systems: Studies in the Process Industry*, Ph.D. Thesis, Eindhoven University of Technology.

MILLER, R. B. (1954) *Psychological Considerations in the Design of Training Equipment*, Wright Air Development Center Report No. 53-136.

MONTAGUE, W. E. (1982) *Is Simulation Fidelity Really the Question?* Report No. NPRDC-TN-82-13. San Diego, CA: Navy Personnel Research and Development Center.

MORAAL, J. & POLL, K. J. (1979) cited by Sanders (1991).

PATRICK, J., HAINES, B., MUNLEY, G. & WALLACE, A. (1989) Transfer of fault-finding between simulated chemical plants. *Human Factors*, **31**(5), 503–18.

PRAETORIUS, N. & DUNCAN, K. D. (1991) Flow representations of plant processes for fault diagnosis. *Behaviour & Information Technology*, **10**(1), 41–52.

RASMUSSEN, J. (1986) *Information Processing and Human–Machine Interaction*. New York: North-Holland.

REASON, J. T. (1990) *Human Error*, Cambridge University Press. Cambridge.

ROLFE, J. M. (1985) *Fidelity and its Impact on the Design, Precurement, Use and Evaluation of Flight Simulation*, Report as part of Defence Fellowship Study, Wolfson College, Cambridge.

ROLFE, J. M. & CARO, P. W (1982) Determining the training effectiveness of flight simulators: some basic issues and practical developments, *Applied Ergonomics*, **13**(4), 243–50.

ROSCOE, S. N. (1971) Incremental transfer effectiveness. *Human Factors*, **13**(6), 561–67.

ROUSE, W. B. (1982) A mixed-fidelity approach to technical training. *Journal of Educational Technology Systems*, **11**(2), 103–15.

SANDERS, A. F. (1991) Simulation as a tool in the measurement of human performance. *Ergonomics*, **34**(8), 995–1025.

SASOU, K., NAGASAKA, A. & YUKIMACHI, T. (1991), Development of the team activity description method (TADEM). In Kumashiro, M. and Megaw, E. D. (eds), *Towards Human Work*. London: Taylor and Francis.

SKLAVER, E. R. (1986) Dynamic simulation in 1991: an Exxon viewpoint, *Proceedings of the 1986 SCS Simulation Conference*, 28–30 July, Reno, NV, USA.

SPURGIN P. (1990) Another view of the state of Human Reliability Assessment (HRA). *Reliability Engineering & Systems Safety*, **29**(3), 365–70.

STAMMERS, R. B. (1979) *Fidelity of Simulation and the Transfer of Training*. Final report to SSRC on Grant 5326, December. Birmingham: Applied Psychology Unit, Aston University.

(1981) Theory and practice in the design of simulators, *PLET*, **18**(2), 67–71.
(1983) Simulators for training. In Kualseth, T. O. (ed.), *Ergonomics of Workstation Design*. London: Butterworth.
(1985) Instructional psychology and the design of training simulators. In Walton, D. G. (ed.), *Simulation for Nuclear Reactor Technology*. Cambridge: Cambridge University Press.
(1986) Psychological aspects of simulator design and use. In Lewins, J. and Becker, M. (eds), *Advances in Nuclear Science and Technology*. New York: Plenum Press.
STANTON, N. A. (1992) *Human Factors Aspects of Alarms in Human Supervisory Control Tasks*. Birmingham: Applied Psychology Group, Aston University.
UJITA, H., FUKUDA, M. & KUBOTA, M. (1989) Plant operator performance evaluation using cognitive process model. Paper presented at the *International Ergonomics Association Triennial Conference*, Paris, France.
WELHAM, D. J. (1986) Technology-based training and simulation. In Labinger, M. and Finch, P. J. (eds), *Technology-Based Training*. Maidenhead: Pergamon Infortech.
WICKENS, C. D. (1992) *Engineering Psychology and Human Performance*. New York: Harper Collins.
YUAN-LING, D. S. (1984) A review of the literature on training simulators: transfer of training and simulator fidelity, Report No. TR-84-1. Atlanta, GA: Georgia Institute of Technology.

PART THREE

Personnel Issues

CHAPTER EIGHT

Assessing nuclear power plant operator variability

YUSHI FUJITA[1] and JODY L. TOQUAM[2]

[1]*Mitsubishi Atomic Power Industries Inc, Tokyo*

[2]*Battelle Human Affairs Research Center*

8.1 Introduction

It has been widely recognised that operators working in a highly automated large-scale human–machine system such as a nuclear power plant (NPP) may experience cognitively demanding situations during emergency in which workload tends to be high and pre-planned operating procedures may not be fully applicable. Unfortunately, this sometimes causes operators to exhibit erroneous behaviour, as has been shown in some disastrous incidents (e.g. Three Mile Island; Chernobyl). To identify effective measures against this inherent human factors deficiency of highly automated large-scale human–machine systems, it is crucial to understand the reasons that cause the misconduct of the operator. However, the reasons are often diverse, and it is sometimes, if not always, difficult to judge whether operators themselves were really responsible, or work environments – either design or organisational – were deficient. Any human performance or behaviour – either normal or erroneous – must however be seen as a manifestation of interactions among human mechanisms, internal factors (e.g. trait, experiences, knowledge, physical and mental state), and external factors (e.g. design, organisation). It is therefore important to understand the nature of human performance or behaviour, not in isolation, but in conjunction with the internal and external factors. This chapter presents some findings concerning the performance characteristics and performance-shaping factors (PSF) of NPP operators. These findings were obtained from a series of experimental studies using a full-scale training simulator.

8.2 Methods of Data Collection

A full-scale simulator was used to collect behavioural data. The simulator was a replica of an 840 MWe, three-loop pressurised water reactor (PWR) used for training utility operators. On the other hand, psychological instruments were used to collect data on PSF. The data were collected in two phases. Those collected in the first phase were used for the analyses of performance characteristics under a high workload situation (Analysis 1) and characteristics of errors of omission and commission (Analysis 2), and the preliminary analysis of PSF (Analysis 3). The data collected in the second phase were used for the detailed analysis of PSF (Analysis 3).

8.2.1 Scenarios

From about thirty scenarios used frequently in training sessions, seven scenarios were selected for the data collection in the first phase, which included four major accident scenarios and three minor incident scenarios. The selection was based on subjective rating by twenty-one utility operators, fifteen instructors and four start-up engineers. In the second stage of the data collection, three scenarios were selected from the seven scenarios used in the first phase.

8.2.2 Subjects

In all, 49 training crews formed by experienced operators were used as subjects in the first data collection. However, not all the crews went through all the scenarios. The number of crews per scenario ranged from 44 to 49. In the second phase, a total of 46 training crews of equal quality were used to collect data. In both cases, the following two training courses were involved:

1 *Normal retraining course* A two-week training course for operators who had already completed the initial training course. Temporal training crews were formed by three trainees from different plants and possibly different utilities. Generally, their level of job experience was the same. Many of the trainees who completed this course are reactor operators at their sites.
2 *Advanced retraining course* A one-week training course for trainees who have already completed the normal retraining. As with the normal retraining courses, temporal training crews were formed by three trainees who have similar levels of job experience (from reactor operators to supervisors).

The three trainees were assigned to the following three job positions (they rotated the positions during the training sessions):

1 *supervisor* an operator responsible for operational decisions;
2 *reactor operator* an operator responsible for the primary systems;
3 *turbine operator* an operator responsible for the secondary systems.

8.2.3 Collection of Behavioural Data

In both normal and advanced retraining courses, behavioural data were collected during the last two or three days, after sufficient familiarisation. Subjects were told that their behaviour would be monitored during the data collection period for the purpose of experiments. Nevertheless, the exact date and time of data collection as well as scenarios to be given were not known to them. Using four video cameras, subjects' behaviour was recorded. Each subject was asked to put on a microphone throughout the data-collection period (2–3 days). Their verbal utterances were recorded on the videotapes. Their control actions, together with process variables and component status information, were recorded by a computerised data logger. In addition, subjects' performance was rated by trained expert raters (i.e. instructors or experienced start-up engineers) through direct observation of subjects' behaviour (Analysis 3 – preliminary analysis), or indirect observation using videotapes (Analysis 3 – detailed analysis). The expert rating used a seven-point subjective performance measure which involved both individual and crew performance factors. The performance factors for individuals included:

- understanding plant status;
- supervising co-ordination;
- communication;
- duty execution;
- supportive activities;
- spirit.

For crew:

- team co-ordination;
- crew performance;
- team spirit.

Inter-rater agreement levels ranged from 0.62 to 0.97 with a mean value of 0.80 (preliminary analysis) and ranged from 0.42 to 0.88 with a mean value of 0.63 (detailed analysis).

8.2.4 Collection of Data on Performance-shaping Factors

A variety of psychological instruments developed and used in the US or other western countries were selected or developed to collect PSF data. Some of the PSF instruments were administered at subjects' home plants before or after the simulator training sessions, and others were administered during the simulator training sessions. The following criteria were used to select the PSF measures:

- Previous research supports their use.
- A Japanese translation was available or little translation was needed.
- Administration was easy and time required was short.
- Sample data showed sound statistical properties (e.g. no ceiling or floor effects).

In a preliminary analysis using the first behavioural data sets, a broad range of PSF were examined to ensure full coverage of PSF that had potential for predicting performance. Later, more specific PSF were examined. Table 8.1 presents a list of the PSF and PSF instruments administered in a detailed analysis using the second behavioural data sets.

8.3 Analysis 1: Performance Characteristics Under a High-workload Situation

Operators are said to experience high workload during plant transient caused by automatic protective or engineering safeguard actions. They are required to confirm that automatic actions are functioning correctly by verifying the status of relevant components. They must verify that the plant process is controlled properly by automatic actions. Using the behavioural data collected from seventeen crews, erroneous tendencies that emerged in these highly procedural verification tasks were analysed. Among the seven scenarios selected for the first data collection, a typical design basis event was used for Analysis 1. It was an accident scenario in which rupture occurs to one of several heat exchangers. Because of pressure imbalance between the primary and secondary sides of the heat exchanger, leakage takes place through the rupture. Subjects follow scenario-based accident procedures that require them to isolate the faulty heat exchanger and balance the pressure mismatch so that the leakage stops. The leak rate was set to be large enough to cause an automatic protective action (reactor trip, RT) and an automatic engineered safeguard action (safety injection, SI).

8.3.1 Data Reduction and Analysis Methods

Subjects are trained to use their fingers to point to monitoring targets (e.g. meters, lamps, annunciator windows) and call out their names. This enabled

Table 8.1 List of performance-shaping factors (PSF) and PSF instruments used in Analysis 3

PSF category	PSF assessed	Instruments
Cognitive ability	■ Perceptual speed ■ Perceptual speed attention ■ Knowledge of job requirements in four areas including basic, design, operations, and others	■ ETS number comparison ■ Battelle visual scanning test ■ Job knowledge test (specially developed)
Personality	■ Adjustment ■ Dependability ■ Validity ■ Stability ■ Impulsive behaviour ■ Tendency to strive for competence in one's work ■ The level of intellectual efficiency attained ■ Tendency to create a favourable impression ■ Potential for becoming an effective leader ■ Tendency to distort one's response intentionally to look good	■ MMPI – Scale 4 ■ MMPI – Scale 7 ■ MMPI – Scale L ■ MMPI – Ego strength ■ MMPI – Impulsiveness ■ MMPI – Academic achievement ■ CPI – Intellectual efficiency ■ CPI – Good impression ■ CPI – Managerial potential ■ Social desirability
Background stress and stress-coping measures	■ Personal background stress ■ Anxiety, depression and feeling of helplessness ■ Extent to which one seeks or avoids information when threatened and one distracts oneself when threatened ■ Perceived stress coping	■ Recent life change questionnaire ■ Psychiatric epidemiology research interview ■ Miller behavioural style scale ■ Cohen's perceived stress scale
Leader behaviour	■ One's perception of one's superior's behaviour (performance-oriented or social) ■ One's perception of one's superior's behaviour (transformational, transactional, . . .)	■ Performance – maintenance ■ Transformational leadership
Background experience	■ Experience and other individual characteristics such as age, physical characteristics, academic background, job experience, and training experience	■ Background questionnaire (specially developed)
Group interactions measures	■ Motivation to work under certain conditions ■ Group cohesiveness ■ Reported stress from one's superior or subordinates	■ Least preferred co-worker scale ■ Group atmosphere ■ Job-related stress

Notes:

1 ETS: Educational Test Service; Minnesota Multiphasic Personality Inventory (MMPI); California Personality Inventory (CPI).

2 In addition to the above-listed PSFs, situational stress factors (e.g., task difficulty) were measured.

the retrieval of monitoring targets from videotapes. Using the data on monitoring actions, the following were evaluated (Fujita, 1992):

- relative level of monitoring tasks (RLMT);
- performance level of monitoring tasks during RT.

RLMT is a measure of relative load of monitoring tasks:

$$\text{RLMT}_i = 100 \times (N\text{-total}_i/N\text{-total})/T_i \qquad [\%\ \text{s}^{-1}]$$

where

RLMT_i = an RLMT for an operational phase i;

$N\text{-total}_i$ = the number of monitoring tasks in operational phase i;

$N\text{-total}$ = the total number of monitoring tasks;

T_i = time required to complete operational phase i.

The following five operational phases were considered:

- Phase 1: from onset of alarms to RT;
- Phase 2: from RT to SI (duration is about 2 min);
- Phase 3: 5 min into SI;
- Phase 4: from 5 min after SI to the reset of SI;
- Phase 5: from the reset of SI to the end of the operation.

The performance level of verification tasks was examined to evaluate if any influences were observed in the busiest operational phase identified by the RLMT measure. As will be shown, the busiest operational phase appeared to be Phase 2. Then, the performance level of verification tasks following RT was examined. From relevant verification items specified in the written procedures, those which were of higher priority were identified by system designers and operational experts. There were nineteen verification items (RT total item group) in which four items were identified as of higher priority (RT higher priority item group). The timing and the level of verification were then compared between the total item group and the higher priority item group. Note that the number of verification items was counted in terms of perceptual units (Swain and Guttmann, 1983). It means that more than one check–read action can be counted as one verification item when it is regarded as a perceptual unit. The grouping was done only when the following were met:

1. Relevant verification actions belong to a conceptually grouped task. For instance, the status verification of three pumps, Pump A1, Pump A2, and Pump A3, that is specified as 'Verify Pump A' in procedures.
2. The type of verification is check-reading (not read-value type).
3. Relevant monitoring devices are grouped.

Table 8.2 RLMTs for five operational phases (Phases 1–5). RLMT is a measure of relative load of monitoring tasks. Phase 2 appears to be the busiest operational phase.

Operational phase	RLMT ($\times 10^{-4}\%\ s^{-1}$)
1	4.42
2	5.90
3	4.52
4	3.70
5	3.59

8.3.2 Results

Table 8.1 shows RLMTs for the five operational phases. It appears that RLMT2 is significantly higher than other RLMTs. The second busiest phase is Phase 3 which is followed by Phase 1. The levels of Phases 4 and 5 are lower. These results indicate that Phase 2 is the busiest operational phase.

Figure 8.1 shows the level and timing of RT verification for the RT total item group and the RT higher priority item group. The level of verification in the total item group appears to be 68.4% showing that the verification was not perfect. In addition, 27.2% was not verified immediately after RT (Phase 2) and left for Phases 3 and 4. Compared with the total item group, the level of verification in the RT higher priority item group is significantly higher, while the timing tends to be delayed to a lesser extent. The level of verification is 92.6%, and only 7.3% was left for Phase 3 alone (not for Phase 4).

For the purpose of comparison, a similar analysis was conducted for the SI verification done in Phase 3, the second busiest operational phase. The total number of verification items for SI was 46, in which 13 items were categorised as of higher priority. Figure 8.2 shows the results. The level of verification is slightly, but statistically significantly, higher in the SI higher priority item group. The tendency to delay the verification of less priority items can also be seen, but again it is not as apparent as it is in the RT data. These results support a previous observation that Phase 1 is significantly busier than other operational phases.

A closer examination of the RT higher priority items which were either delayed or not verified revealed that they are the same item. It was a back-up verification for reactor shut-down equipment. Three out of seventeen crews did not seem to verify it, and another five crews made delayed verifications. All other higher priority items were verified perfectly.

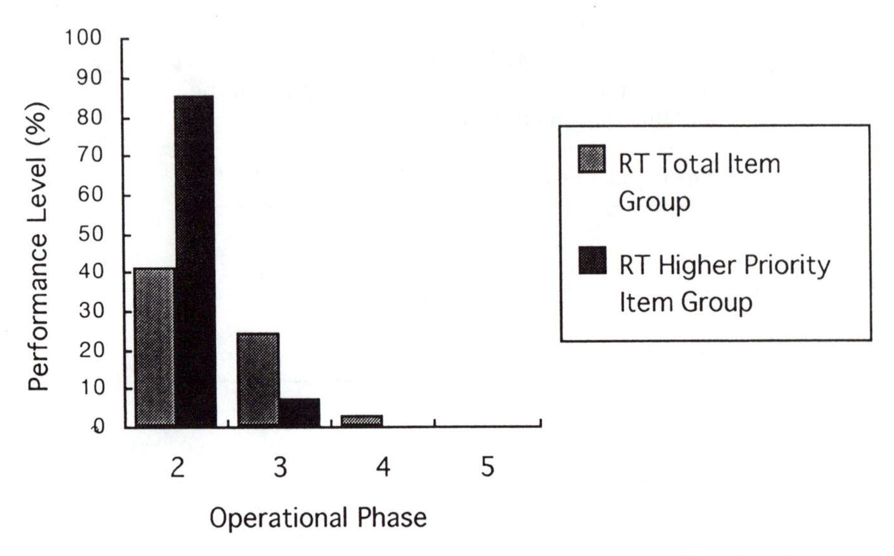

Figure 8.1 Level and timing of verification after reactor trip (RT). The level is higher and the timing is less delayed in the RT higher priority verification item group.

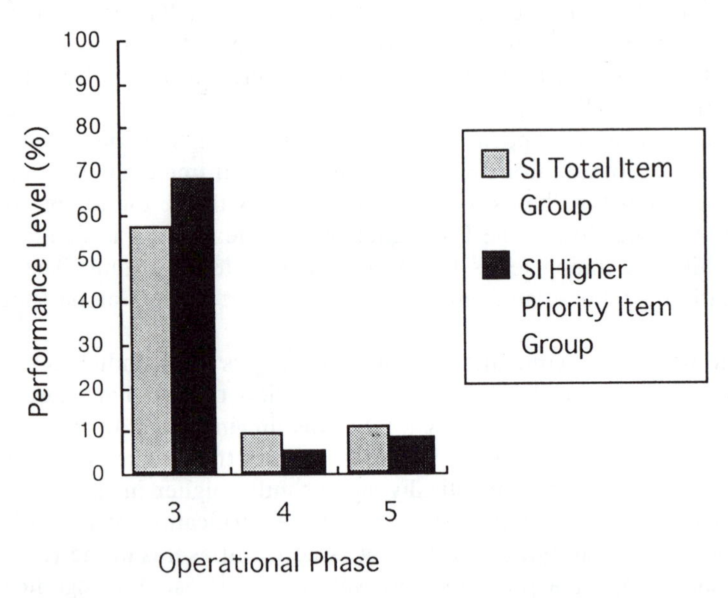

Figure 8.2 Level and timing of verification following a safety injection (SI). It appears that the performance of the SI higher priority verification item group is better than, but not as significant as, the RT higher priority verification item group.

8.3.3 Discussion

The above results suggest that operators are fairly busy immediately after RT. This can be concluded by the facts that (a) RLMT is significantly higher, and (b) priority verification items are attended to a lesser extent in Phase 2. A rationale for the second point is the widely accepted idea that the more important a task is in a dual-task environment, the more it is attended to, causing the performance degradation of less important tasks (Gopher and Donchin, 1986).

An important point is that procedures do not specify any prioritisation scheme. The basis of the prioritisation must, therefore, involve some level of personalisation. In other words, the prioritisation should have emerged as a result of learning through training or other forms of experiences.

It should be noted that the data shown above are not necessarily reflecting 'true' job performance. There was a limitation in the observability of monitoring behaviour. When trainees' behaviour was not clear enough to convince analysts, they judged that the verification was either delayed or omitted. Hence, the data shown above should not be taken as error rates or any sorts of job performance. Nevertheless, the disappearance of clear finger-pointing action and verbal utterances can itself be taken as an indication of tacit performance degradation. It can, therefore, be said that the limitation of observability does not impair the discussion about the personalised prioritisation process.

8.4 Analysis 2: Characteristics of Errors of Omission and Commission

In Analysis 2, erroneous tendencies associated with control actions which included both errors of commissions and errors of omissions (Swain and Guttmann, 1983) were analysed. The scenario used in Analysis 2 was used. The number of control steps involved was 71, and 46 data sets collected in the first stage were used.

8.4.1 Data Reduction and Analysis Methods

Any commissions or omissions were identified and the interpretation of their underlying factors was attempted. In addition, commissions were categorised into the following deviation types:

- *direction* manipulate a control in a wrong direction;
- *selection* select a wrong control;
- *rule violation* violate a predefined control sequence, or preset control conditions;

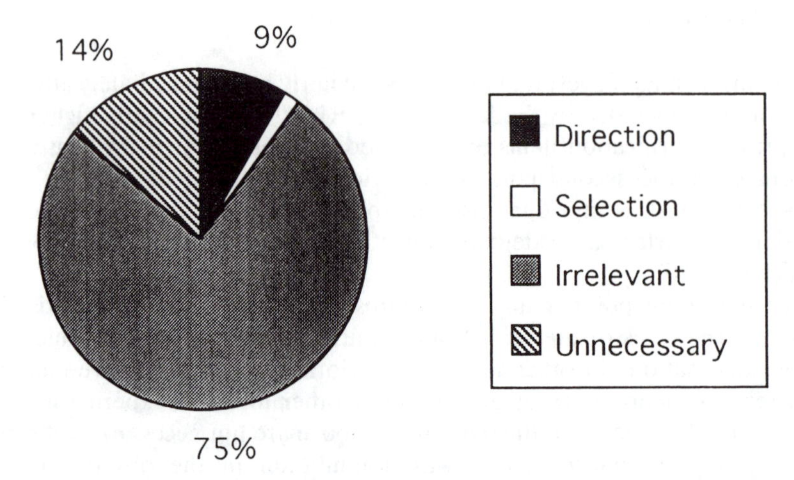

Figure 8.3 Percentage fraction of four types of commissions: direction, selection, irrelevant, and unnecessary. The majority is formed by irrelevant commissions (75%).

- *unnecessary* conduct an action which is not necessarily wrong, but operationally meaningless.

8.4.2 Results

Figure 8.3 presents the percentage fraction of the four deviation types defined for commissions. It appeared that the majority consists of rule violations (75%), and a long way behind in second place comes the commission of unnecessary actions (14%). A closer look at the data revealed that the following are dominant modes of 'violation' which accounted for 86.5% of the commissions of rule violation (65.5% of total commissions):

- violation of control sequence resulting from the omission of conditional control action (62%);
- violation of control sequence resulting from trial-and-error tests (14.5%);
- violation of preset conditions (10%).

The dominant modes of omissions, on the other hand, appeared to be the following, which accounted for 48.5% of total omissions:

- omission of non-urgent control actions (28.5%);
- omission of functionally non-critical control actions (20%).

8.4.3 Discussion

Five control steps are found to have contributed to the commissions of rule violation which are qualitatively similar in their structure: in these control steps, the control action must be preceded by a control action (conditional control action) such as follows:

- start an oil-lift pump before starting the main pump;
- establish instrument air pressure before opening a pneumatic valve.

Usually, the omission of a conditional control action results in an observation of the violation of control sequence, because the omission is recovered immediately after the unsuccessful implementation of the final control action (e.g. the startup of the main pump). This mode of commission indicates a tendency to rush to the final action before satisfying preconditions. This is consistent with a widely accepted view that operators are goal oriented (Rasmussen, 1986). Thanks to interlock design, however, this type of operational violation rarely leads to unfavourable consequences.

Only one control step is found to have contributed to the omissions of non-urgent control actions in which operators are required to zero the control demand to a pneumatic flow control valve. Figure 8.4 presents the control logic for the valve. The valve is designed to be closed automatically after reactor trip. This is accomplished by intercepting the demand with the trip signal. Since the demand is only intercepted, it resumes when the trip signal is reset. This may cause the valve to receive too large a demand abruptly. In order to avoid mechanical damages caused by this possibility, procedures require operators to zero the demand after automatic closure of the valve. However, this is not an urgent task. Operationally, it is acceptable to conduct this operation just before resetting the trip signal. Since operators are in a high workload situation after RT (Analysis 2), it is not surprising that the step is omitted. The nature of this phenomenon is exactly the same as that of the postponement of less priority verification items found in Analysis 1. Although this omission does not cause any immediate consequences, it seems to increase the possibility of damaging the valve because the omission of conditional control action is the cause of the dominant commissions (i.e. commissions of rule violation) as discussed previously.

Several control steps are found to have contributed to the omission of functionally non-critical control actions in which operators are required to isolate a feedwater line leading to a heat exchanger (cf. explanations of the scenario in Analysis 1). The isolation is done with the closure of an isolation valve and two flow control valves located in the up-stream of the isolation valve. Some operators closed only the isolation valve and left the flow control valves open. Functionally, however, the isolation can be accomplished by closing the isolation valve only. It is therefore not surprising that the isolation of flow control valves was omitted.

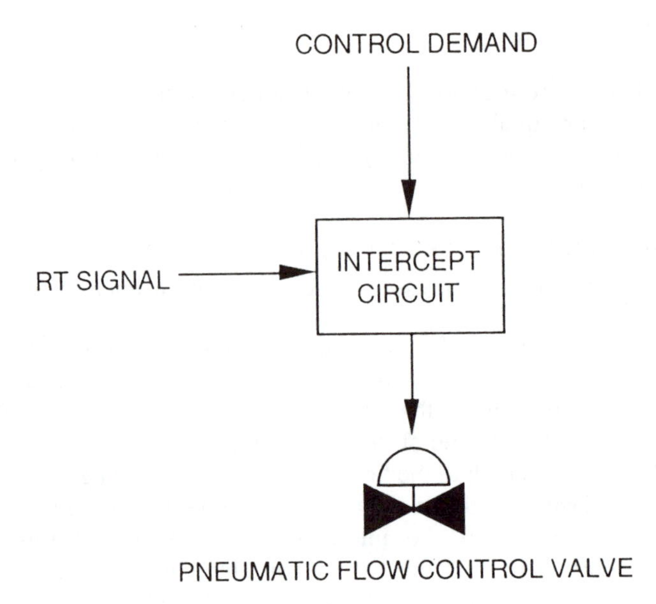

Figure 8.4 Control logic for a pneumatic flow control valve. The reactor trip signal intercepts the control demand, causing the valve to close.

8.5 Analysis 3: Performance-shaping Factors

The analyses of performance-shaping factors (PSFs) were conducted in two steps. In the first step, a broad range of PSF instruments were administered and their correlations with performance examined. From these data a more refined list of useful PSF was identified. Using the refined list of PSF, the PSF battery of tests and questionnaires was refined, and these selected PSF were studied in greater detail in the second step. The behavioural data collected in the first stage and the second stage were used in the first step and the second step, respectively.

8.5.1 Data Reduction and Analysis Methods

The study is based on correlation analyses of performance and PSF data. Those PSFs that correlate significantly with performance are considered to be critical PSF. Judging from the findings from the analysis of various performance measures (Analysis 1), it was decided to use the expert rating data in Analysis 3.

8.5.2 Results

The following results were obtained from the analyses:

1 Previous job experiences and training experiences are significant contributors to performance, which account for up to 20% of performance score variance. Some of the cognitive ability measures (e.g. perceptual speed/attention), job knowledge, personality traits (e.g. social desirability, good impression) and personal stress and coping mechanisms are also significant contributors. The degree to which these PSFs contributed to performance varied by job positions that subjects played during the training sessions.

2 There is a greater tendency for subjects to describe themselves in socially desirable terms, as they move up the job progression ladder in their home plants. In addition, subjects who behave in a socially desirable manner tend to receive higher ratings than those who scored lower on some personality scales (i.e. MMPI Validity Scale, CPI Good Impression Scale, and Socially Desirable Scale) with validity coefficients ranged from 0.15 to 0.46.

8.5.3 Discussion

That previous job experience and training experience are significant contributors is not a surprising result. On the other hand, a finding concerning the correlation between job positions and expert ratings seems to pose an interesting question on the validity of expert rating.

The result that subjects who behave in a socially desirable manner tend to receive higher ratings than those who scored lower on some personality scales contradicts results reported in a similar US study (Dunnette *et al.*, 1981). In the US study, similar personality scale scores correlated negatively with performance scores. This suggests that some cross-cultural differences influence the evaluation of operator performance. The three personality scales which correlate positively with the expert ratings are as follows (Fujita, *et al.* 1992; Toquam and Fujita, 1992):

1 *Validity scale* This is a 15-item MMPI scale which is designed to check whether subjects are reading each item carefully and responding honestly. For those subjects who obtained high scores on this scale (above 7), their personality scores are eliminated from further analyses because they appear to be distorting their responses.

2 *Good impression scale* This is a 38-item CPI scale which measures one's tendency to distort responses intentionally to 'look good' to others.

3 *Socially desirable scale* This is a 13-item personality scale which measures one's tendency to respond in a socially desirable manner.

As evident from these explanations, all these three personality scales indicate a tendency to behave, sometimes by intentionally distorting responses if necessary, in a way that is socially more acceptable. The tendency to present one's self in the most positive manner possible is said to be a natural consequence of the Japanese social system, which is often characterised by collectivism, and is consistent with findings from cross-cultural research on Japanese personality traits (Shigehisa *et al.*, 1985; Triandis, 1989). This finding seems to cast a question on the validity of the expert ratings, or at least to limit their validity.

8.6 General Discussion and Conclusions

From Analyses 1 and 2, the following findings were obtained:

- When overloaded, subjects tend to postpone less important monitoring tasks (Analysis 1).
- Non-urgent or functionally non-critical control actions tend to be omitted (Analysis 2).
- Conditional control actions tend to be omitted, causing the violation of control sequence (Analysis 2).

There seem to be two factors that drive these erroneous tendencies: the goal-oriented nature of humans and the tendency to prioritise tasks. The goal-oriented nature of humans seems to drive the tendency to rush to the final action (the last erroneous tendency). It is considered that the final action should be associated, as the major 'means', with the functional goal more strongly than conditional actions and it is more readily triggered when the goal is evoked. This tendency must become more salient in high workload situations. Conventionally, interlock design is adopted to protect components from this erroneous tendency. As far as operators maintain the goal-oriented approach, this erroneous tendency seems not to be critically important, because it will be recovered immediately by monitoring the component status or the status of the relevant functional goal. If, however, they behave more in a procedural manner and omit the confirmation of functional achievement, this erroneous tendency may trigger a fatal error. The tendency to prioritise tasks seems to drive the first two erroneous tendencies. They appear to become salient especially in high workload situations, acting as a moderating mechanism. Usually, this mechanism enables operators to maintain the workload at a manageable level and to possess a global view on situations. It must be reasonable to assume that only experienced operators can establish a sound basis for the task prioritisation, because it requires a personal build-up process through job experience. However, this superior ability of experienced operators may drive expectation- or assumption-based behaviour which is potentially dangerous. This supports the general psychological view that a mechanism which accomplishes highly efficient reasoning can cause fatal mistakes in some situations. Highly efficient reasoning capabilities are believed to be acquired through adaptation to environments in which humans have to accomplish their goals with their severely limited information-processing capabilities. Such efficient reasoning capabilities are tuned largely to experienced situations, and therefore tend to lose grip when humans are overloaded, especially in novel situations. It seems that process plant operators cannot be immune from this general error-inducing mechanism. It is like a double-edged sword.

The results of correlation analysis between the expert ratings and some personality scales seem to indicate that the expert ratings may be socially distorted. This suggests that one of the determinants of effective performance is to identify the proper behaviour as defined by some socially desirable norm. A problem associated with this type of measure is that the criteria for such measures may differ across sub-cultural groups. In fact, we found from the interviewing of several supervisors that they differed greatly in their views about ways to improve performance. Their responses and suggestions differed greatly across utilities and even within the same plant. This suggests that there are significant variations among sub-cultural groups. The results of previous research in anthropology and cross-cultural psychology suggest that this is not a surprising finding. They suggest that differences within groups are often much greater than differences between cultural groups.

References

DUNNETTE, M. D. *et al.* (1981), *Development and Validation of an Industry-wide Electric Power Plant Operator Selection System*, Report No. 72. Minneapolis, MN: Personnel Decisions Research Institute.

FUJITA, Y. (1992) *Study on The Improvement of Man–Machine Interfaces for Nuclear Power Plants*, unpublished doctoral dissertation, University of Tokyo (in Japanese).

FUJITA, Y., TOQUAM, J. L., WHEELER, W. B., TANI, M. & MO-ORI, T. (1992) Ebunka: Do cultural differences matter? *Proceedings of 1992 IEEE Fifth Conference on Human Factors and Power Plants*, pp. 188–94, IEEE.

GOPHER, D. & DONCHIN, E. (1986) Workload – an examination of the concept. In Boff, K., Kaufman, L. and Thomas, J. (eds), *Handbook of Perception and Human Performance*. New York: Wiley.

RASMUSSEN, J. (1986) *Information Processing and Human Machine Interaction*. Amsterdam: North Holland.

SHIGEHISA, T. *et al.* (1985) *A Comparative Study of Culture: An Introduction to Intercultural Communication*. Tokyo: Kenpakusha (in Japanese).

SWAIN, A. & GUTTMANN, H. E. (1983) *Handbook of Human Reliability Analysis with Emphasis on Nuclear Power Plant Applications*, NRC Report NUREG/CR-1278 (SAND80-0200).

TOQUAM, J. L. & FUJITA, Y. (1992) Individual differences measures: Their correlations in process control occupations in Japan. Presented at the *Society of Industrial and Organizational Psychologists Conference*, Montreal, Canada, May 1.

TRIANDIS, H. C. (1989) The self and social behavior in differing cultural contexts, *Psychological Review*, **96**, 506–20.

CHAPTER NINE

Selecting personnel in the nuclear power industry

NEVILLE STANTON and MELANIE ASHLEIGH

Department of Psychology, University of Southampton

9.1 Introduction

The value of personnel selection is based upon two premises. The first premise is that there are differences between people and that these differences lead to discernible differences in the work that people perform. Unlike hardware components, which tend to be uniform in appearance, reliability, function and performance, human beings are extremely variable (see Chapter 8). The richness of this variation makes the study of people interesting, but can make the task of selecting people to perform specific functions to a required performance standard quite difficult. The infinite variety of manifestations in skill, knowledge, ability, interest, motivation and experience possessed by personnel needs to be matched in an appropriate manner to the tasks they are required to perform. The second premise of personnel selection is that it is entirely possible to measure the differences between candidates on criteria that are relevant to job performance. This can be extended to make predictions about how well each candidate is likely to perform the job for which they are being selected. Personnel selection is about choosing the best person for a particular job. Not necessarily the most qualified or the most experienced, but the most suitable. The extent to which the selection process is carried out is determined by a cost–benefit analysis (although it may not be formalised). For example, the recruitment of an unskilled manual labourer may be on a first-come– first-served basis, whereas the selection of a fighter pilot is through an extensive testing and selection process. The cost of failure in the selection of a labourer is relatively small, whereas it can cost £2.5 million to train a fighter pilot, and normally up to £1.5 million is spent before it is absolutely certain that they will succeed. In between these extremes a clerical worker is likely to go through a less intensive selection procedure than a person recruited into management.

Selection procedures range from a single interview to complete assessment centres which may last a period of several days. Effort expended upon selection pays off most when the demands placed upon the job holder are likely to be high, there are distinguishable differences between candidates, and the selection methods are both reliable and valid (Cook, 1988). These principles will be explored further within the chapter.

A fairly simple and fictitious, although common, recruitment and selection procedure might be as follows:

1 Receive notification of vacancy from budget holder.
2 Draw up job specification in conjunction with departmental staff.
3 Draft advertisement and liaise with budget holder.
4 Place advertisement with local or national paper.
5 Collate information package (including department, district, local area, map, job, job specification, terms and conditions of service).
6 Place on vacancy board.
7 Open file containing advert, information packs, application forms, etc.
8 Send out application forms, information pack, and letter to enquirers.
9 After closing date, photocopy all applications and pass on to manager together with information pack and advise on suitability.
10 Inform shortlisted applicants of time, date, and place of interview, and advise unsuccessful applicants. Contact referees and put shortlisted applicants' forms into a booklet together with job and person specification, terms and conditions, and the information pack, and give to managers.
11 Book room for interview, and inform manager.
12 Two days before interview, telephone applicants who have not confirmed their attendance.
13 Hold interview (welcoming applicants at reception, etc.).
14 Advise successful applicant of offer next day by telephone.
15 Write confirming offer with start date etc. and advise unsuccessful applicants.

This may at first glance appear faultless, but there are some inherent flaws in the procedure. First, there is no way of guaranteeing the quality of the job specification and no mention of a person specification. Second, the recruitment and selection methods are chosen without regard to their suitability for the post. Third, there is no mechanism for validating the recruitment and selection methods. Given these fundamental flaws, it is clear that the recruitment and selection process requires more careful consideration. This chapter will consider the issues of selection that are relevant to the nuclear power industry. First the general process of selection and methods employed will be presented. Second, a review of selection methods used in the nuclear industry be summarised. Finally the implications for the nuclear industry will be evaluated.

9.2 The Selection Process

Selection has been described as a process by many authors (Lewis, 1985; Torrington and Hall, 1987; Cook, 1988; Herriot, 1989a; Smith and Robertson, 1989). To use the process metaphor is very useful, as it implies continuity, adaptation and change, all important facets for a successful selection policy. Torrington and Hall (1987) proposed a systematic methodology for selection given the acronym ARCADIA to describe the seven stages of the selection process:

1 **A**nalysis and specification of personnel requirements.
2 **R**ecruitment of candidates for employment.
3 **C**andidate screening.
4 **A**ssessment of candidates.
5 **D**ecision-making for employment.
6 **I**nduction and placement.
7 **A**ppraisal and evaluation.

The stages may be represented in a process diagram as shown in Figure 9.1.

As shown in Figure 9.1, selection is a closed loop process with two feedback loops. The first feedback loop informs the selectors about the adequacy of the recruitment methods and the second feedback loop informs the selectors about the adequacy of the selection methods and the selection process as a whole. The main stages in the process are identified as: job analysis (where the criteria for selection are identified), recruitment (where the pool of applicants is generated), selection (where the applicants are assessed), decision making (where the resultant data are evaluated and the candidate chosen) and validation (where the recruitment and selection procedure is validated). Herriot (1989b) proposed that selection may be viewed as a social process: it is a reciprocal relationship where both the selector and the candidate have perceptions, and expectations, of each other. The candidate is not a passive actor in the process and has the power to withdraw at any time. Thus, organisations seeking to attract candidates need to realise that any contact they have with the candidate provides the candidate with information about the nature of the organisation.

Prior to entering into the selection process, the organisation has alterna-

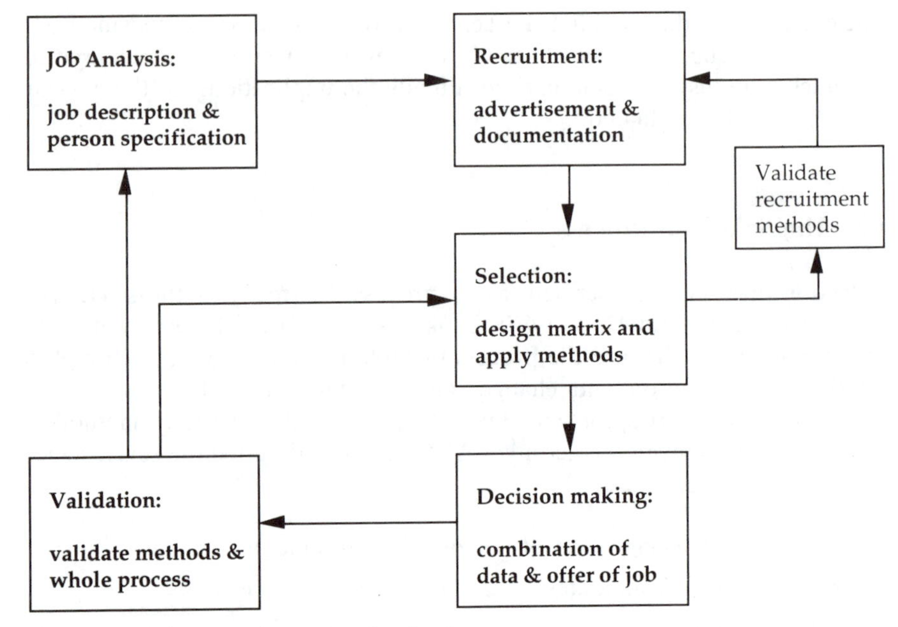

Figure 9.1 The recruitment and selection process.

tives to be considered. Alternatives to selection are job design and training. Before entering into the recruitment and selection procedure, it is first necessary to determine if there is a need to fill the vacancy. There are various alternatives to recruitment, and these are shown in Table 9.1.

Each of these methods has possible drawbacks, making selection a more attractive option in the longer term. As Table 9.2 indicates, job redesign and training are two possible alternatives. An indication of how job design, selection and training (see Stammers, this volume) are interlinked is illustrated in Figure 9.2.

Job redesign may be necessary if adequate performance is not available in the job market. Selection of personnel may have to take the trainability of the candidate into account. Assuming that it is necessary to engage in the recruitment and selection process, the following sections describe the main stages.

9.2.1 Job Analysis

Job analysis is 'the process of collecting and analysing information about the tasks, responsibilities and the context of jobs' (Torrington and Hall, 1987). In essence, the activities of JA involve collecting and organising information about what the job involves and what type of person is needed to perform that work. The main objectives of job analysis are to report this information

Table 9.1 Alternative to recruitment

Alternatives	Possible drawbacks
Reorganise the work	Possible union difficulties
Use overtime	Costly, in long term
Train existing staff	May require negotiation/promotion
Offer bonuses	Costly, in long term
Mechanise	Large capital outlay
Stagger the hours	Possible union difficulties
Make the job part-time	Possible union difficulties
Subcontract the work	Costly in long term
Use an agency	Costly in long term

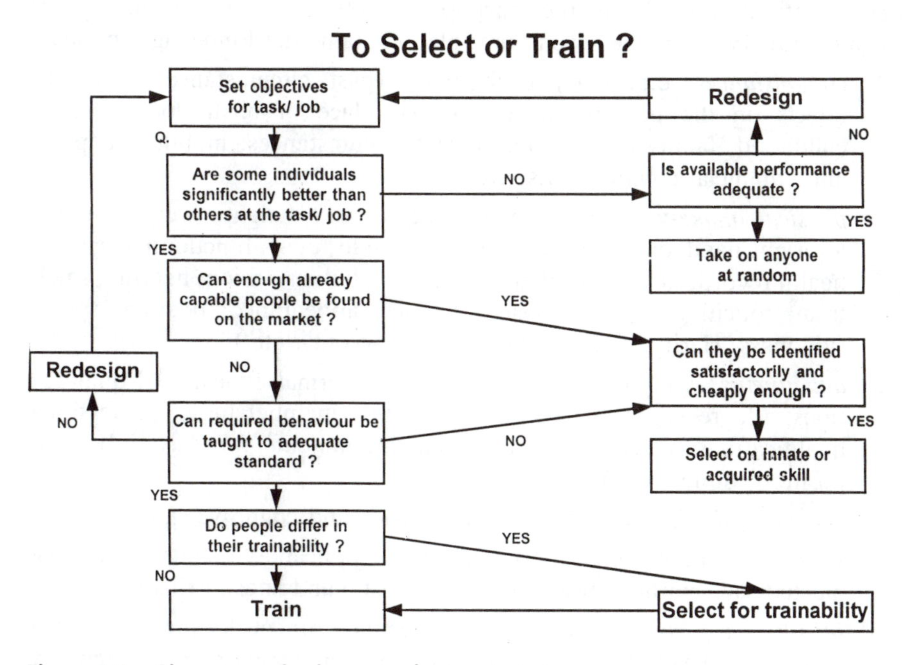

Figure 9.2 Choosing whether to select or train?

in the form of a written job description and person specification. The sort of information to be collected is indicated in Table 9.2.

A more extensive checklist for the purpose of determining job specification in available in Appendix 1. The person specification framework is based upon the influential work of Rodger (1952). The seven-point plan was initially devised to help structure the interview procedure, but it may also be used to help define the person specification. The job specification is used for the purpose of recruitment (i.e. attracting the candidate) whereas the person

Table 9.2 Job specification and person specification

Job specification	Person specification
Job title	Circumstances
Relationship with others	Physical make-up
Job content	Attainments
Main tasks	General aptitudes
Features of the work	Special aptitudes
Working conditions	Interests
Performance standards/objectives	Disposition

specification is used for the purpose of selection (i.e. evaluating the candidate). In brief, the seven-point plan contains the following sections:

1 *circumstances* e.g. early background (past circumstances and their effects on the person's development), place of birth, location(s) in childhood, family occupations, recent circumstances, mobility, dependants, financial commitments, etc.;

2 *physical make-up* e.g. structure (age, sex, height, weight, reach, left/right handedness, disabilities, body build, etc.), functioning (general health record, illness, accidents, allergies, phobias, vision, hearing, smell, taste, touch, etc.) and physical impact (appearance, bearing, speech (loudness, clarity, pitch, pronunciation, accent), etc.);

3 *attainments* e.g. educational attainments (formal educational qualifications, degrees, etc.), specific training (memberships, professional qualifications, etc.) and occupational attainments for each job (title, position, duties, etc.);

4 *general aptitudes* e.g. general intelligence and cognitive skills;

5 *special aptitudes* e.g. occupationally applicable talents (scientific, mechanical, mathematical, practical, artistic and social skills);

6 *interests* e.g. type of interest (intellectual, practical, physical, social, aesthetic), level of interest (self-motivation, vigour, drive, energy, strength of achievement), direction of interest (aims, *ad hoc* motivation, positive and negative reasons) and longer-term aims;

7 *disposition* e.g. self-perception, social relationships, behaviour indicative of self-reliance and dependability.

In comparing individual candidates against a standard person specification: general intelligence, special aptitudes and disposition can be found from tests, physical make-up from appearance, attainments from past records, and interests from questionnaires.

All types of people are involved in collecting the data for job analysis activities: the job incumbent, the supervisor, other members of the role set, technical experts, and resort may also be made to existing written records. Likewise there is a variety of ways in which to collect the information:

- self report
- individual interviews
- other individual methods
  - — checklists
  - — questionnaires
  - — work diaries
- group methods
  - — group interviews
  - — technical conference
  - — mixed conference
- observation
- participation
- existing records.

There are many techniques for analysing this information, including: functional job analysis, the position analysis questionnaire (PAQ), critical incident technique, the results-oriented description, job-reward analysis and task-analysis methodologies (Algera and Greuter, 1989). When collating the job-related information and writing a job description it is necessary to present a standard style, paying attention to possible sources of problems, e.g. sources of error, appropriateness of job descriptions and purposes served by job descriptions. Similarly, when writing a person specification it is recommended that one should avoid overestimating the characteristics required by the job. Rather, one should consider the importance of a candidate's potential and distinguish which characteristics are essential and which are preferred.

9.2.2 Recruitment

Recruitment is 'the activity that generates a pool of applicants, who have the desire to be employed by the organisation, from which those suitable can be selected' (Lewis, 1985). There is a wide variety of recruitment methods from which to choose. As Table 9.3 shows, the choice of the method is largely determined by the type of post being recruited to: management (Mgt), white collar (W.C.) and blue collar (B.C.).

Recruitment, the activity that generates a pool of applicants, is distinct from selection in many ways. If manpower is scarce the emphasis is largely upon *recruitment*, but if manpower is abundant this emphasis shifts to *selection*. For example, in times of staff shortages, employers may attempt to maximise the pool available to all employers, or maximise their own pool

Table 9.3 Recruitment in 350 organisations (from Torrington & Hall, 1987)

Method	No.	Mgt	W.C.	B.C.
Professional and trade journals	238	1	3	11
Internal advertisement	216	2	1	1
National newspapers	208	3	8	12
Local newspapers	129	4	2	2
Consultants	109	5	13	17
PER/job centres	92	6	4	3
Executive search ('head hunters')	83	7	19	—
Visits to Unis/Polys	52	8	11	18
Employment agencies	42	9	6	9
List of job seekers kept	38	10	7	4

at the expense of their competitors. Whereas if manpower is abundant, it may be possible to attract less applicants by not advertising so widely, but this may result in a drop in the quality of applicants. The effects of recruitment and selection strategies upon an organisation may be:

- To attract a scarce skill it may be necessary to increase the salary (putting it out of line with existing salaries).
- If there is an abundance of applicants, recruiters may raise the level of entrance qualifications.

In both of these cases there could be problems arising within the organisation in the longer term. The first case may lead to comparable workers demanding pay parity. The second case may lead to less qualified people worrying about their promotion prospects.

9.2.3 Selection

Selection is 'the activity in which an organisation uses one or more methods to assess individuals with a view to making a decision concerning their suitability to join that organisation, to perform tasks which may or may not be specified' (Torrington and Hall, 1987). There is a variety of selection methods available to organisations, including the following:

- application forms (a form that requires historical information about the candidate to be entered);
- biodata (predicting job performance based upon the candidate's historical information);
- psychological testing (tests of the candidate's cognitive ability and personality);

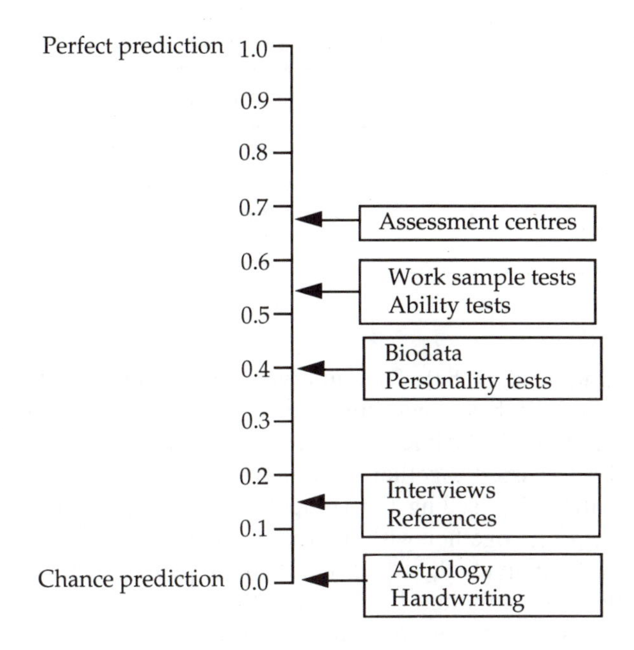

Figure 9.3 Comparison of the accuracy of selection methods (adapted from Shackelton, 1989).

- interviewing (face-to-face discussion of job with candidate);
- work sampling (observing the candidate's work);
- assessment centres (multiple assessment of candidate);
- references (asking third parties to comment on suitability of candidate);
- others (palmistry, horoscopes, handwriting, etc.).

In a summary of the validities of the different methods, Shackelton (1989) presented an indicator of the general performance for comparative purposes, as indicated in Figure 9.3.

As Figure 9.3 illustrates, assessment centres, work samples and ability tests are generally reported to obtain higher validity coefficients than interviews, references and the alternative methods. Biodata and personality tests are generally reported to perform somewhere between these. The interview, psychological testing and assessment centres (which include a discussion on work simulation exercises) will be considered further.

9.2.3.1 *Interviewing*

The interview has been defined as 'a face-to-face interaction between two or more individuals which occurs as a result of an ulterior motive or purpose'

Table 9.4 Characteristics of polls of interview styles

Unstructured	Structured
Unplanned	Systematic
Uncontrolled	Totally controlled
Non-directive	Logically ordered
Commonly disorderly	Facilitates data quantification

(Torrington and Hall, 1987). The interview is the most commonly used selection method, it has great importance in its potential for collecting information (it is very flexible and speedy) and a variety of information sources can be discussed together. The interview also has the potential for giving information, i.e. selling the company to the candidate. It combines both human aspects together with a ritual in personnel selection. Types of interview include: individual, sequential, tandem, panel and group. Interview styles vary along a continuum from unstructured, to focused, to structured. The poles of this continuum are indicated in Table 9.4.

Somewhere between the poles is generally considered to be the best technique, i.e. pursue lines that seem relevant, and ask questions about the application form. Reliability of the interview varies with the amount of structure. It is useful to establish what form the interview is going to take with the respondent in the introductory stages, so that later on this may be referred to if the interview starts to wander off course. Interviewing enables the interviewer to discuss data collected from other sources with the interviewee. Other than verbal information the interview will be packed with non-verbal communication (posture, facial expression, gestures, personal space, etc.).

Rodger's (1952) seven-point plan and Fraser's (1958) five-fold grading are two examples of the 'human attribute classification' approach. The use of such systems in interviews has a number of advantages:

1 The classification can be used as a basis for the middle part of the interview.
2 It divides information into chunks and so encourages the interviewer to make sure that the information is as complete as possible.
3 It sets out the areas of information that are relevant to the employment decision.
4 It encourages interviewers to think systematically about an applicant rather than using an unsubstantiated general impression which may be affected by their own prejudices.
5 These systems emphasise facts and help to prevent concentration on subjective feelings.

The main problem with interviewing as a selection technique is that the physical make-up of the candidate has a very strong influence, although it is of little predictive value. Interviews are also used to make ratings of intelligence, which is a very poor inference. The interview is a discriminatory technique for selection. Most of the psychological evidence suggests that interviews are a bad method of personnel selection. The interview may, however, serve other functions. The thing that influences interviews most is negative information. Positive information is very largely ignored. Also there are contrasting effects with the previous candidate, i.e. interviewers make contrasting judgements rather than absolute judgements. Skilful interviewers are not constrained by rigid rules; they follow up interesting leads. Despite the popularity and widespread use of the interview as a selection technique, there is little in the way of strong evidence for its usefulness in predicting job success. Most of the evidence is equivocal. Reilly and Chao (1982) found predictive validity of unstructured interviews to be low (around $r = 0.19$). However, Weisner and Cronshaw (1988) found predictive validity ($r = 0.62$) and reliability ($r = 0.82$) to be quite high for the structured interview.

9.2.3.2 *Testing*

Psychological testing gives the selection process an overlay of 'scientific respectability'. Testing is used to achieve objectivity by reducing subjectivity (Kline, 1993). This can be achieved only if the test is:

- standardised
- sensitive
- reliable
- valid.

There is a danger of the test 'looking' more accurate than it actually is. Lewis (1985) explains this phenomenon using the analogy of a gun sight: inexperienced users assume that because the gun sight is there, then it must be accurate. However, if they keep missing the target they will realise that it is inaccurate and stop using it until satisfied that it has been corrected. They may, however, justify using it by stating that the target is whatever they hit – this is reckless but could be applied to selection. Selectors must constantly question the validity of their tests, and modify them appropriately. The confidence inspired by tests and their apparent objectivity can be seriously misused.

Testing is a useful selection method when narrow and precise information is required. This is particularly true in the case of ability, but there is a danger that the degree of precision can be overestimated. Surveys have indicated that a selector will be doing quite well if the test used correlates more than $r = 0.4$ with subsequent job performance. Those in favour of tests point out the unreliability of the interview as a predictor of performance and the greater accuracy of test data. Those against the use of tests dislike the objectivity that testing implies, or have difficulty in incorporating test evidence into the rest of the evidence collected.

Those who feel the strongest about tests are the candidates themselves, who usually feel that they can improve their prospects with a 'good interview performance', and that the degree to which they are in control of their destiny is being reduced by a dispassionate routine! The advocates of testing perhaps concentrate too much on the representativeness of the organisation making the right decisions and too little on the importance of the interactive ritual, that is the interview. Nevertheless, tests are useful and have a potential beyond their present level of application (Torrington and Hall, 1987).

Three main types of test measure differences in intelligence, personality and skill.

Tests of intelligence Either candidates regard the idea an an insult, or they rationalise a fear that the answers might not be too good. There is a general debate whether to measure intelligence at a general level (g) or whether specific types of intelligence may be measured. Carroll's (1993) three-stratum theory posits a hierarchy of abilities from fairly specific (at the first stratum) to broad factors (at the second stratum) to highly general (at the third stratum). Whilst the evidence does show quite high correlations from one cognitive ability to another (making the case for g), pragmatics of selection may make the measurement of job-specific abilities more desirable. For a comprehensive review of 73 psychometric instruments, see Bartram *et al.* (1992).

Tests of personality Personality is that which makes one person different from another and includes all the psychological characteristics of an individual. Personality is used to describe the non-cognitive or non-intellectual characteristics of an individual. It refers more to the emotional make-up of a person and is reflected in the style of behaviour rather than the quality of performance. Personality testing is the more glamorous and ephemeral area of psychological testing, but the intuitively appealing nature of the concepts does not guarantee that they are inherently correct. We may all like to believe that success or failure of an individual is due to their personality because it offers a convenient psychological safeguard for us. First it explains why we can't become multi-millionaires, i.e. our lack of success is beyond our control. Second, we are unlikely to be involved in spectacular failure as, say, Lord Raglan's disastrous campaign of the Alma in the Crimean War (1854–6) quoted from Dixon (1976):

> Aristocratic, courteous and aloof, he seemed to display many of the characteristics of the extreme introvert. So distasteful was it to have any direct contact with his fellow men that he could hardly bear to issue an order. . . he issued only one order – to advance until the Alma was crossed.

It would, however, be rather naive to assume that personality is the single dominant characteristic in predicting job performance. In fact, as a predictor of job success, the evidence is equivocal and there is no general consensus on the utility of personality assessment. At one extreme, Blinkhorn and Johnson (1990) claim that an unsuspecting public is being hoodwinked into believing that personality measures are a good deal more predictive than is actually the case. Specifically, they cite the following abuses:

- methodologically flawed development practices
- *a postiori* correlations (i.e. not predicted)
- significance levels barely above the chance effect
- high intercorrelations of factors.

Their strongest claim is that when, say, 20–30 personality scales are used to predict several job performance measures, sufficient correlations will reach significance by chance for the test constructor to claim predictive validity for the measure quite spuriously. Typically, the replication study, using a second independent sample which would indicate the unreliability of the significant findings, is not conducted. Hence, Blinkhorn and Johnson suggest that personality questionnaires are of most use where self-perception is important (e.g. career guidance and counselling), but are of little use in predicting objective performance.

Another type of concern is that personality tests rely upon self-report and subjective evidence, and so responses are subject to various distortions, including the faking of answers. However, once the personality profile is generated, the data may be treated as being entirely objective by other people within the organisation, so that the test taker is unfairly labelled.

The contrary, more optimistic, view of personality assessment is that it is genuinely useful for predicting job performance, provided that specific hypotheses are formulated in advance of data collection. For example, a recent review (Deary and Matthews, 1993) shows that relationships predicted *a priori* tend to be large enough to be practically useful. Hence, Blinkhorn and Johnson's criticisms may be invalid because they do not apply to the better work being done in the field.

Currently, there is some concern amongst practitioners over the professional use of personality measures. Some manuals fail to distinguish adequately between different types of reliability and validity, and fail to point out the basic limitations in the use of the test and in the resultant data. Stanton and Matthews (1995) recommend that personality tests are used with caution. There is never likely to be a situation where they can be used alone. Given the equivocal nature of the information, the combination of personality questionnaires with the interview provides a forum for feedback and discussion of the results. Indeed, there are many reasons why feedback is essential, including legal, ethical and public relations motivations. Stanton and Matthews (1995) caution organisations against using personality questionnaires as a single predictive measure. At best this would constitute a folly and at worst it could lead to a prosecution against the organisation. A recent review of personality tests may be found in Bartram *et al.* (1995).

Tests of skill There is less resistance to the notion of testing skills. Few candidates for a typing post would refuse to take a typing test prior to the interview. The candidates are usually sufficiently confident of their skills to welcome the chance to display them and be approved. Furthermore, the candidate will often be aware of whether they have done well or not. This provides the candidate with a degree of control, whilst they may feel the tester is in control of intelligence or personality tests as they may not understand the evaluation rationale. General tests can be used for vocabulary, spelling, perceptual ability, spatial ability, mechanical ability, and manual dexterity. For a review of 73 psychometric instruments, see Bartram *et al.* (1992).

Summary of tests The aim of selection tests is to obtain some reliable measure of current performance which will predict future performance. A test can be any task that produces a quantifiable output, e.g. questionnaires, rating scales, tests of 'skill' and tests of 'style'. The value of any selection test is a function of the accuracy with which it predicts the criterion, i.e. its forecasting efficiency. The crucial question in test development is that of predictive validity. Related to this is the question of what to use as a criterion. See Aiken (1976) for the history of psychological tests.

Automated testing Bartram and Bayliss (1984) summarised the status of automated testing. In the early days of computer presented tests mainframes were used, but these were expensive and relatively inflexible. The advent of the minicomputer and the cheap personal microcomputer opened up new possibilities for automated testing. Automated testing came to be seen as an increasingly cost-effective alternative to traditional paper-and-pencil techniques. The MICROPAT system developed by Bartram and Dale (1983) to select helicopter pilots for the Army Air Corps provides a complete test-administration operating system, as well as the tests which are dynamic and adaptive. Others (e.g. Vale, 1981) have explored and developed Bayesian adaptive testing techniques where the selection of test items is based on the testee's performance. Most of the early automated tests were semi-computerised. High test–retest reliabilities were found between face-to-face and automated administrations of the Wechler Adult Intelligence Scale (WAIS). The administration costs associated with WAIS testing were found to be considerably less than those for traditional testing. In the area of personnel selection, automated testing frees personnel staff from time-consuming test administration and provides more time for counselling and interviewing. The increasing power and availability of digital computers and on-line computing facilities in the 1970s, along with the development of the microcomputer, led to the complete or partial computerisation of many psychological tests. The general findings in all cases are high test–retest reliabilities and no significant differences between manual and computer administrations for general and specific ability tests.

In general more work has been reported on the automation of personality questionnaires than on ability and intelligence tests, primarily because automation of questionnaire-type test items poses relatively fewer problems. Much of the early work on personality questionnaires focused on automation of scoring and interpretation rather than on the actual administration procedure (i.e. tests presented on paper with boxes to be marked according to the alternative chosen, the marks being subsequently read by machine and marks assigned accordingly). It makes sense to present tests on a computer which can carry out the scoring, rather than using specially designed test-scoring equipment. Another advantage of computer presentation is that it is possible to obtain more than merely test item data. For example, response latencies to each item can be recorded. From its advent, concern has been expressed about the degree of 'rapport' attained with automated testing. The importance of the need to improve the software and hardware ergonomics of the user interfaces has become increasingly apparent as the field of automated testing has expanded.

Adaptive and tailoring testing involves a testing strategy where the test is tailored to the subject during the testing session, so that only the most relevant or informative questions are administered, while items of low discrimination are omitted. The term can also be applied more generally to any interaction where the machine's behaviour is at least partly dependent on the testee's responses. Branching strategies have certain advantages over traditional 'linear' methods of item selection. A greater amount of information about the individual can be collected, for example, information about how the subject goes about solving a problem, and not just whether the correct response is made. Overall testing time may be shorter because items which would have low information value are omitted. In addition, adaptive testing techniques should reduce the test anxiety created by presenting testees with items which are too difficult, and boredom (with consequent 'random' responding) which can arise when items are too easy. In general these factors should enhance the testee's motivation and involvement in the task. Most studies have shown that automated and non-automated versions of the same tests correlate highly. However, caution must be exercised when applying norms originally developed for non-automated tests to their automated versions.

In terms of hardware, future trends seem quite clear. Developments will greatly improve the 'physical ergonomics' of the interface. However, the crucial area for development is in software. Bartram and Bayliss's guiding principle has been that it should be possible for an individual to use an automated system designed especially for him or her, without having to resort to an instruction manual or assistance. In the area of automated testing one is writing software for the most demanding users: the person who may never have used a computer, who has no idea whatsoever about how it works, and who may well be anxious about using one. Some people will see such automation as a threat to their professional expertise. However, automation is unlikely to have a sudden impact on the nature of traditional testing and assessment practices. As changes gradually occur, so the roles of those working in these areas will undergo compensatory changes and grounds for such fears will diminish.

9.2.3.3 *Assessment Centres*

Assessment centres have been popular in the USA since the mid-1960s, and this is echoed by most research and literature on the subject (Feltham, 1989). Assessment centres can be described as a behaviourally based procedure, usually lasting around two days, where groups of managers are assessed using a combination of methods and by a number of assessors. Multiple assessment procedures were first used by German military psychologists during the First World War. The War Office Selection Board developed the technique in Britain during the 1940s, and the US Office of Strategic Services used the method to select spies during the Second World War. The first industrial organisation to adopt assessment centres was the American Telephone and Telegraph Company, which set up its Management Progress study in 1956 (Feltham, 1989). This longitudinal study was the largest and most comprehensive investigation of managerial career development ever undertaken. Its purpose was to attempt to understand what factors are important to the career progress of employees as they moved upwards through the organisational hierarchy.

Assessment centres require a large financial outlay to set up, and continue to be fairly expensive to administer. Assessment centres are most often used in the appraisal of potential supervisors and managers, and it can take up several days of the candidate's time. The advantage of assessment centres for this purpose is that ratings of potential can be assessed on the basis of factors other than current performance. It is well accepted that high performance in a current job does not mean that an individual will be a high performer if promoted to a higher level. It is also increasingly recognised that a moderate performer at one level may perform much better at a higher level. Assessment centres use tests, group exercises, work tasks and interviews to appraise potential.

Instead of assessing individual candidates one at a time through the selection procedure, assessment centres typically form the candidates into groups and assess them in that group. Assessment centres have three essential features in addition to the use of the candidate group:

1 A wide range of techniques is used.
2 Assessment is carried out by a group of judges.
3 The procedure as a whole is validated.

The method works on the principle that candidates should be subjected to a wide range of techniques or methods (the more the better!). Individually, these may have a limited but positive validity, whereas collectively they may produce data that can be converted into a useful predictor of job performance. A typical procedure may include the following methods:

Conventional selection methods These include the use of biographical questionnaires, interviews, psychological tests (e.g. ability and personality questionnaires).

Group discussion This is a technique whereby the candidates sit down together and are given a topic to discuss. The assessors, who may not appoint a leader to the group, remain outside, observing the ensuing interaction. This appears to be an effective procedure for measuring such dimensions as oral expression, interpersonal skills, and influence. The topics should be carefully chosen to maximise the interaction between group members. Each should meet the following requirements:

- It must not require specialist knowledge that favours individuals in the group.
- It should relate to an issue about which candidates are likely to have strong views.

Early consensus of opinion is unlikely. The discussion can be assessed in an unstructured way with assessors being asked to make any comment they wish on the group member's behaviour that they have observed, e.g.:

- Does the group member get his ideas across?
- Does he or she appear to influence other members' thinking?
- Does he or she dictate the direction that the discussion should go?

A highly structured classification system can be used. One long-established approach utilises the following categories:

- shows solidarity
- shows tension
- agrees
- gives suggestions
- gives opinions
- gives orientation
- asks for orientation
- asks for opinions
- asks for suggestions
- disagrees
- shows antagonism.

Group members can be compared by counting the number of times they exhibit behaviour under each of these categories.

Work simulation exercises At the macro-level business games are used. Often the candidate group is divided into smaller competing teams. These games, which can be computer based, can indicate to observers leadership ability in a business context and a more general business sense. There can be problems of insufficient information, and candidates must state their assumptions.

At the micro-level, a common technique is the 'in-tray exercise'. It involves group members individually being confronted with the same in-tray containing a pile of simulated management problems in the form of paperwork which the candidates have to deal with. How they perform is observed, and group members are compared. These exercises should be designed so that they can be objectively marked by assessors. This technique is a useful way of seeing how candidates will perform under job pressure and how rational their decision-making ability is.

Summary of the assessment centre The assessment centre contains both the statistical devices of tests and the clinical judgement method. Typically there is one assessor to three candidates. The role of the assessors is to reach some consensus on each candidate's rating against the job requirement. The whole procedure should be validated as a single method. Documented predictive validities have not been overwhelmingly high, but some assessment centre predictions have given cause for optimism (Ungerson, 1974).

The AC has a very high face validity. It looks like the sort of procedure that would sort out who are going to make good managers, and who are not. This *may* be misleading! Its main drawback is its cost, which limits it to large and wealthy employers. Disregarding the cost, the assessment centre is the best selection method, and may be used when:

- selecting for a key position (decision must be correct);
- there is a large throughput of candidates (this spreads the cost).

There is the inherent danger that it might be used as a 'sledgehammer to crack a nut', i.e. the amount of resources put into the assessment centre cannot be justified in terms of the effect it will make in terms of job performance differences. More recently, utility analysis techniques have been considered to determine the cost-effectiveness of selection methods (Boudreau, 1983; Cook, 1988).

9.2.3.4 *The Assessment Matrix*

It is important, prior to applying the selection methods to candidate, to determine which job-related criteria are being measured by which selection

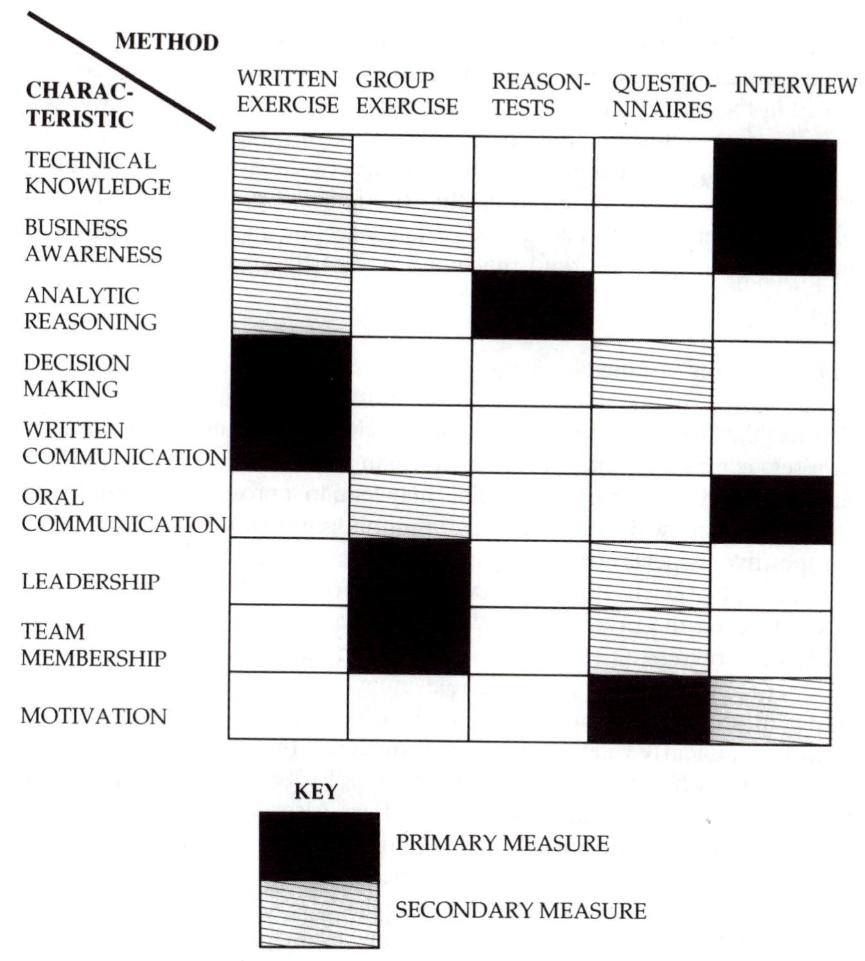

Figure 9.4 The assessment matrix.

method. Obviously, there will be some overlap, but this can be used to confirm the data gathered from the other source. The assessment matrix provides an approach for mapping methods of assessment onto the characteristics needed for the job (as identified in the job analysis). Figure 9.4 illustrates what this might look like.

As indicated in Figure 9.4, most methods are able to collect data about more than one characteristic. For example, the interview collects data on technical knowledge, business awareness, oral communication and motivation. In addition, the methods can be used as primary (the main source of that information) or secondary (additional sources) measures. The advantage of collecting the same information from more than one source is that it allows the internal validity of the selection methods to be evaluated. If, for example, the data on leadership gathered by the group exercise and questionnaire were in concordance, then this would increase the confidence which could be placed in these data. However, if these two sources were in conflict, caution with the data would be advised.

In summary, the assessment matrix improves the systematic structure of the selection process, by ensuring that the focus of data collection is upon job-relevant criteria. It also ensures that selection methods are chosen to measure specific criteria, and that none is superfluous to the process.

9.2.4 Decision Making

Making the decision about the appropriateness of the candidate often requires compiling data from more than one source, e.g. references, interviews and test performance. This may lead to a problem of combination. There are many ways of doing this, multiple regression being one. One of the positive aspects of multiple regression is that the predictors are independent of each other; however, if they overlap (i.e. predict the same criterion) then the predictive quality is not so good.

Multiple regression (MR) is an extension of correlation, which considers each of the predictors to assess the relevant weightings, and arrives at a single predictor score. This often looks impressive, but MR assumes that each predictor is linearly related to job performance – that is if there is a positive relationship between part of the selection and subsequent criterion performance, then the more that exists the better. Also, for example, very high intellectual ability will not necessarily provide very high job performance, which is what MR would suggest. MR operates a 'central trade-off position'. This means it will assume that a very high score on one of the measures compensates for a less than adequate score on one of the others.

Given the problem of the quantification of the predictors, how can one quantify predictors into data? Tests are normally designed to produce scores, but interviews, application forms, references are not. There are four levels of scaling:

1 *nominal scale* numbers are used as labels;
2 *ordinal scale* numbers refer to some degree of magnitude;
3 *interval scale* uses units that represent equal intervals;
4 *ratio scale* where zero equals the absence of that which is being measured.

The rating scales used as a measurement of interview performance are often *not* interval scales, but nevertheless find themselves being built into multiple regression equations. It is possible to reduce the number of decisions the interviewer has to make by having few well-defined traits assessed on a numerical scale, i.e. intelligence, sociability, personal warmth, etc. Often the justification for using a test is that there is 'evidence' that they work, meaning that those that do well in the test also do well in the job. A typing test may show if an individual is a good typist, but it says nothing of what that person will contribute towards office harmony. Traditionally there are two major criticisms of statistical validity:

1 Correlation coefficients are never very high ($r = 0.4$ is considered good).
2 The methods are too situationally specific.

However, both can be attributed to the poor application of statistical methods:

- *Sampling error* Various kinds of sampling error can occur. The failure to use an adequate sample of applicants on which to base the statistical analysis. For example, the sample size is too small, or used on job incumbents rather than applicants. In addition, no one knows what happens to those candidates who are rejected. The implication of these errors is that absolute confidence may not be placed in the accuracy of selection decisions, as some information is missing. However, the alternative of taking on all applicants as a means of validating selection procedures is often not practical.
- *Job performance measurement error* This is the failure to allow for the unreliability of crude measures of job performance. It has been suggested that human judgement is not as reliable as statistical techniques. However, this has been challenged. Some evidence suggests that in fact judgements for managerial and supervisory staff are surprisingly successful. Even when rating a scale such as 'sociability' the raters are using their own judgement, and making a complex decision based upon a variety of sources of information.

Clearly statistical prediction is essential to selection, but not at the expense of 'clinical judgement' – largely because it is too difficult in practice to unravel the two approaches. A systematic attempt to bring the art and science of selection together is to be found in the use of assessment centres. Selection is concerned with prediction. Which candidate will be best at doing the job. In order to predict, selection practices need to be able to measure certain relevant aspects to those seeking employment. These measures must be both stable (that is reliable) and accurate (that is valid). One view is that this can be achieved scientifically by using well-established statistical methods, especially the coefficient of correlation. Another very different approach is to consider the assessment of human potential to be too complex for statistical analysis. It requires human judgement – this being an art which should be developed.

9.2.5 Validation

Selection ratings for each individual can be compared with eventual performance over a variety of time periods. The comparison can also be to compare different selection criteria. Large discrepancies would require further investigation. There are big differences between individuals and jobs, and in selection matching or mismatching may occur. Mismatching may be reflected in many ways:

- poor job performance;
- reduced job longevity;
- lateness and absenteeism;
- number and severity of loss-time incidents and accidents.

There are two types of factors associated with differences in behaviour; these are quantitative and qualitative. Examples of quantitative include job performance (quantity, number of errors, accuracy), tenure on the job, number of absences, attitudes (as scored by an attitude scale), time required to complete a job, energy expenditure and job satisfaction (as measured by a job satisfaction questionnaire). Examples of qualitative include standard of work (quality) and behavioural measures of work (e.g. good/ indifferent/bad). Job performance varies between employees (and within employees). Reasons associated with these differences include variables such as individual variables (e.g. attitudes, personality characteristics, physical characteristics, interests, motivation, age, sex, education and experience), situational variables, job and working conditions variables (e.g. methods of work, design and conditions of work, work space and arrangement, physical work environment) and organisational and social variables (e.g. the character of the organisation, type of training and supervision received, types of incentives, the 'social' environment, union relations, etc.).

Thus there are many different types of variables that can influence the work performance or behaviour of people. The practice of matching people to jobs may regard the job as fixed. In selecting people who do not already have the necessary skills, the selector would have to focus on identifying individuals who have the potential for learning that particular job. The other strategy is to design the job to fit the people available, i.e. job simplification. More recently work design has gone into job enrichment to make the work more intrinsically satisfying, and therefore motivate employees to perform better. Most behavioural research is aimed at making some form of prediction, and in selection this is predicting future behaviour in the job from current behaviour during the selection process.

Selection ratings for each individual can be compared with eventual performance over a variety of time periods. The comparison can also be to compare different selection criteria. Large discrepancies would require further investigation. Selection techniques attempt to determine from the behaviour at the interview and in the tests what future behaviour is likely to be, and does this match the organisation's needs? The longer term the prediction, the more unreliable it is likely to be (e.g. consider the unreliability of the weather forecast!). Risky though it may be, selection does require prediction, and that prediction must be reliable and valid. The use of statistics in selection may help to make decisions more objective. Anything that can be dealt with in a quantifiable way (i.e. test results) can be compared to subsequent job performance to see if there is a valid relationship. To assess the relationship 'the aspect' would need to be considered as one variable and a measure of job performance as another. The normal way of statistically examining the degree of association between any two variables is to use the coefficient of correlation (r), where:

- The nearer r is to $+1$ the more highly positive is the relationship.
- The nearer r is to -1 the more highly negative is the relationship.
- A coefficient of $r = 0$ indicates no relationship.

If, for example, employees are given a test on appointment, and this is compared to subsequent job performance, the degree of relationship can be mapped onto a graph. The test should not only show a good correlation, but the correlation should be statistically significant at the 5% level or below (i.e. less than a 5% probability of the event occurring by chance). Statistical analysis of the relationship between the single selection measures and job performance (bivariate approach) can be refined further to justify its use. In reality most selection procedures do not use single selection methods, and some of the procedures used are not easy to quantify.

9.3 Implications for the Nuclear Industry

In a comprehensive review of selection of process operators, Williams and Taylor (1993) proposed that the lack of research into the subject was due to three main reasons. First, it is a complex topic to study, there are many variables to be investigated and it is difficult to conduct research in a controlled manner. Second, characteristics possessed by the individual may be considered to have less impact upon performance than factors in the environment, such as work and equipment design. Third, past research studies into selection show that the predictive power is quite poor. This becomes a circular argument; because it is a difficult area to study and people perceive that it is less important than other areas little research is done, so the predictive power of the methods is not optimised. A more optimistic perspective would be to try to get to grips with the topic to demonstrate that the present position can be improved substantially. The overview of the selection process presented in section two indicates how this may be undertaken, by systematically addressing elements within the selection process.

In their review, Williams and Taylor (1993) report that their survey of companies, both in the UK and internationally, reported similar findings to that of the IAEA (1984) survey. All the organisations surveyed were fairly pragmatic about their selection process and all employed a systematic approach, having many stages in common. They were also well informed about selection methods, tending to rely upon the more traditional approaches (e.g. interviews, technical tests, cognitive ability tests, and oral examinations). Williams and Taylor (1993) point out that the selection process was mainly used in a negative way (to filter out unsuitable applicants) rather than as a positive exercise (to identify ideal candidates) and that formal validation studies are rarely undertaken. They suggested that there is scope for improving current selection practices, including:

- improving the selection framework;
- basing the job analysis upon a functional analysis of the operator's task;
- using the functional analysis to identify selection attributes;
- measuring job performance;
- conducting validation studies.

Each of these points is consistent with the selection process model presented in Figure 9.1. The review by Williams and Taylor (1993) was principally concerned with selection of process operators. It may be interesting to consider differences in the selection of other personnel. In a survey of management selection in Britain, Robertson and Makin (1986) found that references and interviews are ubiquitous methods. This finding was further supported by Shackelton and Newell's study (1991). It would be interesting to compare these findings with those methods utilised in the nuclear industry to answer the following questions:

- Are different methods used to select engineering and management personnel?
- What range of methods is used?
- How effective are the methods in practice?

To answer these questions, an international survey of selection methods used in the nuclear industry was conducted; 34 companies in 13 countries (England, Scotland, Wales, France, Italy, Germany, Sweden, Switzerland, USA, Canada, Argentina, Japan and Russia) participated. A comparison with previous surveys is shown in Table 9.5.

Table 9.5 shows the percentages of organisations reporting on the use of selection methods. The first column indicated the selection methods used. The second column shows the data reported by Robertson and Makin (R&M) for the selection of UK managers (the sample taken from the Times 1000 index). The third and fourth columns show the data reported by Shackelton

Table 9.5 Comparison of surveys on selection methods

Method	R&M (1986) Managers	S&N (1991) UK Managers	S&N (1991) FR Managers	S&A (1995) Managers	S&A (1995) Engineers
Interview	99	100	100	100	100
Reference	99.3	73.9	11.3	95.2	86.6
Personality test	35.6	9.6	17	85.7	65.2
Cognitive test	29.2	69.9	48.9	76.2	60.2
Handwriting	27.3	2.6	77	5.8	2.9
Astrology	0	0	0	0	0
Biodata	5.8	19.1	3.8	2.9	5.8
Assessment centre	21.4	58.9	18.8	8.7	8.7

and Newell (S&N) on methods used to select managers in the UK and France respectively (the sample taken from the Times 1000 index and Les 200 Premières Groupes de Echos). The fifth and sixth columns show the data from Stanton and Ashleigh's (S&A) study of managers and engineers in the nuclear industry (for a further breakdown of these findings, see Appendix 2).

As Table 9.5 shows, interviews and references remain popular approaches, despite the equivocal evidence on their validity (Herriot, 1989a). There is a dramatic increase in the use of personality tests, despite the controversy surrounding their use in the selection context (Blinkhorn and Johnson, 1990). Cognitive tests remain popular; we feel that this is justified in terms of their predictive validity (Bethell-Fox, 1989). There is little surprise in the low use of the alternative methods (handwriting analysis and astrology). Biodata also appear to be unpopular. It is a little surprising, however, to see that assessment centres are not utilised more. The predictive validity of this combined approach is very respectable (Feltham, 1989). We also asked respondents of our survey to report the performance of their selection measures. In general, they reported that the performance of traditional approaches (interviews and references) was good. Psychological tests (personality and cognitive tests) performed above average. However, reports of handwriting analysis and biography were very poor. There were insufficient data on the performance of assessment centres to make a generalisation. It is highly likely that these reports were subjective in nature, not based upon formal validation studies, and therefore should be treated with caution.

9.4 Conclusions

In conclusion, there is much that can be done to improve existing selection practices in the nuclear industry. This applied to the selection of all personnel, but perhaps process operators are a special case because of licensing requirements. The selection of methods to evaluate personnel should depend upon:

- selection criteria of the post to be filled;
- acceptability and appropriateness of the methods;
- abilities of the staff involved in the selection process;
- cost–benefit analysis of methods;
- accuracy of methods;
- legality of methods.

Establishing the validity of the selection process make good commercial sense, especially when one considers the costs associated with poor selection decisions, for example:

- poor productivity of inappropriate personnel;
- knock-on effects of inappropriate personnel;
- costs of carrying a less effective employee;
- waste in training;
- cost of re-recruitment and selection.

We therefore argue that the organisation needs to pay careful attention to all of the stages in the selection process from developing appropriate selection criteria, to choosing recruitment techniques, to choosing selection methods, to making the selection decision, to validating the process.

References

AIKEN, L. R. (1976) *Psychological Testing and Assessment*. Boston: Allyn & Bacon.

ALGERA, J. & GREUTER, A. M. (1989) Job analysis for personnel selection. In Smith, M. and Robertson, I. (eds), *Advances in Selection and Assessment*. Chichester: Wiley.

BARTRAM, D., ANDERSON, N., KELLETT, D., LINDLEY, P. & ROBERTSON, I. (1995) *Review of Personality Assessment Instruments (Level B) for use in Occupational Settings*. Leicester: BPS Books.

BARTRAM, D. & BAYLISS, R. (1984) Automated testing: past present and future, *Journal of Occupational Psychology*, **57**(3), 221–37.

BARTRAM, D. & DALE, H. C. A. (1983) *Micropat Version 3.0: A Description of the Fully Automated Personnel Selection System Being Developed for the Army Air Corps*, Ergonomics Research Report, University of Hull, Report ERG/Y6536/83/7.

BARTRAM, D., LINDLEY, P., FOSTER, J. & MARSHALL, L. (1992) *Review of Psychometric Tests (Level A) for Assessment in Vocational Training*. Leicester: BPS Books.

BETHELL-FOX, C. E. (1989) Psychological testing. In Herriot, P. (ed), *Assessment and Selection in Organisations*. Chichester: Wiley.

BLINKHORN, S. & JOHNSON, C. (1990) The insignificance of personality testing. *Nature*, **348**, 671–2.

BOUDREAU, J. W. (1983) Effects of employee flows on utility analysis of human resource productivity-improvement programs. *Journal of Applied Psychology*, **68**, 396–406.

CARROLL, J. B. (1993) *Human Cognitive Abilities*. Cambridge: Cambridge University Press.

COOK, M. (1988) *Personnel Selection and Productivity*. Chichester: Wiley.

DEARY, I. J. & MATTHEWS, G. (1993) Personality traits are alive and well. *The Psychologist*, **6**(7), 299–311.

DIXON, N. F. (1976) *On the psychology of military incompetence*, London: Cape.

FELTHAM, R. T. (1989) Assessment centres. In Herriot, P. (ed.), *Assessment and Selection in Organisation*. Chichester: Wiley.

FRASER, J. M. (1958) *A Handbook of Employment Interviewing*. London: Macdonald & Evans.

HERRIOT, P. (1989a) *Assessment and Selection in Organisations*. Chichester: Wiley.

(1989b) Selection as a social process. In Smith, M. and Robertson, I. (eds), *Advances in Selection and Assessment*. Chichester: Wiley.

IAEA (1984) *Qualification of Nuclear Power Plant Operations Personnel: A Guidebook*, International Atomic Energy Agency, Vienna, 1984 (Technical reports series no. 242). STI/DOC/10/242, 290 pp.

KLINE, P. (1993) *The Handbook of Psychological Testing*. London: Routledge.

LEWIS, C. (1985) *Employee Selection*. London: Hutchinson.

REILLY, R. R. & CHAO, G. T. (1982) Validity and fairness of some alternative employee selection procedures. *Personnel Psychology*, **35**, 1–62.

ROBERTSON, I. T. & MAKIN, P. J. (1986) Management selection in Britain: a survey and critique. *Journal of Occupational Psychology*, **59**(1), 45–57.

RODGER, A. (1952) *The Seven Point Plan*. London: National Institute of Psychology.

SHACKELTON, V. (1989) *How to Pick People for Jobs*. London: Fontana.

SHACKELTON, V. & NEWELL, S. (1991) Management selection: a comparative survey of methods used in top British and French companies. *Journal of Occupational Psychology*, **64**(1), 23–36.

SMITH, M. & ROBERTSON, I. (1989) *Advances in Selection and Assessment*. Chichester: Wiley.

STANTON, N. A. & MATTHEWS, G. (1995) Twenty-one of personality: an alternative solution for the occupational personality questionnaire. *Journal of Managerial Development*, **14**(7), 68–77.

TORRINGTON, D. & HALL, L. (1987) *Personnel management: a new approach*. Englewood Cliffs, NJ: Prentice Hall.

UNGERSON, B. (1974) Assessment centres: a review of research findings. *Personnel Review*, **3**, 4–13.

VALE, C. D. (1981) Design and implementation of a micro-computer based adaptive testing system. *Behaviour Research Methods and Instrumentation*, **13**, 399–406.

WEISNER, W. H. & CRONSHAW, S. F. (1988) A meta-analytic investigation of the impact of interview format and degree of structure on the validity of the

employment interview. *Journal of Occupational Psychology*, **61**(4), 275–90.
WILLIAMS, J. C. & TAYLOR, S. E. (1993) *A Review of the State of Knowledge and of Current Practice in Selection Techniques for Process Operators*, HSE Contract Research Report No. 58/1993. London: HMSO.

APPENDIX 1

Job analysis information checklist

(Requirements: essential/desirable/undesirable)

Job identity
- Job title
- Department
- Location
- Number of incumbents

Work performed
- Duties/responsibilities performed
- How they are performed
- Why they are performed
- Frequency and scope of particular duties (main duties + minor duties)
- Performance standards (criteria for evaluation)
- Difficult/distasteful duties
- Features of the work
- Features of the workplace (physical/social)

Knowledge required
- Areas of knowledge
- General disciplines
- Specialised expertise
- Formal education (how much)
- Experience (how long)

Skills required
- Mental
- Physical
- Interpersonal

Physical demands
- Exertion (availability of support equipment)
- Motion
- Environment (heat, cold, humidity, noise)
- Hazards
- Exposure to unpleasant conditions

Special demands
- Work hours
- Travel
- Isolation

Sources of workers
- From other jobs
- Job postings
- Apprenticeship programmes
- Other companies
- Hands-on training

Accountability
- Equipment value
- Assets
- Budgets and expenditures
- Outside relations

Personal requirements
- Attainments
- Attitudes
- Physical characteristics
- Personality characteristics
- Interest
- External restraints

APPENDIX 2

Review of selection methods used in the nuclear industry

The sample for this table is taken from 13 countries and 34 participant companies (responses are in percentages). The use of methods of selection is analysed with respect to types of employees (engineers and managers).

		1 never	2 <half	3 half	4 >half	5 always
Method of selection						
Interviews	Managers	—	9.52	4.76	9.52	76.19
	Engineers	—	4.35	13.04	4.35	78.26
Application forms	Managers	4.76	9.52	—	14.92	71.43
	Engineers	4.35	—	—	13.04	82.61
References	Managers	14.29	9.52	9.52	9.52	57.14
	Engineers	13.04	17.39	8.7	8.7	52.17
Personality tests	Managers	14.29	33.33	14.29	—	38.1
	Engineers	34.78	34.78	4.35	8.7	52.17
Cognitive tests	Managers	23.81	28.57	9.52	—	38.1
	Engineers	39.13	34.78	4.35	4.35	17.39
No. of interviews						
One only	Managers	28.57	—	19.05	33.33	19.05
	Engineers	21.74	—	21.74	38.78	21.74
More than one	Managers	19.05	28.57	14.29	4.76	33.33
	Engineers	26.09	34.78	13.04	4.35	21.74
Participation						
Personnel department	Managers	4.76	19.05	19.05	9.52	47.62
	Engineers	17.39	8.7	13.04	4.35	56.52
Line management	Managers	—	—	4.76	14.29	80.95
	Engineers	—	4.35	4.35	—	91.39
Internal specialist	Managers	28.57	23.81	14.29	9.25	23.81
	Engineers	34.78	21.74	13.04	13.04	17.39
External specialist	Managers	71.43	19.05	—	4.76	4.76
	Engineers	91.3	8.7	—	—	—
Type						
One-to-one interview	Managers	60	20	—	20	—
	Engineers	50	33.33	—	16.67	—
2–3 interviews	Managers	20	—	—	40	40
	Engineers	—	16.67	—	50	33.33
Panel	Managers	60	20	—	20	—
	Engineers	33.33	50	—	16.67	—
Other methods						
Handwriting analysis	Managers	94.2	2.9	2.9	—	—
	Engineers	97.1	—	2.9	—	—
Astrology	Managers	—	—	—	—	—
	Engineers	—	—	—	—	—
Biography	Managers	97.1	—	—	—	2.9
	Engineers	94.2	—	—	—	5.8
Assessment centres	Managers	91.2	5.8	3	—	—
	Engineers	91.2	5.8	—	3	—

CHAPTER TEN

Training issues

ROBERT B. STAMMERS
Aston University, Birmingham

10.1 Introduction

Unlike many industries, there is little need to call for training to be taken seriously in nuclear power operations. It has a long-established precedent in most countries, with regulatory bodies exerting control on the nature and frequency of training. It is an industry that has made major investment in training facilities, with large-scale control room simulators being to the forefront.

All industries have made more use of automation: over the last decades there has been a considerable impact on the skill needs of system operators. This in turn has affected the training needs of those operators. The major change has been the increased utilisation of human cognitive skills rather than perceptual or motor activity alone. The cognitive skills include monitoring for events or for system values going out of tolerance, fault diagnosis from displays and search within the system displays for faulty items. In turn there is the need to correct faults, to reconfigure the system to bypass faulty elements or to run the system in a degraded mode to achieve the task objective in an alternative way.

In automated plant, the operator needs to be able to understand how the system under control functions, and this must include the automated features. The operators need to have a model of the total situation, built up through this knowledge and via the currently displayed information.

Systems like this give rise to difficulties for operators in initially learning and subsequently practising the required knowledge and skill. Problems stem from the inherent reliability of automation. This means that there are limited opportunities for training in the on-job situation. Whilst procedures can be learned in isolation, they will be practised rarely, if at all, in their real context. When problems occur, it may be envisaged that the operator's competency will be at high levels of readiness. From what is known about skill retention (discussed below), we should not expect this simplified model of training to work too well. The opportunity to try out in a concrete way what has been learned in the abstract is the key training problem. From what has been said above, the limited opportunities for practice in this context give rise to the need for simulators.

To this can be added that the tasks may have to be performed under stressful conditions, where the operators may be under scrutiny for their actions or where doing the wrong thing may make the situation worse. The build-up of skill through practice opportunities is the key to combating performance-limiting effects of stressors. These examples of training issues need to be put into a wider training context.

10.2 Training Systems

A training system for a task of any complexity needs to have a number of important characteristics. The primary one is that training should be based on a detailed task analysis of what will be required of the operator in meeting the demands of real task context. Training should not be based on expert opinion only, reliance should not be put on trainees informally learning what is required of them, nor should it be expected that their inherent intelligence or previous learning will help them 'work out' what is required (Stammers and Shepherd, 1995).

Another characteristic of training in this context is that it goes on over a long period of time, moving through incremental stages of difficulty. As proficiency grows, the trainee should be challenged by scenarios of increasing complexity. At any stage of training there will be a mix of *instruction* (concerned with learning factual knowledge) and *practice* (opportunities for the development of skill). The best mixture of theory and practice to bring about the appropriate cognitive performance needs to be carefully determined (Stammers, 1995).

As has been alluded to already, training in this context is difficult to envisage without some form of simulation. A range of simulations can be justified on the basis of a number of factors. Some reasons for using simulations have to do with the nature of the systems to be learned about, i.e. their inherent complexity. Other reasons concern the costs of using complex systems for training purposes only. These types of factors can be termed 'system' factors. Different types of reasons come from the difficulties of learning about rare events and complex control systems in the on-job situation, i.e. 'learning' factors. However, the most common reasons for justifying simulation arise from an interaction of system and learning variables. For example, some systems cannot allow operators to make errors (system factors), yet such errors are characteristic of the learner (learning factors). It is important to consider a range of *simulations* and to put an emphasis on the effective learning rather than to use the more restrictive term, *simulator*, which focuses on the equipment used and its functional realism (Stammers, 1986).

In the design of simulations, the focus should be on making them into effective training devices, rather than to concentrate on replicating the real situation. There are a number of important training facilities that need to be built in, such as tailored performance measurement and feedback capabilities, part-task practice facilities, graded difficulty and slow and fast time running, etc.

10.3 Utilising Training Resources

It is essential to view a training system as part of the larger system which it serves. Training research focuses on learning efficiency and effectiveness (Patrick, 1992). This is an area of key importance, but to build really effective training facilities it is essential that the broader system is taken into account.

The 'Taskmaster' model of training functions was developed as an aid to training policy decision making (Shepherd *et al.*, 1992). It is based on a task analysis of the functions of training management. Following the approach of hierarchical task analysis (HTA) (Shepherd, 1989), the highest-level operations for the overall task of 'Manage and deliver training for the organisation' are (Figure 10.1):

1 promote and optimise training function;
2 design and develop training;
3 install new training;
4 provide training delivery.

The fourth operation illustrates the need to see that everyday training activities are carried out. Operations 2 and 3 are executed when a new training project is initiated. The first operation is of central importance and suggests how the effective training manager can be proactive, rather than reactive to changes in the external and internal environment of the

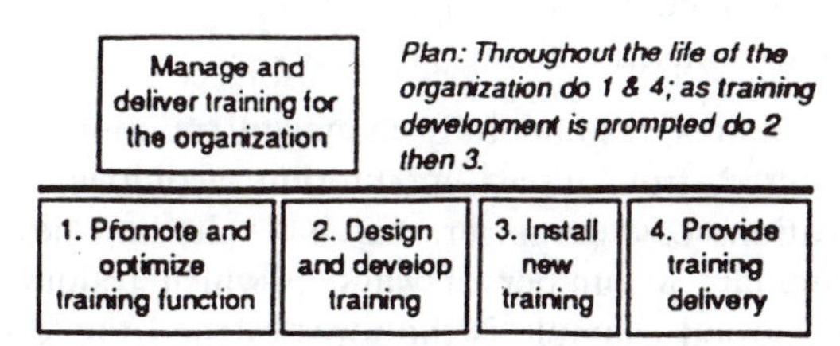

Figure 10.1 Functions in 'manage and deliver training for the organisation'.

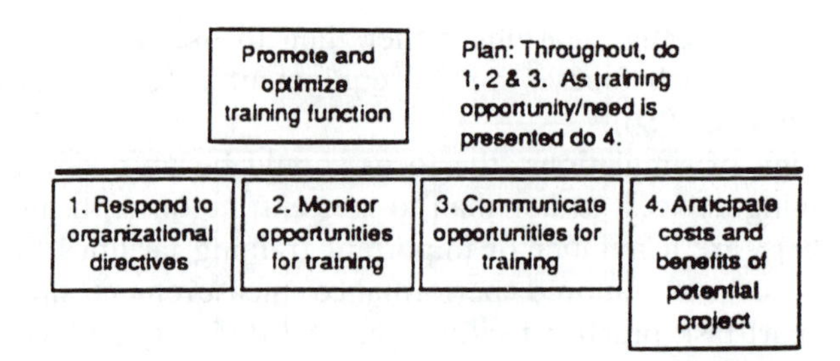

Figure 10.2 Functions in 'promote and optimise training function'.

organisation. This operation can be further decomposed thus (Figure 10.2):

1 Promote and optimise training function:
- 1.1 respond to organisational directives;
- 1.2 monitor opportunities for training;
- 1.3 communicate opportunities for training;
- 1.4 anticipate costs and benefits of a potential training project.

Operations 1.1, 1.3 and 1.4 make reference to how the manager can make the organisation more aware of the potential of training. There is a need to be aware of the organisation's overall strategy and a need to see how training can contribute to those goals. Operation 1.2 concerns a different type of activity. It envisages the training manager having responsibility for monitoring for outside events and for opportunities within the organisation to use developments in technology and training research. This operation can be broken down further (Figure 10.3):

1.2 Monitor opportunities for training:
- 1.2.1 monitor potential demand, e.g. new systems under development;
- 1.2.2 monitor problems associated with incidents and observed drifts in operational achievements;
- 1.2.3 monitor opportunities to extend training involvement in on-going projects;
- 1.2.4 monitor the emergence of new training technologies;
- 1.2.5 monitor general human resource environment;
- 1.2.6 monitor unused training resources.

It can be seen that a resourcefully managed training function can use information collected from these monitoring activities to bring about important applications of emerging research results and new technological innovations. There are a number of ways in which training research can influence policy: a good example is the topic of the long-term retention of training, outlined below.

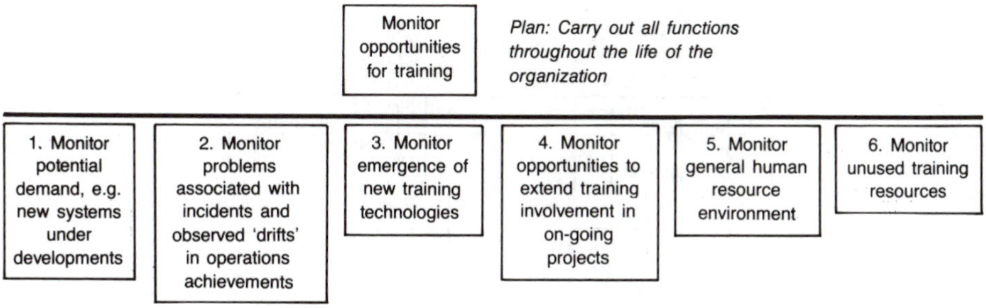

Monitor opportunities for training

Plan: Carry out all functions throughout the life of the organization

1. Monitor potential demand, e.g. new systems under developments

2. Monitor problems associated with incidents and observed 'drifts' in operations achievements

3. Monitor emergence of new training technologies

4. Monitor opportunities to extend training involvement in on-going projects

5. Monitor general human resource environment

6. Monitor unused training resources

Figure 10.3 Functions concerned with 'monitor opportunities for training'.

10.4 Retention

As discussed above, operators may work with a system for prolonged periods of time and not have to execute certain key tasks. These tasks can be routine, but rarely performed, tasks or they can arise from faults occurring in the system. Concern over situations like these has been around for some time and literature reviews have been carried out in a number of contexts e.g.:

1 aircraft control (Naylor and Briggs, 1961; Prophet, 1976);
2 spaceship control (Gardlin and Sitterley, 1972);
3 military skills (Schendal *et al.*, 1978; Hagman and Rose, 1983);
4 training for the unemployed (Annett, 1979);
5 process control operations (Stammers, 1981).

In these contexts, a common theme is that training that is given at one point in time may not be regularly practised under normal conditions. Whilst there are some uncertainties in the literature a number of important generalisations emerge.

It is clear that skill loss occurs, and can be at a significant level. It can also occur after quite short periods of non-practice. As an example, Cotterman and Wood (1967) found with a simulated spacecraft control task, a fall in performance with only a week without practice. Using a range of military tasks, Vineberg (1975) found skill losses of between 18% and 26%. Studies by Glendon and colleagues (e.g. Glendon *et al.*, 1988) have found large losses in cardiopulmonary resuscitation (CPR) skills. In one study, fewer than 20% of 124 first aiders reached a score of 75% six months after training. It is possible to say that the tasks used in these studies are different to those found in the nuclear industry. However, there are now sufficient data from a number of studies to give rise to concern. A number of factors influence the scale of the skill loss, i.e. the amount of original training, the nature of the original training and the amount of retraining received (Hagman and Rose, 1983).

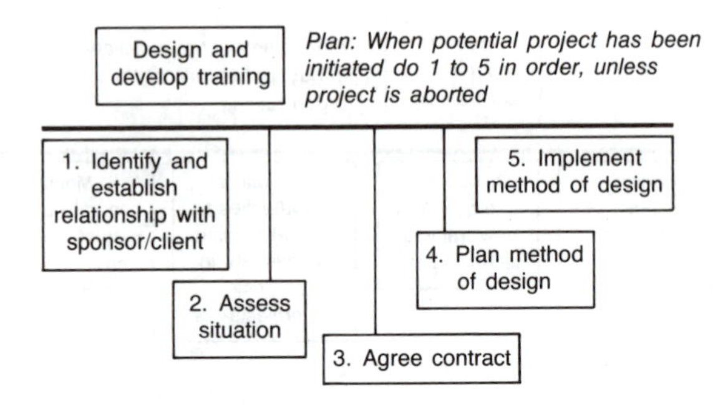

Figure 10.4 Functions in 'design and develop training'.

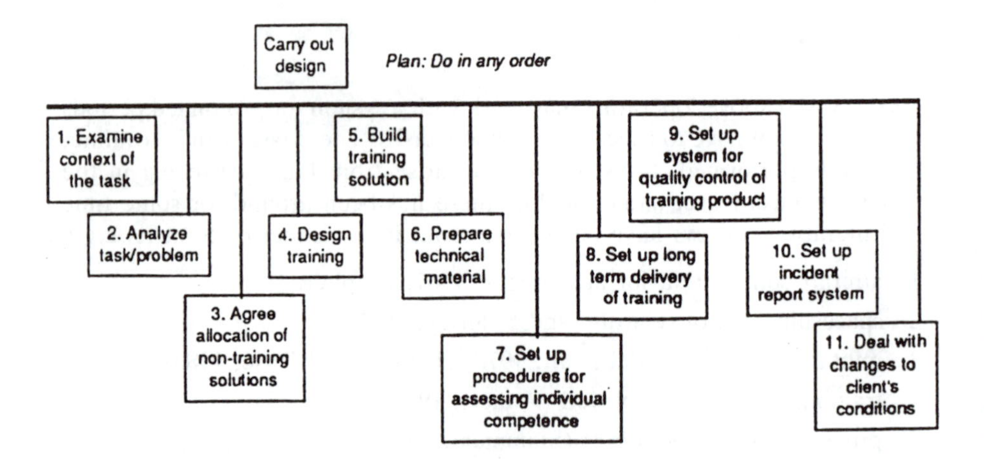

Figure 10.5 Functions in 'carry out design'.

On a more positive side there are established training activities that can help to reduce this loss. The research results generally support the finding that 'overtraining' can result in skill loss. This involves taking training past the achievement of normal criterion performance. Rehearsal training is also very important. Even if skill loss is substantial, relearning to previous criteria can occur quickly. In some cases it has found that relearning can be achieved with reduced realism simulations (Gardlin and Sitterley, 1972).

Further analysis is illustrated in Figures 10.4 and 10.5.

10.5 Conclusions

An effective training function needs to be sensitive to the organisational context in which it finds itself. At the same time it must be able to introduce innovations and new approaches and be able to respond to changes in the organisation's role brought about by external events and internal policy changes. To convince people of the need for change, it must collect evidence from a variety of sources and marshal them in order to release sufficient funds to be able to meet its objectives. In turn it must be able to justify those costs.

References

ANNETT, J. (1979) Memory for skill. In Gruneberg, M. M. and Morris, P. E. (eds), *Applied Problems in Memory*, pp. 215–47. London: Academic Press.

COTTERMAN, T. E. & WOOD, M. E. (1967) *Retention of Simulated Lunar Landing Mission Skills: A Test of Pilot Reliability*, Rep. No. AMRL-TR-66-222 (AD-817232), Wright-Patterson Air Force Base: Aerospace Medical Research Laboratories.

GARDLIN, G. R. & SITTERLEY, T. E. (1972) *Degradation of Learned Skills: A Review and Annotated Bibliography*, Rep. No. D180-15080-1 (N73-10152), Seattle, WA: Boeing Co.

GLENDON, A. I., MCKENNA, S. P., HUNT, K. & BLAYLOCK, S. S. (1988) Variables affecting cardiopulmonary resuscitation skill decay. *Journal of Occupational Psychology*, **61**, 243–55.

HAGMAN, J. D. & ROSE, A. M. (1983) Retention of military tasks: a review. *Human Factors*, **25**, 199–213.

NAYLOR, J. C. & BRIGGS, G. E. (1961) *Long-term Retention of Learned Skills: A Review of the Literature*, Rep. No. ASD-TR-61-390 (AD-267043), Wright-Patterson Air Force Base, Ohio: Aeromedical Systems Division.

PATRICK, J. (1992) *Training: Research and Practice*. London: Academic Press.

PROPHET, W. W. (1976) *Long-term Retention of Flying Skills: A Review of the Literature*, Rep. No. HumRRO-FR-ED (P)-76-35 (AD-A036077), Alexandria, VA: Human Resources Research Organization.

SCHENDAL, J. D., SHIELDS, J. L. & KATZ, M. S. (1978) *Retention of Military Skills: Review*, Tech. Paper 313 (AD-A061338), Alexandria, VA: U.S. Army Research Institute for the Behavioral and Social Sciences.

SHEPHERD, A. (1989) Analysis and training in information technology tasks. In Diaper, D. (ed.), *Task Analysis for Human–Computer Interaction*, pp. 15–55. Chichester: Ellis Horwood.

SHEPHERD, A., STAMMERS, R. B. & KELLY, M. (1992) Taskmaster – a model of training functions to support the organization of effective training provision. In *Proceedings of the International Training Equipment Conference and Exhibition*, pp. 328–32. Luxembourg: International Training Equipment Conference and Exhibition.

STAMMERS, R. B. (1981) *Skill Retention and Control Room Operator Competency*, Report No. 19, Karlstad, Sweden: Ergonområd AB.

(1986) Psychological aspects of simulator design and use. In Lewins, J. and Becker, M. (eds), *Advances in Nuclear Science and Technology*: Volume 17, *Simulators for Nuclear Power*, pp. 117–32. New York: Plenum.

(1995) Training and the acquisition of knowledge and skill. In Warr, P. B. (ed.), *Psychology at Work*, 4th edn. London: Penguin.

STAMMERS, R. B. & SHEPHERD, A. (1995) Task analysis. In Wilson, J. R. and Corlett, E. N. (eds), *Evaluation of Human Work*, 2nd edn. London: Taylor and Francis.

VINEBERG, R. (1975) *A study of the retention of skills and knowledge acquired in basic training*, Rep. No. HumRRO-TR-75-10 (AD-A012678), Alexandria, VA: Human Resources Research Organization.

CHAPTER ELEVEN

Team performance: communication, co-ordination, co-operation and control

NEVILLE STANTON
Department of Psychology, University of Southampton

11.1 Introduction

Teamwork is a central feature of control-room operations, yet there appear to have been few attempts to review existing literature on team working and apply it to nuclear control rooms. Undoubtedly, attempting to identify and analyse all of the factors that could influence team performance is a considerable undertaking. Some researchers have taken on this task, however (see Chapter 8). One such attempt to identify and classify the factors that affect team performance was devised by Foushee and Helmreich (1988). They identified three main factors: inputs, process and outputs. These are illustrated in Figure 11.1.

'Inputs' identify those factors which are brought to the team situation: the characteristics of the individuals within the team, the characteristics of the team and the characteristics of the environment. 'Process' refers to those factors which relate to the functioning of the team. 'Outcomes' refers to the performance of the team in terms of the task and team processes.

Another attempt to analyse team performance focuses on the constituents of the team processes (in aviation) in terms of communication, team building, workload management and proficiency (Cooper *et al.*, 1980). These four factors were proposed as basic team skill areas; a team may be identified as positive or negative on any of the sub-scales (see Figure 11.2).

Whilst some of the sub-scales are obviously related to the context of aviation, the general areas are more generally applicable to team performance in nuclear power plants.

A growing interest in the ideas behind the socio-technical approach to system evaluation is evident in more recent research. The socio-technical

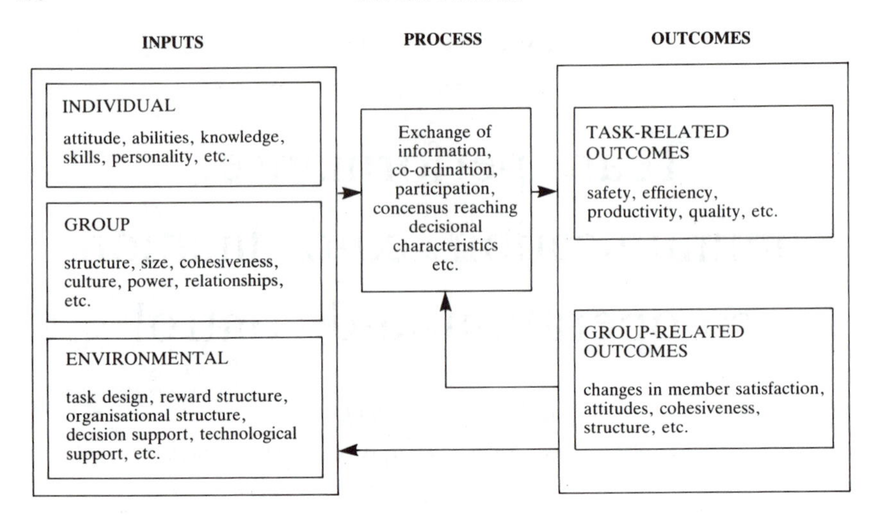

Figure 11.1 Factors affecting team performance (from Foushee and Helmreich, 1988).

approach identifies two major sub-systems, the technical system (the hardware and software), and the social system (engineers, control room staff, management, etc.). This analysis reflects the realisation that both these systems need to be considered. As Woods (1987) writes, nuclear power plants are not just technical systems, nor are they just human systems. In particular, it is the nature of the interaction between the social and technical sub-systems that provides the greatest insight into performance problems. Hettenhaus (1992) considers that a social-technical approach should address:

- the technical process;
- the interaction of operators with the technical process;
- the interaction of people with each other;
- the boundaries of management within the system.

He suggested that a fundamental principle to the approach is that the operators be in control of the technology which is employed to enable them to work at a higher level. The systems ideas can be traced back to the pioneering work of von Bertalanffy (1950) who introduced general systems theory (GST). Originally applied to biological systems, GST was used to describe the relationships between molecules, organs and beings. Subsequently, GST has been applied to describe organisations (Lockett and Spear, 1980) as a system in dynamic terms:

- Activity results from continual adaptation of the system components.
- Changes in one component affect other components and the whole system.

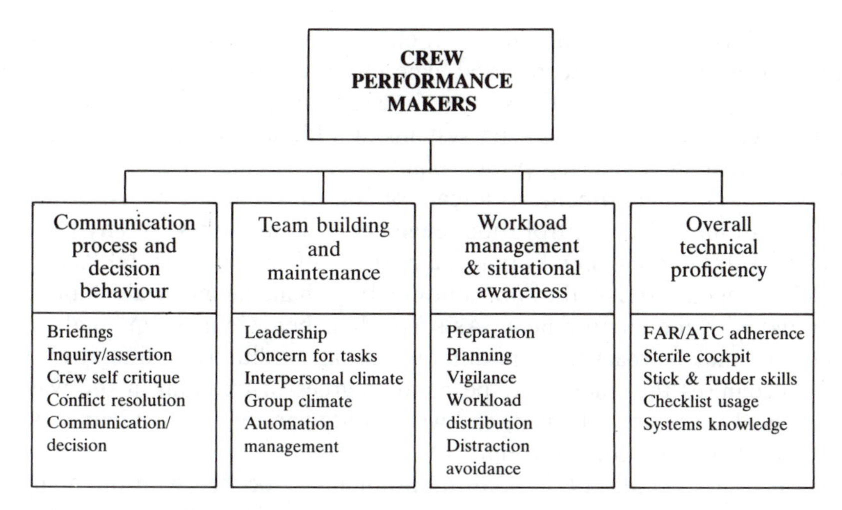

Figure 11.2 Factors relating to team processes.

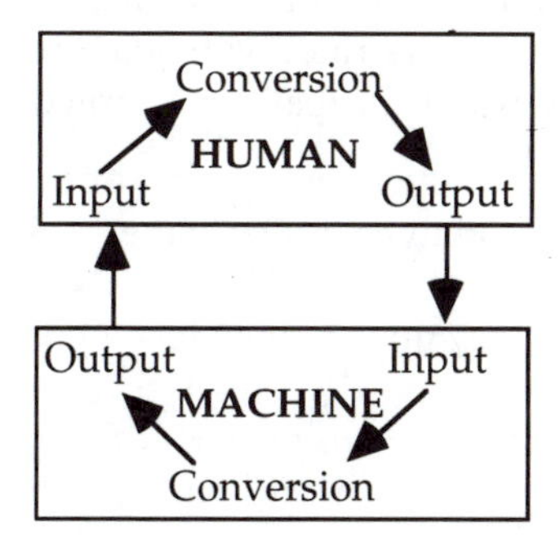

Figure 11.3 Basic building-block of systems analysis.

- Systems components become linked by exchanges (inputs and outputs).
- Within a component an internal conversion occurs.
- Each type of input has a corresponding output.
- Errors become apparent at boundaries between components.

The input–conversion–output cycle is the basic building-block of a socio-technical system (see Figure 11.3). Three main concepts dominate systems theory: interaction and interrelatedness, levels of analysis and equifinality.

A system is defined as a '. . . set of interrelated elements, each of which is related directly or indirectly to every other element. . .' (Ackoff and Emery, 1972). This emphasises the point that system elements cannot be construed as separate, self-contained, elements. All elements of the system are interrelated and interconnected. Therefore, changes in one element will have an effect upon the entire system. The system can be analysed at various levels, i.e. broken down into sub-systems in a hierarchical manner, e.g. the social system in a nuclear power plant could be analysed as the whole subsystem, as a department, as a team of control room engineers, or at the level of an individual control room engineer. The usefulness of any level of analysis will largely depend upon the purpose of the analysis. NPPs are by definition open systems, they are continually exchanging materials (input and outputs) with the environment. As such, their behaviour is equifinal: their final state may be reached from different initial conditions in different ways. This recognises that there may be many degrees of freedom in the way in which goal states are achieved, and why a wide variety of behaviour may be observed from one control room to the next.

Montgomery *et al*. (1991) note the problems associated with the analysis of team performance. They compared the use of behaviourally anchored rating scales (BARS) and behavioural frequency rating scales (BFRS) by training instructors at a nuclear power plant (for a review of training issues, see Chapter 10). The instructors rated videotape scenarios of control-room performance. The study was principally concerned with six dimensions of performance:

1 communications;
2 openness;
3 task co-ordination;
4 team spirit;
5 maintaining task focus;
6 adaptability.

The results favour the use of BARS for assessing team performance, as the data were generally more stable. It was thought by Montgomery *et al*. (1991) that BFRS imposed too high a demand on the raters, which resulted in the poorer performance of the scaling technique.

Methods of evaluating team performance need to be able to record, code and analyse as many facets of the team members' behaviour as possible. Past efforts in this area have considered individual behavioural activities, physical location of the individual and verbal communication. Other aspects of behaviour could also be captured, such as non-verbal communication (Argle, 1972), concurrent verbal protocol (i.e. think-aloud techniques, e.g. Bainbridge, 1984), and collective group meta-information processing (by applying Rasmussen's SRK framework at the group, rather than the individual level). The team could also be involved in the analysis of the data by playing back videotape and information presented to them during the simulated event. They could be asked to justify their decisions, indicate what

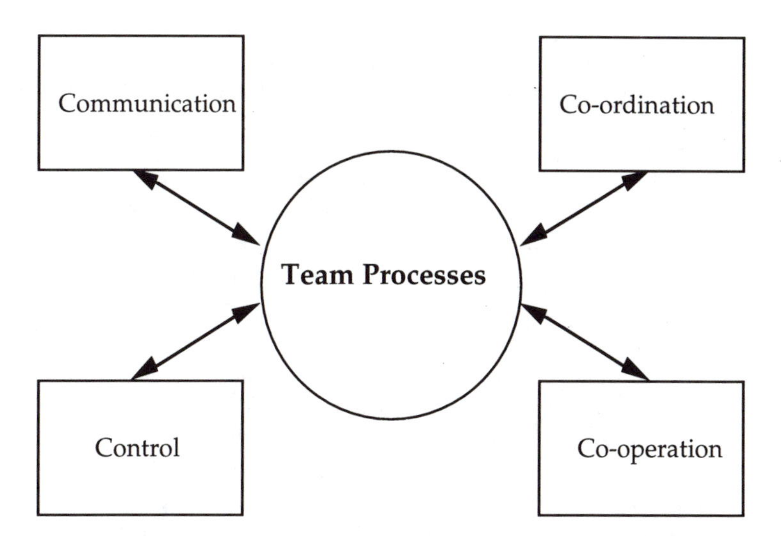

Figure 11.4 The four Cs in team processes.

led them to develop their hypotheses, and where they sought information from. The use of eye measuring equipment may help facilitate this latter suggestion.

In an attempt to reconcile the various attempts to analyse team performance, this review focuses on the core dynamics of teamwork. Drawing from a wide variety of teamworking situations, the review will present the research under four major aspects of team process, as shown in Figure 11.4.

11.2 Team Communication

The relationship between team communication and team performance has been investigated in the aviation industry, where Kanki *et al.* (1989) report that teams '. . . communicating in highly standard, hence predictable, ways were better able to co-ordinate their task, whereas [teams] characterised by multiple, non-standard communication profiles were less effective in their performance.' There are many similarities between teams in the cockpit and teams in NPP; typically both are small, highly structured, highly co-ordinated and required to process vast amounts of information. This apparent similarity justifies the use of research data from the domain of aviation to inform NPP operations.

Kawano *et al.* (1991), in the analysis of NPP operators, found that communication differences affected team performance in simulation studies. They noted four major crew communication styles as indicated in the following figure. The communication styles were (Figure 11.5):

- top down;

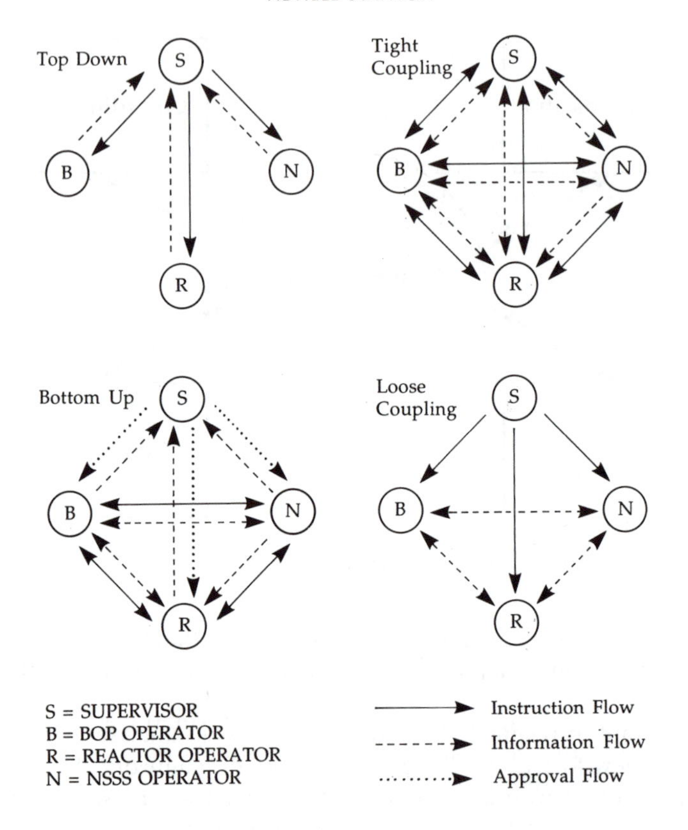

Figure 11.5 Crew communication type.

- tight coupling;
- bottom up;
- loose coupling.

Kawano *et al.* (1991) indicate how team performance may be affected by communication styles. In general, it seems that tight coupling and bottom up lead to superior performance to loose coupling and top down. This is due, in part, to reduction in bureaucracy and, in part, improved communication channels providing the possibility for feedback.

Coury and Terranova (1991) note that few studies of human performance in complex systems consider team decision making. They propose three elements of collaborative decision making:

1 decision making by individuals and teams;
2 the characteristics of the user interface;
3 aspects of communication outside the displays.

These elements contain various attributes that are illustrated in Table 11.1.

Table 11.1 Elements of collaborative decision making

Decision making:
- composition of mental models
- role of mental models in decision making
- decision processes and strategies
- individual and team mental models
- effects of risk and uncertainty

User interface:
- characteristics of display formats
- effect of displays on mental models
- navigating complex data structures
- information management

Communication:
- content of communication
- communication skill and ability
- role responsibility
- effect of display formats on communication

Coury and Terranova suggest that collaborative decision making occurs at a higher level than individual decision making would (see Figure 11.6).

This is due to the 'team mental model' resulting as a collaboration of individual mental models. This could mean that the 'team mental model' has fewer consistencies and flaws than any individual person's understanding. It would certainly help explain, at least in part, the superior team performance reported by Hooper *et al.* (1991).

The nature of the communication content is considered to be a better predictor of operator workload than measures of knowledge and skill (Terranova *et al.*, 1991). Hooper *et al.* noted that individuals were more concerned with managing joint decision-making effort and co-ordinating the interaction with each other. These differences are illustrated in Table 11.2.

In summary, it seems that team communication is more effective where standard communication practices are adopted, team communication is tightly coupled, decision making is collaborative and communication includes co-ordinating activities.

11.3 Team Co-ordination

Team co-ordination and organisation may also have a direct relationship with team performance. Co-ordination refers to the formal structural aspects of the team: how tasks, responsibilities and lines of communication are assigned.

Table 11.2 Categories of communication content

Individuals	Teams
Goal/decision	Co-ordinating teams' activities
Plan of action/specification	Giving confirmation
Uncertainty/reflection	Requesting feedback or approval
Reaction	Seeking information
Observation/description	Offering a suggestion
Expectation/hypothesis	Giving a command
Justification/explanation	Giving negative feedback/reprimand
Restatement/confirmation	Giving positive feedback/motivation

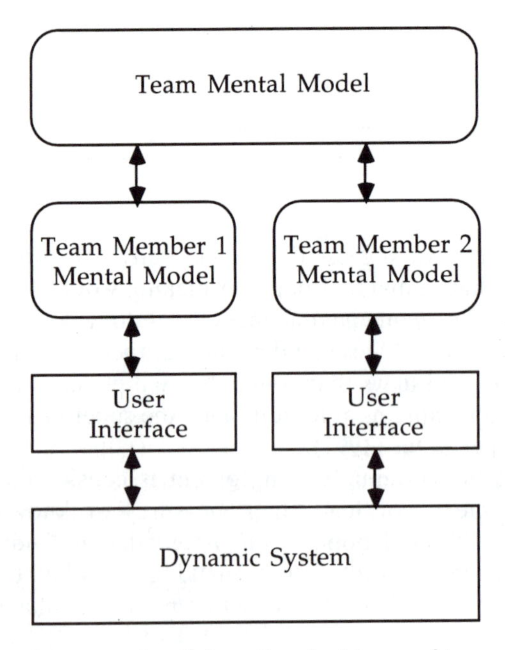

Figure 11.6 The elements of collaborative decision making.

Stammers and Hallam (1985) report a series of case studies investigating task demands and team organisation in control room (air traffic and emergency services) domains. They suggest that optimal task separation will largely depend upon the operational system. Stammers and Hallam (1985) identify two main team organisation methods as vertical and horizontal. In vertical organisation, sets of tasks are allocated to one operator. Horizontal organisation means that any of the operators can carry out the tasks. These two principal methods of team design are illustrated in Table 11.3.

Stammers and Hallam (1985) assert that the introduction of another

Table 11.3 Forms of team organisation with two operators and three tasks

	Operator 1	Operator 2
Vertical (i)	AB	C
Vertical (ii)	A	BC
Horizontal	0.5 (ABC)	0.5 (ABC)

Table 11.4 Forms of team organisation with three operators and three tasks

	Operator 1	Operator 2	Operator 3
Vertical	A	B	C
Horizontal	0.3 (ABC)	0.3 (ABC)	0.3 (ABC)
Hybrid (vertical–horizontal)	A	0.5 (BC)	0.5 (BC)
Hybrid (horizontal–vertical)	0.5 A	0.5 A	BC

individual into the control team extends the list of possible permutations and configurations for team organisation. Some of the alternatives are illustrated in Table 11.4.

More complex allocations may exist in operational environments, but these figures serve to illustrate the different types of team organisation that could exist. From their studies, Stammers and Hallam (1985) were able to draw seven main conclusions regarding the design of team organisation. These are as follows:

1 Some initial assessment should be made of the complexity of the tasks in the system.
2 Task organisation, the nature and degree of interaction between tasks, should also be assessed.
3 Sets of tasks can then be provisionally allocated to team members, with attempts to balance the demands imposed by task complexity.
4 The degree of intra-task organisation should be maximised for individuals, i.e. they should be given tasks that easily link together.
5 At the same time, attempts should be made to minimise the amount of inter-task organisation or communication demand between operators.
6 Various mixes of vertical or horizontal team structures are likely to be possible and should be assessed.
7 It is also important to try to build in flexibility, to allow team organisation to vary under changing task demands.

Although these points are aimed at team design, or redesign, they are also pertinent for team analysis.

Singleton (1989) suggests that inadequate team organisation may manifest particular behavioural symptoms, such as:

- contention between individuals or sections, particularly about demarcation;
- excessive reliance on human memory;
- secretiveness about progress to reduce the possibility of interference from others;
- inadequate leadership;
- poor communications;
- inconsistencies in style of problem solving.

In summary, it seems that effective team organisation largely depends upon task complexity and the flexibility of individual team members to adapt to situational changes.

11.4 Team Control

Control in team performance may refer to the control over team members in performance of the task by the team leader. Analysis of the literature suggests that over-controlling team leaders seems to result in performance problems for teams.

Green (1990) noted from Confidential Human Incident Reporting Programme (CHIRP) reports that breakdowns in team performance can result from the conformity and compliance. For example, the subordinate's reluctance to question the superior's actions, even when matters are abnormal. Two illustrations from CHIRP make this point clear:

> The captain had misread his altimeter and thought he was 100 feet lower than he was. I believe the main factor involved here was my reluctance to correct the captain. This captain is very 'approachable' and I had no reason to hold back. It is just a bad habit that I think a lot of co-pilots have of double-checking everything before we say anything to the captain.

In the second illustration, Green (1990) describes a helicopter accident which was manned by two captains.

> . . . the handling pilot suffered a temporary incapacitation as he came to the hover before landing on an oil rig. The aircraft descended rapidly, and the cockpit recorder shows that although the non-handling pilot appreciated that matters were extremely abnormal, he initially tried to gain the other pilot's attention by asking what the matter was, and actually took control too late to prevent the aircraft from hitting the sea.

These examples illustrate that the situation has to become extremely hazardous before the non-handling pilot (i.e. the pilot who is not flying the aircraft) is willing to intervene. Even then, it may be too late to prevent an accident. Such reluctance to intervene is not beneficial to optimum team performance and, as the second example shows, is not simply restricted to rank seniority differences.

In many respects, most of the issues pertinent here have been covered by previous sections. However, the issue of compliance to 'authority' is worthy of considering separately. In some classic research performed in the early 1960s, Milgram was able to demonstrate the power of authority in gaining compliance from subordinate figures. The implications of this work suggest that individuals will perform acts when they are able to assign responsibility to another that they would not do under 'normal' circumstances.

To put these observations into context, there may be power differentials of members of the team. This may lead to destructive conflict or blind compliance. Physical power refers to the use of physical threat to coerce behaviour. This is normally referred to as 'the power of last resort' and is unlikely to be observed within the control-room context. Resource power, however, emphasises the power base: i.e. the person is in control of resources that are desired by others. Resource power (sometimes referred to as 'reward' power) may mean that the superordinate may be able to sanction pay, promotion or shift allocation if compliance is not forthcoming. The type of power that is associated with a formal role is termed 'position' power. This type of power is underwritten by the organisation, and provides the occupant with information, right of access and the right to organise. Expert power, however, is a power base that requires no sanctions, because individuals are willing to comply on the basis that the expert has greater knowledge and/or experience in the undertaking than themselves. Personal power resides in the individual and their personality or charisma. It may be enhanced by position or expert status, but compliance is gained by popularity rather than respect. Negative power is the capacity to stop things from happening: to delay, distort or disrupt. This may occur in times of low morale, stress and frustration. As has been suggested, the leader may possess one or more of these attributes of power. Indeed, those subordinated to the official group leader (in terms of position of power) may possess other types of power (e.g. expert, personal or negative) which could lead to some clash of interests within the group. This could lead to suboptimal performance by the group to the detriment of the system as a whole.

The style of leadership adopted may be selected from many alternatives including:

- autocratic;
- paternalistic;
- consultative;
- participative;
- or a combination of the above.

Determining which is appropriate depends upon the characteristics of the leader and features of the leadership situation, such as:

1 Characteristics of subordinates:
   (a) ability knowledge experience training;
   (b) need for independence;
   (c) professional orientation;
   (d) indifference towards organisational rewards.
2 Characteristics of the task:
   (a) unambiguous knowledge experience training;
   (b) methodologically invariant;
   (c) provides its own feedback concerning accomplishment;
   (d) intrinsically satisfying.
3 Characteristics of the organisation:
   (a) formalisation;
   (b) inflexibility;
   (c) highly specified and active advisory and staff functions;
   (d) closely knit cohesive work group;
   (e) organisational rewards not within the leader's control;
   (f) spatial distance between superior and subordinates.

Reason (1987) in an analysis of the nuclear incident at Chernobyl (26 April 1986), suggested that a psychological explanation for the antecedents of the event could, in part, be explained by 'groupthink' (Janis, 1972; Janis and Mann, 1977). Groupthink is a phenomenon that has been observed amongst small, cohesive, elite groups. Janis and Mann (1977) describe several classic historical cases which illustrate defects in group decision making. Common to all of these cases is a concurrence-seeking tendency, where the group develop rationalisations to support their illusions about the invulnerability of their position. Janis and Mann (1977) listed eight major symptoms that characterise groupthink:

1 an illusion of invulnerability, shared by most or all of the group members, which creates excessive optimism and encourages taking extreme risks;
2 collective efforts to rationalise in order to discount warnings which might lead team members to reconsider their assumptions before they recommit themselves to their past policy decisions;
3 an unquestioning belief in the group's inherent morality, inclining the members to ignore the ethical or moral consequences of their decisions;
4 stereotyped views of rivals and enemies as too evil to warrant genuine attempts to negotiate, or as too weak or stupid to counter whatever risky attempts are made to defeat their purposes;
5 direct pressure on any member who expresses strong arguments against any of the group's stereotypes, illusions, or commitments, making clear

that such dissent is contrary to what is expected of all loyal members;

6 self-censorship of deviations from the apparent group consensus, reflecting each member's inclination to minimise to themselves the importance of their doubts and counter-arguments;
7 a shared illusion of unanimity, partly resulting from this self-censorship and augmented by the false assumption that silence implies consent;
8 the emergence of self-appointed 'mindguards' – members who protect the group from adverse information that might shatter their shared complacency about the effectiveness and morality of their decisions.

Figure 11.7 illustrates the antecedents and symptoms of groupthink.

Reason (1987) argues that five of the eight symptoms could reasonably be ascribed to Chernobyl operators: their actions were consistent with an 'illusion of invulnerability'; it is likely that they 'rationalised away' any worries (or warnings); their single-minded pursuit of repeated testing implied 'an unswerving belief in the rightness of actions'; they clearly 'underestimated the opposition' in this case the system's intolerance of being operated in the forbidden reduced power zone; finally if any one operator experienced any doubts they were probably 'self-censored' before they were voiced (paraphrased from Reason, 1987). Reason suggests that the Chernobyl operators may not be unique in the nuclear power industry. From a systems perspective, bearing in mind the principle of equifinality, it is entirely possible that other operators in other stations could find themselves in a similar situation.

Janis (1972) as well as defining the criteria which identify this syndrome, has also proposed a number of guidelines on how to discourage team members from conforming uncritically with the leader's position or with an emerging group consensus in an effort to preserve group harmony. These guidelines are applied in the context of a team responding to a control-room emergency. It is argued that in order to improve decision-making in a simulated emergency, the team leader should follow the seven leadership practices listed below:

1 Present the problem to the group at the outset in a neutral manner, without pushing for any preferred solution. This will encourage an atmosphere of open enquiry so that members explore a wide range of alternative solutions.
2 Encourage the group to air objections and doubts by accepting criticism of his own judgement in a manner that shows that members are not expected to soft-pedal their disagreements.
3 Arrange for a sizeable block of time to be spent by a member surveying critical information and constructing alternative explanations of the problem.

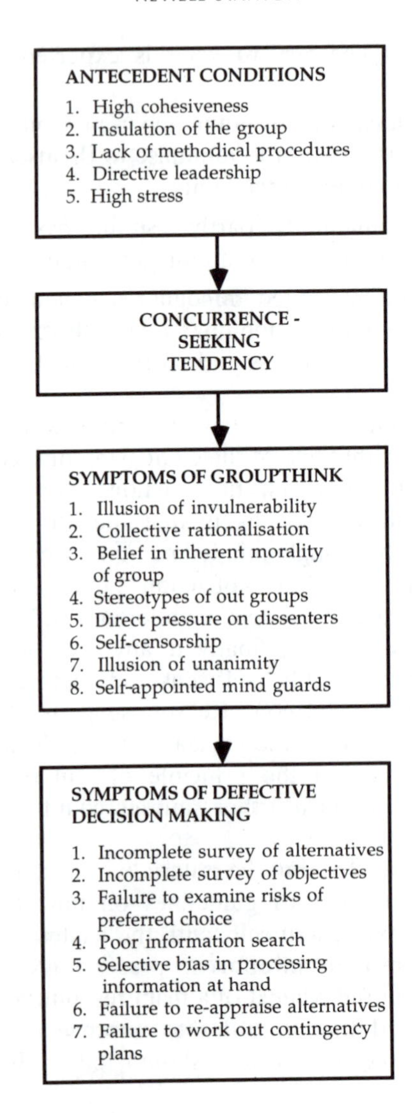

Figure 11.7 Analysis of groupthink.

4 Invite one or more outside experts within the organisation who are not core members of the particular team to attend the training exercise and encourage them to challenge the views of core members.

5 Assign the role of 'devil's advocate' to a member and give him an unambiguous assignment to present the arguments of the rejected alternatives as cleverly and convincingly as possible, so as to challenge the preferred solution of the group.

6 Divide the group into two separate groups whenever possible to discuss

the pros and cons of each solution and then bring the subgroups together to explore their differences.

7 Hold a 'second chance' discussion (after the group has reached a preliminary consensus and before a commitment is made to a particular solution) at which members are asked to express as vividly as they can all their doubts and to rethink the entire issue before making a definitive choice.

Janis and Mann (1977) suggest that not all cohesive groups suffer from 'groupthink' even though they may display its symptoms from time to time, if the group is:

- open to outside evaluations;
- subject to periodical critical inquiry;
- run by a non-authoritarian leader.

In summary, an effective leader is neither universally autocratic nor universally participative. Rather they should utilise the approach in response to the demands of the situation. Further, a group is probably capable of making a better decision than any individual in the group making the decision alone.

11.5 Team Co-operation

The degree to which team members co-operate is a pertinent issue in team performance. This issue relates to all aspects of team processes, from the initial development of the team through to crisis management by the team.

It is suggested that the group will not develop into an effective working unit until after they have passed through the initial stages of group formation. Tuckmann (1965) proposed four stages of group development. The leader must understand what stage the group is at in its development.

- *Forming* The beginning of the group development. An orientation stage, where people are looking for guidance. Marked by feelings of anxiety, members may be overly polite, look to the chairman for cues.
- *Storming* People begin to challenge their initial assumptions about the group. May be marked by power struggles, open conflict, anger, frustration.
- *Norming* People recognise that storming is taking them nowhere and adopt norms that are acceptable to the group. This is marked by increased group cohesion.
- *Performing* The group is now in a position to work, and people are given jobs in the group, and these roles are recognised by other group members.

The survival of the group is dependent on the members of that group adhering to the norms (i.e. codes of behaviour) that have been established. Enforcement is accomplished by a process of:

- *Education* In the early initiation phase of acceptance to group membership, the 'right' way of behaving is presented to mould the new member.
- *Surveillance* This is performed to detect deviance from group norms. It is a function of each group member, who adhere because group norms are essential to group survival.
- *Warning* This is issued when a deviation from group norms is detected. Progression from supportive and friendly contact to threatening behaviour occurs if a group member fails to comply.
- *Sanctions* These are the actual disciplinary actions imposed only when it is clear that the deviant is not going to change behaviour to that approved of by the group. Ostracism from the group is the ultimate sanction.

Hooper *et al.* (1991) analysed team performance in simulated processing plant tasks. Essentially, the 'operators' were required to perform, monitor, diagnose and control actions to meet production requirements and minimise deviation from set points. If a failure was detected, simulated repairs were carried out. The results show that in all cases team performance was superior to that of individuals. Teams used fewer control actions and took less time to diagnose faults than individuals. As Hooper *et al.* (1991) concluded: '. . . teams were more efficient, exhibited better control over the process and made better use of the displayed information.'

Group decision making has certain advantages over decisions made by individuals, for instance:

- It brings multiple knowledge and skill.
- It expedites acceptance.
- It can result in higher-quality decisions.
- It increases commitment.

However, there are some potential disadvantages associated with decisions by groups, for example:

- It requires more time.
- It can ignore individual expertise.
- It encourages riskier decisions.
- It creates the possibility of 'groupthink'.

Wickens (1991) notes that the high cognitive demand placed upon the team in the control room of a nuclear power plant is in part due to the limited time window within which to establish the appropriate procedure to restore plant safety once the event has been detected. He calls the change in activity, from routine functioning to responding to abnormal or emergency conditions, 'teams in transition'. He notes certain features that impact on the transition process:

- time constraints;
- structure of the event;
- environment;
- risk;
- organisational structure.

Wickens (1991) suggests that although there is little in the technical literature regarding the facets of team performance in workload transitions, some tentative conclusions can be drawn. These conclusions are reproduced below:

1 The workload transition produced by a crisis may often be described as a qualitative as much as a quantitative shift; because in the pre-transition period many environments impose high workload associated with vigilance and monitoring, anxiety, and planning.
2 Sleep loss and sleep disruption form a major impediment to the ability of teams to function in a crisis. Yet often insufficient and inadequate steps are taken by operators (and management) to ensure that disruption does not occur.
3 There are many activities that can be carried out during the pre-transition period, which can facilitate the post-transition adjustment. The most important relate to planning future action possibilities, and maintaining adequate situation awareness.
4 Communications, and the way in which effective communications may be mediated by personality types, and expected functions within the team appears to play a strong role in crisis management.
5 Training is obviously relevant for several reasons: knowing how to respond in emergency and developing automaticity of actions make a difference in the response to crisis. However, an unresolved issue remains. This is the extent to which training should focus on automatising routines, or on higher-level knowledge-based thinking that can allow the routines to be altered if they are not appropriate. Another issue is the extent to which stress management can be effectively trained.
6 Stress at some level, from boredom, to anxiety, to arousal, to panic, is a regular companion of teams in workload transition. The research data-base in the area remains sparse.

In summary, given sufficient maturity (in terms of the group's formation), teams are likely to perform better than individuals. However, as Wickens (1991) notes, there is plenty of scope for further research.

11.6 Conclusions

The literature reviewed suggests that team performance is 'greater than the sum of its parts', i.e. investigating and evaluating a team of operators requires more than adding analysis of single operators together. This means that when individuals are brought together to perform a co-operative task, the group processes go beyond the explanation of individual information-processing activities. Phenomena such as 'groupthink' suggest that group behaviour requires special consideration and analysis. Quite clearly, groups of people behave differently than do individuals. Some researchers have chosen to describe this phenomenon as a team mental model, collaborative decision making, collective cognition, and so on. This emphasises the higher-order nature of the individuals working together.

References

ACKOFF, R. L. & EMERY, F. E. (1972), *On Purposeful Systems*. London: Tavistock.

ARGLE, M. (1972) *The Psychology of Interpersonal Behaviour*. Harmondsworth: Penguin.

BAINBRIDGE, L. (1984) Diagnostic skill in process operation, *International Conference on Occupational Ergonomics*, 7–9 May, Toronto.

BERTALANFFY, L. VON (1950) The theory of open systems in physics and biology. *Science*, **111**, 23–39.

COOPER, G. E., WHITE, M. D. & LAUBER, J. K. (1980) *Resource Management on the Flightdeck*, NASA/Industry Workshop, NASA CP-2120, NASA: USA.

COURY, B. J. & TERRANOVA, M. (1991) Collaborative decision making in dynamic systems, *Proceedings of the Human Factors Society 35th Annual Meeting*, 2–6 September, 944–8, Santa Monica, CA: HFS.

FOUSHEE, H. C. & HELMREICH, R. L. (1988) Group interaction and flight crew performance. In *Human Factors in Aviation*. New York: Academic Press.

GREEN, R. (1990) Human error on the flight deck. In Broadbent, D. E., Baddeley, A. D. and Reason, J. T. (eds), *Human Factors in Hazardous Situations*. Oxford: Clarendon Press.

HETTENHAUS, J. R. (1992) Changing the way people work: a sociotechnical approach to computer-integrated manufacturing in a process industry. In Kragt, H. (ed.), *Enhancing Industrial Performance*. London: Taylor and Francis.

HOOPER, K. N., COURY, B. J. & TERRANOVA, M. (1991) Team performance in dynamic systems, *Proceedings of the Human Factors Society 35th Annual Meeting*, 2–6 September, 949–53, Santa Monica, CA: HFS.

JANIS, I. J. (1972) *Victims of Groupthink*. Boston: Houghton Mifflin.

JANIS, I. J. & MANN, L. (1977) *Decision Making: a Psychological Analysis of Conflict, Choice and Commitment*. New York: Free Press.

KANKI, B. G., LOZITO, S. & FOUSHEE, H. C. (1989) Communication indices of crew coordination. *Aviation, Space and Environmental Medicine*, January, 56–60.

KAWANO, R., FUJIIE, M., UJITA, H., KUBOTA, R., YOSHIMURA, S. & OHTSUKA,

T. (1991), Plant operators behaviour in emergency situations by using training simulators. In Quéinnec, Y. and Daniellou, F. (eds), *Designing for Everyone*. London: Taylor and Francis.

LOCKETT, M. & SPEAR, R. (1980) *Organisations as Systems*. Milton Keynes: The Open University Press.

MONTGOMERY, J., GADDY, C. & TOQUAM, J. (1991) Team interaction skills evaluation criteria for nuclear power plant control room operators, *Proceedings of the Human Factors Society 35th Annual Meeting*, 2–6 September, 918–22, Santa Monica, CA: HFS.

REASON, J. T. (1987) The Chernobyl errors, *Bulletin of The British Psychological Society*, **40**, 201–6.

SINGLETON, W. T. (1989) *The Mind at Work*. Cambridge: Cambridge University Press.

STAMMERS, R. B. & HALLAM, J. (1985) Task allocation and the balance of load in the multiman–machine system: some case studies. *Applied Ergonomics*, **16**, 251–7.

TERRANOVA, M., HOLLY, D., COURY, B. J. & HOOPER, K. N. (1991) Individual and team communication in a dynamic task, *Proceedings of the Human Factors Society 35th Annual Meeting*, pp. 954–8, Santa Monica, CA: HFS.

TUCKMANN, B. (1965) Development sequence in small groups, *Psychological Bulletin*, **63**, 384–99.

WICKENS, C. D. (1991) Teams in transition, *Proceedings of the Human Factors Society 35th Annual Meeting*, 2–6 September, Santa Monica, CA: HFS.

WOODS, D. D. (1987) Technology alone is not enough: reducing the potential for disaster in risky technologies. *Human Reliability in Nuclear Power*. London: IBC.

CHAPTER TWELVE

Human failure in the control of nuclear power stations: temporal logic of occurrence and alternating work times

MARC DOREL

Université de Toulouse-Le Mirail, Toulouse

As far as the control of relatively complex industrial facilities is concerned, there is no doubt that man is indispensable. In evolving situations, his capacities for analysis, diagnosis, forecasting and decision are required for correcting defects, but also for adjusting recommendations. Automatons do not have these capabilities.

The latter are capable of tireless execution of complex and lengthy sequences of actions. On the one hand they simultaneously and continuously identify all of the values taken by variables, while on the other hand reacting immediately and reliably. Real-time measurements are made possible through the use of several sensors combined with significant computing power.

However, automatons do not think about the immediate and later consequences of the actions undertaken, nor are they capable of evaluating them thoroughly. Finally, in spite of having reliable storage memory, they do not construct individual experiences similar to those of operators.

In spite of the fact that man's usefulness has been recognised, there are differences concerning the way in which he is taken into account. Indeed the integration of the human factor in the running of industrial facilities has many aspects whose respective significance is underestimated to varying degrees.

From this point of view, it is necessary to acknowledge that the weakness of systems is less due to man himself than to the fact that he has not been sufficiently considered. In spite of the difficulties raised by this approach, man should be globally taken into account by identifying all of the factors involved in potential problems. This is why, if individuals are considered only in terms of human weakness, the fundamental subject for reflection will be bypassed, if not obfuscated, and the question of systems weakness will in no way be resolved.

Man plays a key role in the overall reliability of any given industrial facility. This idea is mentioned, among other authors, by Faverge (1970), Leplat (1985) and Poyet (1990). Nevertheless man, while contributing to reliability, is also a factor of unreliability. Execution of a task is sometimes obstructed by human failure.

12.1 Origins and Objectives of the Work

Over the last few years the question of human reliability has become an important preoccupation of ergonomics. In the main, research has developed within a cognitive framework which aims to identify the failures or biased procedures of the human operator when supervising complex systems (e.g. Leplat, 1985; Rasmussen *et al.*, 1987; Reason, 1990). However, the error itself has not been considered with respect to the operator's temporal dimension.

To cite only one example, admittedly an extreme case, but nevertheless indicative of the importance of this question, it should be observed that Reason (1987), in his analysis of the Chernobyl accident, identified a set of cognitive, metacognitive and social factors, but did not attribute any particular meaning to the moment at which the malfunction occurred: 0 h 20 for the beginning of the critical period, 1 h 24 at the time of the explosion.

Nevertheless, the time at which human failures occur is of interest, both theoretical and applied, and should be taken into account.

12.1.1 Theoretical Interest

In the many cases of alternating shift times, when work is done irrespective of the temporal structure of the human organism, a disturbance in the rhythms of biological functions may be observed, by comparing them with the normal circadian fluctuations for the levels of activation for these functions (e.g. Reinberg, 1971, 1979).

In theoretical terms, our previous research mainly dealt with the relation[1] between the moment at which the work was done and the individual and collective organisation of the supervising activity in a control room.

Dorel (1978, 1983) as well as Dorel and Quéinnec (1980) have shown that operators, without doing it badly, do not always perform a task in the same way. They maintain a circadian rhythm during their activities, with a high amplitude. This rhythmicality is nevertheless modulated in amplitude, by the position of each shift within the complete cycle of which the 3×8 hours shift plan consists, but also by the presence or absence of a colleague in the control room.

It has thus been clearly shown that an activity such as supervising, with a large cognitive aspect, depends on biological and social determinants which are continuously interacting (Dorel, 1983).

Other work (Terssac *et al.*, 1983) has also emphasised the systematic occurrence of changes in the actual activity, between the morning, afternoon and night shifts.

In practical terms, these few observations raise questions concerning technical and human reliability, particularly with alternating work times (Quéinnec and Terssac, 1987). However, the costs, incurred by an organisation of tasks which does not take socially and biologically penalizing moments into account, are often ignored.[2]

This cost category includes the consequences of incidents caused on the one hand by the specificity of work in industrial plants and, in particular, by human failures which are considered without really taking work times into account.

12.1.2 Application

In a company any integration of knowledge affects the way in which the company is organised and also its development.

It is legitimate, when studying specific cases of work undertaken in nuclear power stations, to identify, on the occurrence of problems, the more or less probable effect of certain choices, no matter on which levels they may have occurred. In this respect, the studies which have been done provide useful information for any attempt to reduce factors leading to difficulties and problems.

Generally speaking, they lead to identification of critical points which may increase the awareness of the work group. Our main concern in isolating a temporal factor in the production of certain events is indeed to emphasise:

- the existence, within 24 h, of moments which reduce concentration, i.e. the reception of signals as well as the responses to them (e.g. Marquié and Quéinnec, 1987), thus limiting focused attention and maintained supervising;
- the not insignificant probability of human failures occurring at those moments;
- the specific reasons for deciding and acting, at those times, by concentrating mainly on the sub-task[3] being undertaken, in other words with as little diversion of attention as possible towards several concomitant sub-tasks,

some of which have already been begun, while others have been planned.

12.1.3 Objective

In the framework of this investigation, an attempt was made to determine links between:

- on the one hand, the occurrence of human failures partially responsible for triggering incidents;
- and on the other, the moments at which these failures occur.

Several levels have been considered:

- the work shift time during the 24 h period;
- the succession of three shifts per team;
- the reactor's operating phase.

Our intention here is not to identify, nor *a fortiori* to quantify the elementary factors involved in incidents. The factors concerned may be related in particular, to the design of the work station, or to the form and content of operating documents, or even to the way in which the work in the different departments is organised, while not forgetting the complexity of the task nor the training and skill of the operators.

We do not intend either to characterise the nature of human failures nor to analyse the cognitive mechanisms brought into play. Our concern is rather to reveal the existence of a temporal logic which might help to explain the conditions under which a human failure or several, non-random failures involved in triggering an incident, occur at a given moment in time.

12.2 Methodology

The objective which we are trying to achieve requires that the following rules be complied with:

- the choice of work situations in which the operators relieve each other 24 h out of 24;
- the analysis of archives which can be used to identify human failures involved in incidents;
- the identification of temporal variables related to the human failures.

According to the above rules, research has been conducted in three nuclear power stations[4] producing electricity: PS1, PS2 and PS3.

12.2.1 Identification of Failures

Failure is taken to mean any difference from the operating standard[5] (human and/or technical), which, at any given moment in time, either immediately or through a delayed reaction, jeopardises the required operation.

Human failures are identified by analysing the reports on significant incidents,[6] covering the years from 1981 to 1989. The incidents in question occurred during exploitation of the facilities, whether or not the reactor was powered up, and irrespective of the type of task being undertaken, the work station, its position on the site, the operators concerned, the failure risk factors or the nature and mechanisms of the human or technical failures. They are not derived from a research programme with an experimental approach consisting of simulated situations, under conditions far removed from natural conditions.[7]

Without ignoring the possible existence of various potential sources of difficulty, an incident may result from human failure or from the combination of several human failures, or from one or several technical failures, or from both types of failures. The human failures recorded are therefore not necessarily esteemed to be preponderant, or the first or last in a series of failures, directly responsible for the incident. They might be previous failures, or even follow and thus be independent of the incident.

The entanglement of different elements, as well as the rare information given by a report, or perhaps the wealth of information given by another report, implies an analysis whose duration cannot be defined ahead of time: from one to two hours up to ten or more. In addition, on three distinct occasions separated by several weeks, each incident report resulted in three in-depth analyses, in order to guarantee a systematic application of the same analytical criteria.

Finally, during collection of data the author was able to benefit from direct conversations with the engineers and control crew shift bosses whether concerning technical details or replies to questions raised during reading of the reports. Thus, to the extent that the information acquired was sufficient, each failure, identified after thorough and repeated reading of the data, was classified:

- either as a technical failure;
- or as a human failure, imputed to a member of the control department;
- or as a human failure, imputed to a member of another department or an outside company carrying out work in the station.

12.2.2 Determining the Dates and Times at which Human Failures Occur

The subject under analysis means that it is indispensable to take into account the dates and times of occurrence of factors involved in producing a given incident, no matter what the time interval between each human failure and the event. It also requires consideration of the dates and times of human failures which might have occurred following the incident.

Each report indicates the date and time of the event. Then the chronological sequence of the incident is described, using the automatic log,[8] and mentioning any information considered to be relevant, while including comments.

A posteriori, it is often more difficult to isolate and record the date and time of occurrence of a human failure. The complexity of the exercise increases when the comments, while sparing individuals' feelings, raises the problem of several human failures. An extended and renewed investigation mainly serves to maintain the trustworthiness and the reliability of the analysis.

On the other hand, a precise case would be one in which an incident occurs immediately or almost immediately after, for example, a wrong action.[9] The date and time of the action thus correspond to the date and time of the incident.

12.2.3 Determination of the *n*th Successive Manning of the Shift when Human Failure Occurs

The date and time defined above determine the shift in question: morning, afternoon or night, for which a trace has been found in the documents[10] containing the work activity reports which have to be made by members of the crews controlling nuclear power stations.

These documents can be examined to reconstruct the shift rotation at the time, and especially to identify with certainty a specific shift in question in relation to the sequence of shifts. The official shift crew is less significant than the operators who were actually present and registered in the logs for the days preceding and following the chosen date. They are the significant marker for locating the shift within the work rotation, and indicate exactly the number of successive times that this same shift was occupied by them, before and following the incident.

This reconstruction requires a systematic analysis of the documents established for several weeks before and after the moment at which the incident occurred.

12.2.4 Selection of Human Failures

The analysis led to the definition of three independent populations of human failures (one per power station) which strictly satisfy the following selection criteria.

12.2.4.1 *Reactor Operating Phase*

The human failures in question take place:

- after the loading of nuclear fuel into the reactor,[11] from the beginning of start-up procedures for operation, up to the next planned refuelling (this interval corresponds to the reactor's operating phase);
- during several successive operating phases after fuelling: six phases from 1981 to 1988 for the PS1 station, six phases from 1982 to 1989 for the PS2 station, five phases from 1982 to 1987 for the PS3 station.

12.2.4.2 *Person Responsible for the Failure*

This refers only to human failures caused by people who are part of the crews responsible for supervising the power stations.[12]

12.2.4.3 *Crew Shift Rotation Plan*

The succession of shifts, performed by members of the shift crew involved in the failure, respects the planned rotation for the seven-day period during which the failure occurred and also for the week preceding it and the week following it.[13]

12.3 Results: Human Failure and Organisational Aspects

The results presented here have thus been developed from 110 human failures, i.e.: 41 for the PS1 facility, 38 for the PS2 facility, 31 for the PS3 facility including 17 from 1982 to 1984 for one type of rotation and 14 from 1985 to 1987 for a second type of rotation. Let us consider human failures:

- first by focusing on work time aspects, more specifically by identifying the three types of shift manned during a 24 h period (i.e. morning, afternoon, night), but also by identifying the time of occurrence within each shift;
- and then later by considering the more specific organisational aspects of nuclear power stations consisting of the stage reached in the reactor's operating phase, but also of the type of rotation in question.

In order to avoid reducing contradictory evolutions, or even masking possible differences between nuclear facilities, the data collected are analysed for each facility.

To the extent that the variables manipulated are nominal variables, the statistical hypothesis of independence, between the ⟨power station⟩ variable and the temporal variables taken one by one, is validated by the chi-square test. Significantly, each application of the test indicates that there is no relation between the ⟨power station⟩ variable and the temporal variable in question. In other words, the two variables do not affect each other. The probability p that the chi-square reaches or exceeds a certain value, given the independence hypothesis, is shown in the figure which corresponds to the temporal variable evaluated.

Under these conditions, the analogous trends observed justify the analysis in terms of mean results. Many of the figures shown are based on data from the three facilities PS1, PS2 and PS3. The deviation index mentioned represents the standard error for the mean (it does not occur when its value is zero or close to zero). Some of the data are only to be found in one or two of the facilities. These cases are discussed in detail in the comments.

Concerning the Y-axis, the relative frequency (F) appears to be in proportion to the size of the population of human failures, when the results concern only one facility. Otherwise, when two or three facilities are considered, the relative frequency (F) is the mean of the proportions with respect to the populations corresponding to the failures.

12.3.1 Shifts and Work Times

The distinction made between the three consecutive work shift times, covering a period of 24 hours, shows that the human failures occur with a frequency which is almost twice as great during the night shift as it is during the afternoon shift (Figure 12.1). However, the night shift is not the only critical work shift period. In fact the morning shift shows a concentration of almost as many failures, to the extent that a difference appears between the morning and afternoon shifts ($p = 0.035$). Generally speaking, with the three shift periods being alternatively occupied, the frequency for the night shift decreases for the afternoon shift.

The interpretation of this fact in terms of a more or less great human cost, thus provoking human failures, is based on accepted knowledge of chronobiology.

An intra-shift breakdown enables a more precise analysis during mornings, afternoons and nights. Dividing each shift into three equal time periods also makes it possible to isolate the beginning-[14] and end-[15]of-shift effects which add to the consequences of individuals' operating capacities, which vary over the 24 h (Figure 12.2).

For the morning and the afternoon, it would appear that the time periods at the beginning and end of the shift are, just as equally, even more critical than the middle period. In other words, depending on the mean trend in evidence, there is no difference between the beginning and the end; only the median part has significantly fewer failures.

The beginning–end-of-shift phenomenon is also to be found in the interval [21 h; 6 h]. On the other hand, with the night effect, the median part cannot

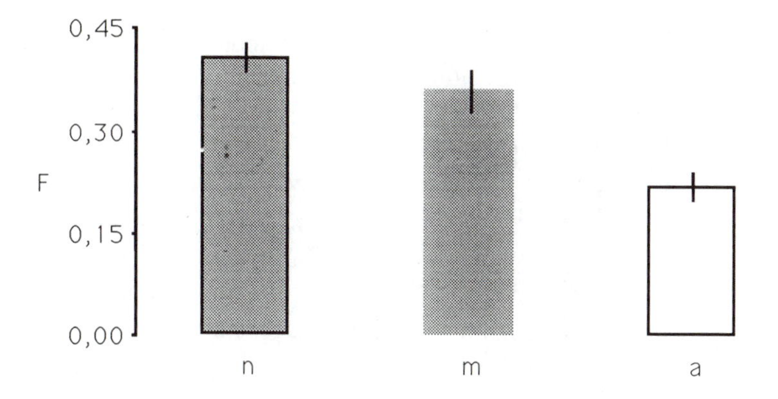

Figure 12.1 Frequency per shift.
The ⟨Shift⟩ and ⟨Power station⟩ variables are independent: $p = 0.81$.
F(*Y*-axis) is the mean relative frequency of human failures.
The *X*-axis shows the successive night (n), morning (m) and afternoon (a) shifts.

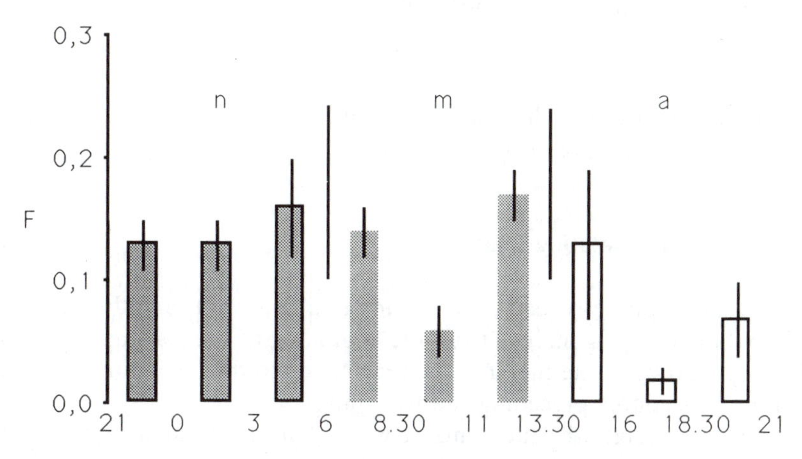

Figure 12.2 Frequency at beginning, middle and end of shift.
F(*Y*-axis) is the mean relative frequency of human failures.
The *X*-axis represents, for each of the shifts (changing at 6 h, 13 h 30, and 21 h), three successive time periods of equal length.

be distinguished from either the beginning or the end of the shift. Consequently the failures are distributed according to mean frequencies which are more or less the same throughout the three different parts of the night shift, as it has been divided up.

Even though the moments which are least vulnerable to failures are only to be found, on the one hand from 8 h 30 to 11 h, and on the other from 16 h to 18 h, the type of shift does not significantly affect the intra-shift distribution

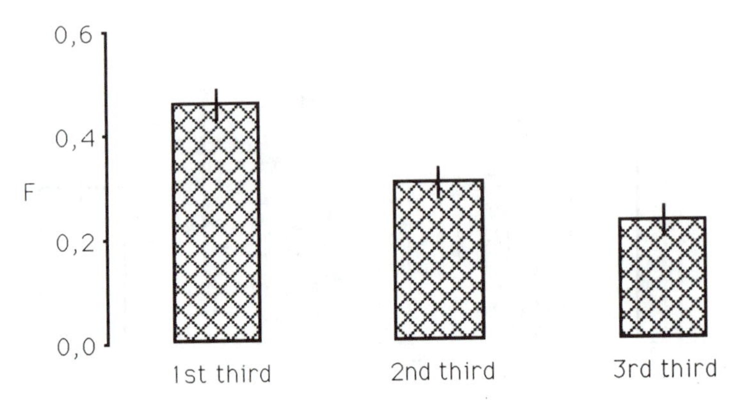

Figure 12.3 Frequency during operating phase.
The ⟨Operating phase⟩ and ⟨Power station⟩ variables are independent: $p = 0.58$.
F(Y-axis) is the mean relative frequency of human failures.
The X-axis represents the current stage of the reactor's operating phase. The phase is divided into three equal and successive intervals showing the beginning (1st third), middle (2nd third) and end (3rd third).

($p = 0.19$). Finally, the distribution of human failures is not uniform and varies over the 24 h period ($p = 0.033$).

12.3.2 Reactor Operating Phase

The operating phase spreads over an interval of several months defined by the number of days gone by, from the beginning of the start-up procedures following loading of the fuel into the reactor, up to the beginning of the next stopping procedures preceding the refuelling.

Let us now relate the moments at which failures occur to the operating phase. The arbitrary division of the phase into three successive intervals of equal length reveals that there is a lower frequency of human failures towards the end of the phase (Figure 12.3).

Moreover, the frequency per shift within each third of a phase is shown in Figure 12.4. The mean trend shown in Figure 12.1 is again confirmed.

More precisely, the mean distribution of human failures between the three types of shift (Figure 12.1) is only the reproduction of an identical phenomenon present in each third of the reactor's operating phase ($p = 0.91$). In other words, the current stage of the phase does not determine whether failures occur rather more in a certain type of shift than in another. The overall decrease, observed from the beginning to the end of the operating phase (Figure 12.3, $p = 0.017$), is the result of a decrease for each of the three types of shift.

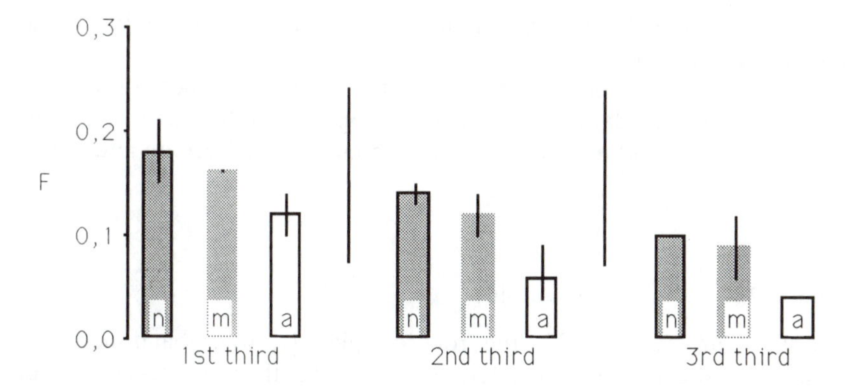

Figure 12.4 Frequency per shift during phase.
F(*Y*-axis) is the mean relative frequency of human failures.
The *X*-axis shows the beginning (1st third), middle (2nd third) and end (3rd third) of the reactor's operating phase. Night (n), morning (m) and afternoon (a) are shown for each of the three thirds.

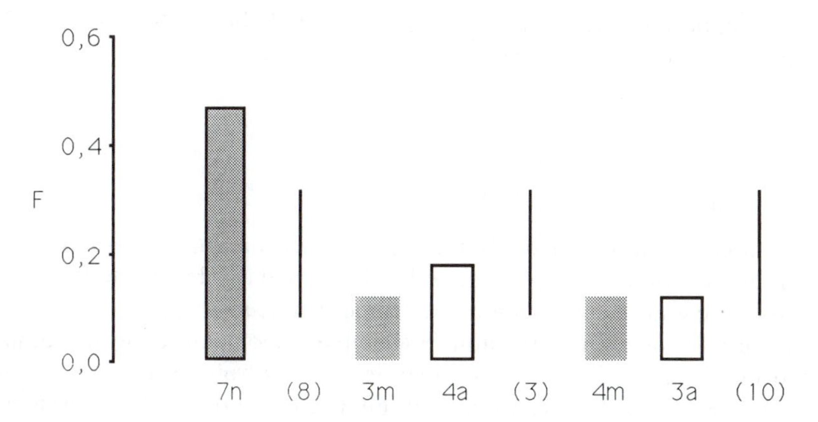

Figure 12.5 Frequency throughout rotation [I].
F(*Y*-axis) is the relative frequency of human failures.
The *X*-axis shows the successive shifts and rest periods.

12.3.3 Shift Rotation Plan for Control Crews

Three work plans for three shifts in a 24 h period will be discussed. Each of their 42-day cycles consists of three periods of seven consecutive working days.

Rest days (r) are inserted in between these periods, and added to the former are work days whose so-called normal working time may be compared with government office hours. Nevertheless, up to the limit of a previous number of overtime hours, a certain number of due days will possibly replace these work days (d).[16]

In facility PS3, from 1982 to 1984, three successive periods of seven consecutive working days took place according to the following rotation plan [I], in which m, a and n respectively denote morning, afternoon and night:

		7n	(4r and 4d)
3m	4a		(3r)
4m	3a		(6r and 4d)

In the same facility PS3, the rotation plan [I] was abandoned[17] and replaced by the rotation plan [II] in 1985. The three successive periods became:

2m	3a	2n	(5r and 4d)
3m	2a	2n	(3r)
2m	2a	3n	(5r and 4d)

The rotation plan [III] concerns the two other facilities PS1 and PS2. The three successive periods were:

3m	2a	2n	(5r)
2m	3a	2n	(5r)
2m	2a	3n	(6r and 5d)

For each of the two plans [II] and [III] based on quick alternation, the order of the three successive periods of seven days determined on each occasion the end of the rotation with three consecutive nights.

We are looking for a relation between the moments at which human failures occurred and the rotation plan for the control crews.

The focus on the type of rotation in force is based on the hypothesis that the distribution of human failures varies according to the place of the morning (m), afternoon (a) and night (n) shifts in the rotation cycle.

12.3.3.1 *Slow Alternation (Rotation [I])*

The rotation plan [I] for the PS3 facility consisted of seven consecutive long night shifts (nine hours), which implied a high frequency of human failures (Figure 12.5).

Indeed, of the failures recorded during the successive manning of night, morning and afternoon shifts, almost half of them occur during the night.

An examination of their distribution between the first and seventh nights, i.e. from Friday night (Fr) to the following Thursday night (Th), reveals a concentration of failures mainly during the second half of the week (Figure 12.6).

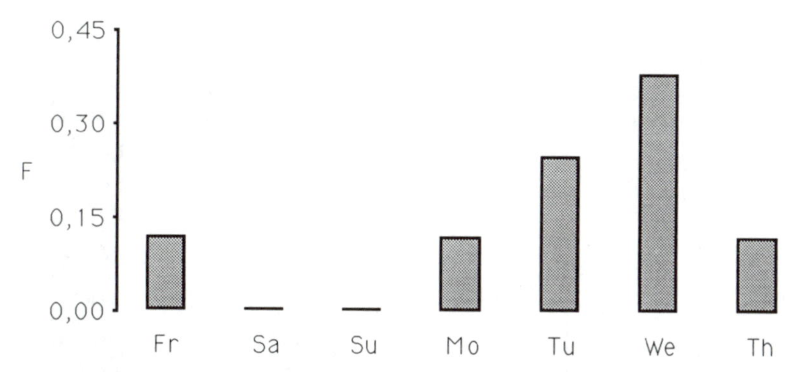

Figure 12.6 Frequency for consecutive nights.
F(*Y*-axis) is the relative frequency of human failures.
The *X*-axis shows the consecutive nights in the rotation [I].

12.3.3.2 *Quick Alternation (Rotations [II] and [III])*

In this case the analysis focuses on human failures per period, by grouping together the seven shifts, which occurred between several consecutive rest days, which might have been mixed with one or several working days within so-called normal working hours (cf. above). In each seven-day period, the three successive identical shifts define the type of period: 3a, 3m or 3n, for periods consisting respectively of three successive afternoons, three successive mornings or three successive nights.

The overall distinction made between the three types of period emphasises clearly an average difference between each of these types (Figure 12.7, $p = 0.001$). In a non-random distribution, the third period (type 3n), which includes the three nights at the end of the week and at the end of the rotation, accounts in particular for half of the human failures.

An examination of the successive shifts throughout the plans [II] and [III] reveals that the difference in relation to the two previous periods 3m and 3a has a basis in fact towards the end of the rotation, mainly during the three last night shifts (Figure 12.8 for the PS1 and PS2 facilities; Figure 12.9 for the PS3 facility) but also, to a lesser extent, during the last two morning shifts. Indeed there are significant differences between the observed and the expected distribution frequencies throughout the rotation (Figure 12.8, $p = 0.002$).

By comparing the frequency of human failures throughout the two successive types of rotation for the same PS3 facility (reminder: plan [II] replaced plan [I]) it should be noted (Figure 12.9) that the trend towards a higher frequency remains true during the night shifts at the end of the cycle:

- in spite of having abandoned rotation [I] with its slow alternation (see Figure 12.5), by adopting rotation [II] with quick alternation;

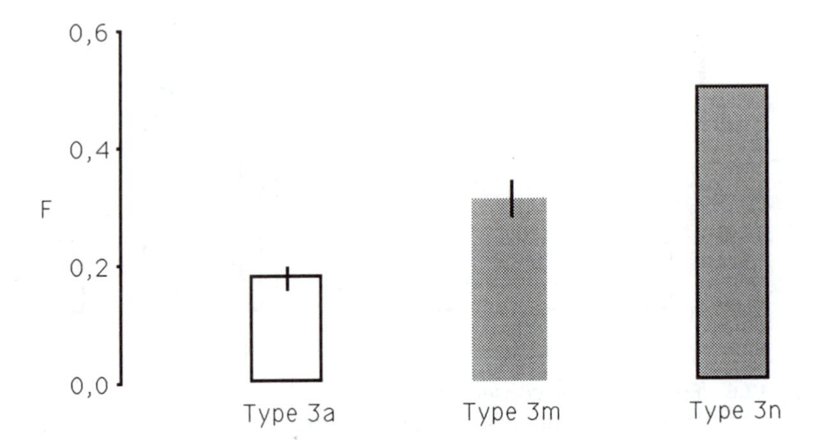

Figure 12.7 Frequency per type of period.
The variables ⟨Type of period⟩ and ⟨Power station⟩ are independent: $p = 0.81$.
F(*Y*-axis) is the mean relative frequency of human failures.
Seven consecutive work days make up a period grouping mornings (m), afternoons (a) and nights (n).
Along the *X*-axis, the three periods are characterised by their type.

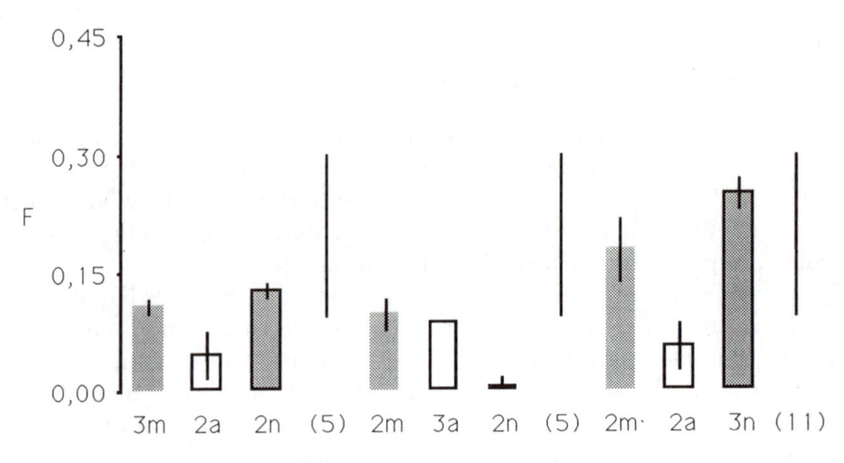

Figure 12.8 Frequency throughout rotation [III].
F(*Y*-axis) is the mean relative frequency of human failures.
The *X*-axis represents the successive work shifts and rest periods.

- and contrary to a hypothesis which might result from this: the hypothesis that there are fewer human failures in the case of a reduced number of consecutively worked nights, no matter what their order in the rotation.

To the extent that work periods include seven consecutive days, the

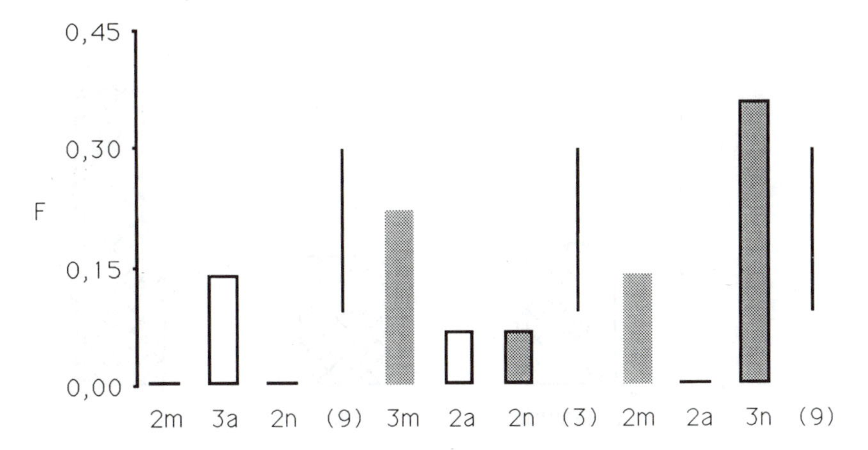

Figure 12.9 Frequency throughout rotation [II].
F(*Y*-axis) is the relative frequency of human failures.
The *X*-axis represents the successive work shifts and rest periods.

analysis deals, finally, with human failures which take place during the week.

The rotations divide the week into three groups by means of the successive shifts manned. The rotation for the PS1 and PS2 facilities combines Monday with Tuesday, Wednesday with Thursday, Friday with Saturday and Sunday, while that for the PS3 facility combines Tuesday with Wednesday, Thursday with Friday, Saturday with Sunday and Monday. With a one-day offset, each group previously defined nevertheless characterises a beginning, and respectively a middle and an end of week.

The week in question is the calendar week in which the human failures occurred during the weeks related to the rotations. In fact the latter started alternatively with one of the three groups, previously defined, in order to ensure a balance between the crews subjected to the rotations.

The distribution yielded by the analysis shows a higher mean frequency for the three days at the end of the week. The beginning and the middle only account, with four days, for half of the human failures (Figure 12.10). In spite of that, the distribution of failures is not significantly different from the theoretical distribution ($p = 0.43$). In other words, the observed frequencies do not generally contradict the expected occurrence frequencies directly related to the amount of days at the beginning, in the middle and at the end of the week.

Nevertheless, the frequency of human failures per type of shift depends on the whole on the manning of the shift at the beginning, in the middle or at the end of the week ($p = 0.001$). It should be noted in particular that the high frequency at the end of the week is largely due to the night shifts (Figure 12.11).

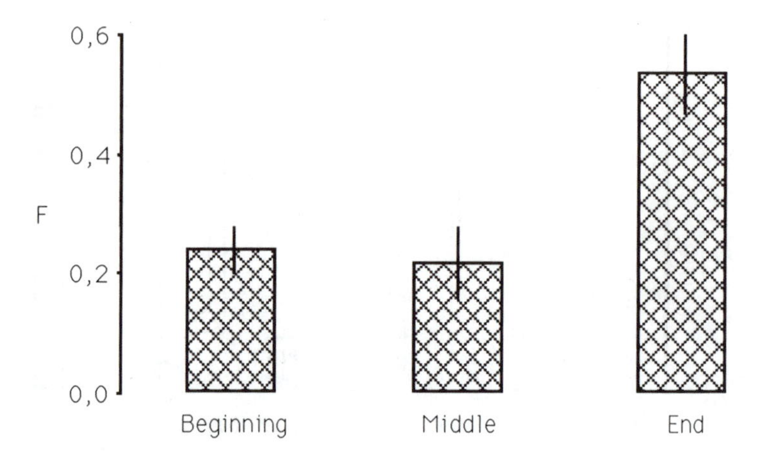

Figure 12.10 Frequency during the week.
The ⟨Week⟩ and ⟨Power station⟩ variables are independent: $p = 0.11$.
F(*Y*-axis) is the mean relative frequency of human failures.
Along the *X*-axis, the week is divided into three successive groups of two days (Beginning), two days (Middle), and three days (End).

To sum up, the general distribution of human failures, observed for the three types of shift (see Figure 12.1), is a result of the similar trends recorded throughout the reactor's operating phase (see Figure 12.4), but is nevertheless not systematically the same throughout the week (Figure 12.11).

12.3.4 Disturbed Rotation

Temporary changes, for one or several individuals, made to the schedule for handling the sequence of rest periods and the three work periods, are due to organisational necessities.

The interfering variable resulting from this aspect, which is likely to affect the occurrence of human failures, is neutralised in the developed analysis (cf. 12.2.4.3).

As a complement to this analysis, and as an exploratory investigation in the two facilities PS1 and PS2, let us consider, under the same conditions, and using the same methodology, the occurrence of human failures in a succession of work shifts governed mainly by operating necessities and the concern to guarantee the required number of individuals for the successive shift crews.

Analysis of the data led to identification of twenty human failures (i.e. fourteen for the PS1 facility and six for the PS2 facility) which occurred in a rotation with a disorganised and irregular schedule, before and after the day on which the incident occurred. A random sequencing of shifts and rest

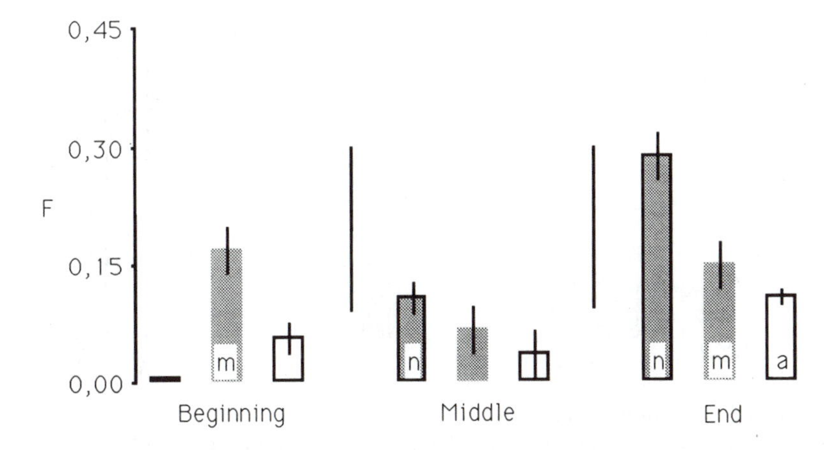

Figure 12.11 Frequency per shift during the week.
F(*Y*-axis) is the mean relative frequency of human failures.
Along the *X*-axis, the week is divided into three parts: Beginning, Middle, End.
Nights (n), mornings (m) and afternoons (a) are distinguished.

periods was defined, the seven-day work periods (cf. 12.3.3) were shortened or lengthened, and the make-up of the crews[18] was changed.

Attention should be paid to the probable effects of a distinct disorganisation of the schedule sequencing the work and rest periods. These effects are caused as much by changes in the make-up of crews as by anarchistic application of the work shift rotation schedule. They are also due to the addition of extra work shifts, which, in a repetitive fashion, reduce the length of rest periods.

Thus the variability in the order in which work and rest alternate, but also the irregularity in the length of consecutive work day sequences, are likely to temporarily and unfavourably transform the nature and status of the afternoon shift, a relatively privileged work shift (particularly in terms of physiological cost) with respect to the night and morning shifts.

Figures 12.12–12.14 illustrate the disturbance referred to. It should be noted, as opposed to Figure 12.1, which is related to a set and regular work rotation, that the human failures occur more frequently[19] in the afternoon than at night or in the morning (Figure 12.12, $p = 0.02$). Moreover, the distribution between the beginning, the middle and the end of the afternoon shift tends to become uniform while at the same time reflecting a strong deviation (Figure 12.13). This trend is also to be found from the beginning to the end of the week (Figure 12.14, $p = 0.74$).

In conclusion, in spite of the reduced size of the population considered, the first observations confirm that it is worth while focusing on the effect of the disorganised and anarchistic succession of work shifts on the distribution of human failures (intra-shift and inter-shifts, during the week, . . .).

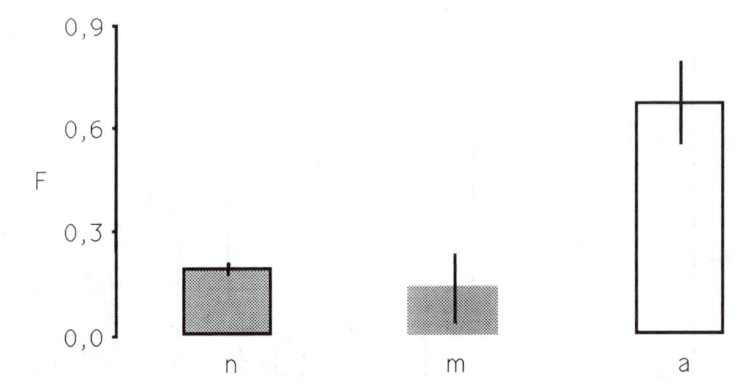

Figure 12.12 Frequency per shift.
The ⟨Shift⟩ and ⟨Power station⟩ variables are independent: $p = 0.28$.
F(*Y*-axis) is the mean relative frequency of human failures.
The *X*-axis shows the successive shifts for night (n), morning (m) and afternoon (a).

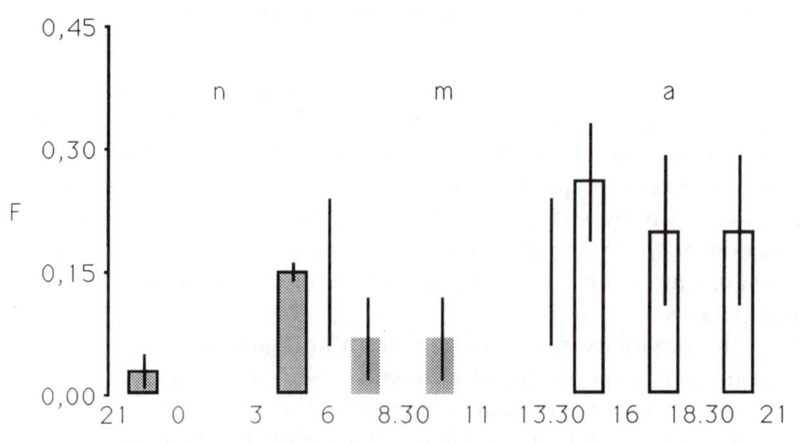

Figure 12.13 Frequency at beginning, middle and end of shift.
F(*Y*-axis) is the mean relative frequency of human failures.
The *X*-axis shows a representation of three successive work times of equal length, for each of the shifts (changing at 6 h, 13 h 30, and 21 h).

12.4 Discussion: Human Failure and Optional Choices for Transforming Work Situations

The use of the expression 'human failure' is due to the fact that the concept of human error has caused a passionate controversy. Indeed while this concept is such that it has caused a scientific controversy, it has also provoked energetic discussion amongst a large majority of workers in industrial facilities, who do not appreciate its connotation.

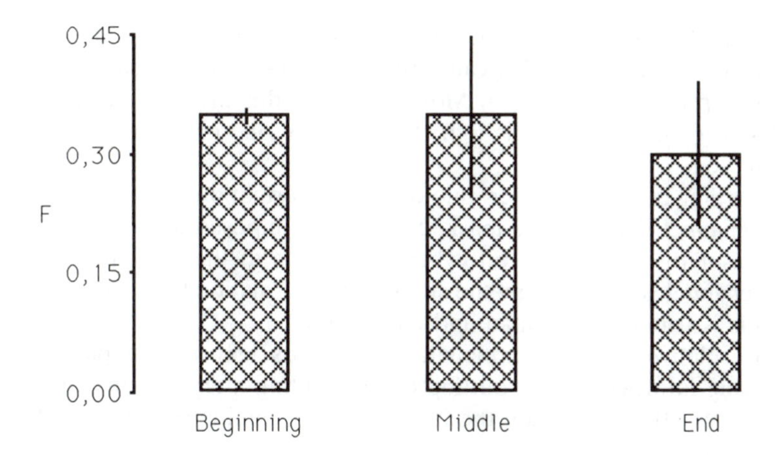

Figure 12.14 Frequency during the week.
The ⟨Week⟩ and ⟨Power station⟩ variables are independent: $p = 0.39$.
F(*Y*-axis) is the mean relative frequency of human failures.
Along the *X*-axis, the week is divided into three successive groups of two days (Beginning), two days (Middle), and three days (End).

While of a heuristic nature, the concept of human error nevertheless does not explain anything. It has been widely used by several authors who have helped define and classify errors, or who have tried to identify the underlying processes which led to them, or have contributed to error management and prevention. More particularly, used by the general public, it tends to mask the complexity of the factors which actually have to be faced, due to the apparent simplicity of the cause identified (man), and to a universally negative connotation which has been attributed to it.

A more subtle interpretation of error is thus necessary.

12.4.1 A Way of Approaching Error

The different fields of study of the cognitive functioning of man have led to different conceptions of what an error is. From the end of the nineteenth century to the development of recent schools in cognitive sciences, several papers have marked progress in the study of human error and shed light on the way in which it is approached (e.g. Reason, 1990).

The profusion of models used reveals both the extent of interest in the subject and its complexity. Thus no one model gives full satisfaction. In other words, none of them has succeeded in explaining all types of errors and even less the variations within a given type. An example of this is given below by considering two models of information processing on which several papers have been based. These two models generate two approaches to error.

Thus, according to the information theory proposed by Shannon and Weaver (1949) an underlying cause of error is the limited capacity of human beings to process information. More precisely, this theory presupposes that the basis for human error is that there is only one channel for processing information.

According to cybernetic terminology, the information or signal transmitted in the channel may then be masked by a superimposed disturbance (noise). The reasoning logic based on cybernetics leads to the hypothesis that the single channel is saturated by the amount of information to be processed, and that this is true both for man and computers.

This relatively simple explanation of errors ignores the possibility of processing information by splitting or distributing attention, in other words by selective attention, when the information is coming simultaneously from several sources.

The influence of the computerised model for programmed solving of logical and mathematical problems (Newell and Simon, 1972) has led to a different conception of error. During intellectual activity, the error occurs at a moment of reasoning which aims to solve one or several difficulties or one or several defined and limited problems. It occurs when the programmed logic, in other words the method used to achieve the set objectives and the problem's solution, is either abandoned or faulty.

Whether the information processed is or is not symbolic, a parallel is once again made between man and the computer. However, it is well known that the computer does not process information with respect to its recorded knowledge of the world. More precisely, it does not process information by giving it meaning, as does man, since he takes account of the context in which the information exists.

In the case of the control of a given industrial facility, the mechanism being considered would appear not to recognise the existence of a process of successive adjustments made by man as he interacts with the chronological states of the facility, for instance during a transient problem situation. Since the mechanism does not take into account man's extremely flexible automatic processing of information, which calls on parallel circuits, it does not integrate the attempts made to regulate, or to restore control, which, triggered by the error, enable man to surmount it and often to resolve it or eliminate it.

12.4.2 Error in a Sequence

Behind the idea of error there is often another underlying idea which likens the error to the initial cause of the incident or accident. This judgement raises the question of the lack of understanding of the phenomenon. In fact various factors, to different and varying degrees, play a role in causing problems and the human error which lead to an incident.

It is not rare for certain characteristics of work situations to turn out to be so many potential problem-causing factors for the control of a given industrial facility, in spite of the compensatory role of the experience of individuals involved. For example, difficulties might possibly be generated or exacerbated by:

- the topography of the industrial facility, or the marking of equipment, or the availability and presentation of information to be processed, or operating instructions which are unduly complicated, if not unsuitable for a specific state of the facility (technical level);
- a lack of experience concerning the operation of part of the facility (level of skill);
- alternating work times, or a temporary change in the composition of a work crew (organisational level).

In that it is tied on the one hand to several risk factors, and on the other that it sometimes leads to incidents, human error is a symptom within a system. In these terms, it is obviously worth studying. Decisions taken for the purpose of improving systems show the extent to which it has been taken into account.

Let us now consider some improvements which have been carried out in nuclear power stations, as practical solutions developed for the purpose of reducing and preventing the occurrence of symptoms.

This is the case for several training actions, whether for basic training or for complementary training following a specific event, or training mainly aimed at the mastery of problematic situations which arise during the control of facilities.

Any improvement in the content of control documents, and the way in which this is presented, is also due to the symptoms having been taken into account. This involves facilitating the required control both in normal operating situations and during incidents. The aim is also to guarantee the good condition of the equipment tested.

The same is true for any measure but also for any individual behaviour which helps to maintain or further reciprocal control, when performing either control operations or maintenance operations.

In passing we should mention actions concerning the man–machine interface, with the reorganisation of control rooms, mainly designed as an aid to control by means of a more organised presentation of the information on the control boards, control consoles and monitor screens.

A concrete answer is also to be found in the investigation and identification of premises and equipment which is not in the control room (Leckner *et al.*, 1987). Another concerns the automation of certain tasks, and another the development of planning tools which integrate the powering-down constraints for loading nuclear fuel into the reactor.

Finally, the procedures for transmitting relevant information during a change in shift crews, or again the procedures for coordinating respective tasks between the shift crew and both the chemistry and maintenance departments, and especially the way in which operational dialogues are conducted between members of a given crew, are all examples of solutions which are renewed daily in an attempt to limit the renewal of old symptoms.

12.4.3 Supervising and Failure, Alternating Times and Watchfulness

Human failure is anchored in mental work. It is an expression of a fleeting and unusual behaviour in man. It is part of the processes of acquiring information, decision making, and the actions for controlling and maintaining the facility, all of which cannot be separated. It enters into the strategies for supervising the facility, in other words the way in which the activity of the members of the control and maintenance crews is organised in time and space. Within a complex facility, this activity is generally split up into specialised tasks.

An incident may be described as a malfunction for which it is difficult to estimate a cost. For obvious economic reasons, attention is always paid to the event and its consequences, but those factors which appear to be only indirectly profitable or not of immediate interest or which are even not particularly clear, are often neglected.

This is also true for the moment at which human failure occurs, which apparently is not generally considered to be a problem, even though there is a not insignificant probability that human failures will occur at given moments affecting the watchfulness, in other words obstructing the reception of signals and responses to them. Indeed, the momentary functional capacities of the human organism depend on daily rhythms based on the change from day to night.

Consequently, operative procedures and strategies chosen vary more or less, over the 24 h, according to the changes in the normal psychophysiological capacities of the individuals performing a task at unfavourable moments.

Moreover the operative procedures and strategies adopted may also vary somewhat in response to changes in the nature and type of work required.

These few reflections should partly guide the joint responses to questions of reliability in complex facilities and those concerning the safety of the people working in them. During the last decade, certain catastrophic accidents said to have been caused by human errors bear witness to this.

12.4.4 Conditions of Individual Work Activity

The control and maintenance of a complex industrial facility include a large number of sub-tasks. This is why the effect (its measurement and the subsequent results) of successive shifts should be relativised somewhat. It is effectively a result of the work activity, but also of all of the conditions in which this activity takes place. In other words, the effects of the temporal conditions under which sub-tasks take place are modulated and given meaning just as much by the individual work activity as by all the other conditions of organisation and performance, such as: the nature and content of sub-tasks, and the available means and methods for completing them.

To be precise, in a given context, any possible choice for improving temporal working conditions can only lead to significant improvements:

- if the work requirements, in other words the sub-tasks to be performed and the organisational and physical conditions under which they are performed, have previously been taken into account in a complete and thorough fashion;
- but also if the operative methods and procedures, as well as the mental processes brought into play by individuals, are taken into account, in other words by taking into account the actual work activities which take place in a given context;
- and finally, if the effects of these actual activities are taken into account in terms of physical and mental workloads, and of immediate and delayed fatigue.

When the task requires being active at night time, it is of fundamental importance to integrate the characteristics of individuals, in particular the functional requirements of the human organism alternately penalised by activity in the morning and then during the night. Expressed in another way, it is necessary to accept that it is not possible to avoid requiring individuals, who are all different from each other, to sleep or rest during the day and work during the night, with varying frequency according to the needs of public service.

New temporal work conditions are a result of organisational choices. Any arrangement may be preferred to another one. On the one hand, it will have to involve a compromise between contradictory positive and negative implications (e.g. concerning night work: Quéinnec *et al.*, 1987). On the other hand, this compromise should be acceptable to everyone, but in particular it should be open-ended and hence reviewed regularly according to a predefined schedule.

It would be particularly worth while if this compromise were to limit the annual time of exposure to the intrinsic risks in the alternation of successive work shifts (in terms of physiological, psychological, family-related and social effects).

12.4.5 Malfunctioning and Organisational Choices

The characterisation of problematic events in terms of the temporal dimension, and the attempt to relate this to other extra-temporal variables to understand better the meaning, leads to a more accurate representation of reality, both for decision makers and for all levels of management, no matter to what extent each of the members of the group is involved in the choices.

For example, this could help:

- to arouse interest in the implementation of complementary tools (e.g. observation report grids) making it ultimately possible, from several points of view, to enrich analysis of the behaviour of the man–machine system in problem situations;
- to confirm organisational choices;
- to throw light on worthwhile directions by contributing to reflection on organisational arrangements (including technical aspects) likely to reduce malfunction risks even further in tomorrow's facilities.

This last possibility will first of all be illustrated by the technical orientation of two types of arrangement resulting from two different approaches: respectively the preventing of failures and the limiting of their effects. Finally the main development will involve options for discussion on the temporal organisation of work according to alternating times.

12.4.5.1 *Possible Technical Solutions*

In an attempt to deal with reliability problems, our work has focused on the failures observed in the members of facility control crews, for the purpose of ultimately preventing human failures. From this point of view, in other words with the intent of eliminating problem-causing factors, one might envisage a diagnosis and forecasting aid which would integrate several factors involved in the triggering of incidents and which would reinforce anticipation of certain critical phases in the genesis of problematic events.

Moreover, knowing that a given sub-task, currently being undertaken, is generally managed simultaneously with other sub-tasks (started or planned), the intellectual processing activity, concentrated on the current sub-tasks, is of particular interest during diagnosis, decisions and actions taking place at those times. It is by means of this processing activity in particular that:

- certain actions are envisaged;
- the reactions of the facility and the results of the actions to be performed are anticipated;
- these actions are performed;[20]
- correct execution of the order given by the action is controlled.

Under conditions of simultaneous management of several sub-tasks, the possible reliance on automation of certain complex and lengthy sub-tasks would help diminish the scattering of attention. In other words, the potential capabilities of automatons would enable an increased focus of the operators on the decisions for which they are responsible, i.e. on the actions entrusted to them and controlled by them. From this point of view, organisation of a harmonious balance between sub-tasks performed by automatons and sub-tasks reserved for operators would increase the efficiency of their execution.

From the same point of view, let us consider the existence of alarms which appear upon withdrawal of equipment from activity or again during maintenance work on equipment which has not been withdrawn from operation. The possible elimination of these alarms would be helpful for the operators in the control room. In other words, it would not activate intellectual activity for processing useless signals and would facilitate the focusing of attention on more relevant information. Indeed, for the purpose of creating a more suitable balance, between on the one hand the information displayed in the control room and on the other the state of the equipment, the signals referred to should never have been assigned alarm status when the operating of the facility was being designed.

A complementary approach to the preventive one is to be found in the idea of technical mechanisms for limiting the effects of human failures, in other words in the idea of failure tolerance, developed in particular in the field of reliability, availability, maintainability and safety (RAMS) for industrial computing systems (e.g. Arlat *et al.*, 1995).

Thus human failures which occur during maintenance operations in facilities could be critically examined. In a complementary way, it would also be interesting to focus on all problem-causing factors in the framework of the maintenance organisation.

Finally, let us imagine that technical devices are envisaged, which would integrate mechanisms capable of tolerating human failures. It would be particularly interesting to implement several previously defined tolerance procedures, respectively compatible with a given state of the technical systems, and which would not disturb the programmed and harmonised sequence of sub-tasks performed for maintenance but also for controlling the facilities.

12.4.5.2 *Sequencing and Programming Work and Rest Times*

The alternating rotation of shifts increases the number of individuals subject to unusual work times. Thus, a given group of individuals will not have to continually bear the inconveniences of night or morning work. The inconveniences are shared between a greater number of individuals who are alternately subjected to uncomfortable and less uncomfortable work times.

Night and morning shifts inevitably shorten the daily sleep time and will probably have, for instance, a negative effect on watchfulness and/or health, and/or the individuals' private lives, in terms of inconvenience and risks. Nevertheless, while it is strongly penalised on certain days, the family and social life is however made easier at other times.

In short there is no good solution for organising alternating work times. Any solution can only be a compromise between advantages and disadvantages. Both of these will be discussed briefly.

Generally at least equal to two or three, the number of successive mornings or nights is never reduced to one. However, such a temporal organisation could limit the accumulation of fatigue during morning and night shifts. It would limit the cost for the organism, which is in particular related to the repeated adjustments of some of its biological rhythms. However, frequent working of a different type of shift to the preceding one would risk a daily adjustment of personal, family and social time, which would be more difficult.

Consequently, the sequence of work and rest times should be envisaged by consulting those who already live with this sequence. It would be a good idea, in particular, to achieve a balance between the regular distribution and concentration of rest. Indeed, a normal or reduced number of consecutive working days favours more frequent recuperation, particularly with respect to possible lack of sleep, and thus enables avoiding an accumulation of fatigue. But of course in this case, personal, family and social time would be distributed and shortened.

On the other hand a great number of consecutive working days or even the repeated reduction of rest days, but also a time interval reduced from one shift to the next, will later lead to a longer rest period (or several successive periods) enabling the individual to rediscover a normal life rhythm, to relax and to forget somewhat his professional activity, or even sometimes to devote himself to another activity. But accumulated fatigue makes recovery more difficult.

Between two successive periods of several rest days, there is generally a sequence of consecutive working days. Between the end of each shift and the start of the next consecutive shift, there is an interval on which the individual's aptitude for recovery directly depends. The more this interval is reduced in time, the more the individual's resources are mobilised. Recovery is thus limited to the same extent as are later reliability and performance. At the same time the number of meals normally taken decreases and family and social life is restricted.

Moreover if a reduced recovery interval follows a night shift, this will have more negative effects on sleep and health. In this case, the consequences will be even greater when the night shift has already been preceded by a reduced recovery interval which itself was preceded by a night shift which had prevented the individual from sleeping the night before.

When recovery time is shortened between two work shifts, and furthermore when part of it is taken up by a more or less long round trip between the work site and the recovery place, then two questions are raised. The first is whether a sufficient degree of watchfulness is being maintained and the second involves the concomitant risks run during the trip.

In addition, the beginning and end of shift times, respectively for the morning and the night shifts, are directly related to sleep disturbances, affecting both quantity and quality of course, but also, in particular, watchfulness and reliability. Thus the later the shift starts in the morning or the earlier it ends in the night, the lesser will be the corresponding disturbance. Likewise the earlier individuals get up in the morning or the later they go to bed in the night, the greater will be the effects on health.

Under these conditions, in order to facilitate recovery, the time interval between the end of a morning or afternoon shift and the beginning of the following shift, when it is of a different type, must be at least 24 consecutive hours. After the night shift, a recovery period should include at least two whole nights.

Finally, it is extremely difficult to maintain a satisfactory period of sleep and, at the same time, to take regular meals at home, except in the case of a palliative consisting of unequal shift durations, or possibly of work times which are periodically changed (and this would cause other problems).

Whether they are equal or unequal, shift durations are defined on the basis of choices which could not be qualified as a necessity, no matter what the type of work or the desires of various people. The longest duration of one of the shifts should be compensated by the shortest duration of a shift which is considered to be more uncomfortable. Thus it would be understandable to lengthen the night period in order to shorten the morning period. It would make much less sense to shorten the afternoon also, unless this is done for reasons other than striving for a normal psychophysiological balance.

The choice of four work shifts for a 24 h period, instead of three, for example, would also encourage limiting the duration of one, two or three initial shifts. However, this means the number of times the individual is required to go to work increases unless the weekly work duration is decreased.

As for the conditions (times and places) in which meals are taken, these are defined by the time constraints of the beginning and end of the shift:

- either before and after work times (to the extent that meal times at home are not too disturbed, it would be a good thing if this were the most frequent solution since family and social life and also the individual's biological rhythms would be the least disturbed);
- or during work times, under conditions which are generally less favourable, unless provision is made for meal times and a suitably equipped dining room, and hence the possibility of taking normal meals.

Thus, on the one hand work would not stop the individual from eating, and on the other comfort and hygiene would be improved, while a less unbalanced diet would help limit the risks of various later disorders.

12.4.5.3 *Pause Times, Number of Workers, Number of Crews*

The question of the duration of a shift as well as the type of shift, but also of their sequencing, both lead to the question of pauses, whether or not these have been planned in the work organisation.

For instance, when normal fatigue is combined with the effects of lack of sleep during the night or morning shifts, a pause, taken on the individual's initiative (according to need), to the extent that the work undertaken allows this, enables him to relax a little, to rest or to eat, thus helping to maintain an acceptable level of human reliability. Indeed:

- any individual's level of activity can never be constantly maintained at maximum efficiency (under the hypothesis that a maximum might exist and could be measured and known);
- each individual's performance may vary between the beginning and end of a given shift.

Thus, in the case of certain sub-tasks, if the need for a pause coincides with the impossibility of taking it, this would mean that the individual must remain active and try to maintain his level of watchfulness and concentration. Consequently, and at the very least, this would increase the accumulated fatigue and hence the human cost intrinsic in alternating work times.

Any decision concerning the number of workers would of course have an effect, in particular, on the way in which pauses are taken and also the way in which absent workers are replaced.

As far as pause times are concerned, when the working mode is continuous, an insufficient number of workers would involve the risk:

- of having to transfer sub-tasks to other crew members;
- or at the very least of causing a delay in the work activity, which would ultimately lead to a later accumulation of sub-tasks.

The management of planned and unplanned absences, as well as the replacement procedures, also depends on the number of people in the crews. The lower the number of workers or crews, the more there is a risk of reducing the recovery interval between two work shifts and the more likely a rest period is to be interrupted. On the other hand, having an available reserve of skilled workers, subject to specific conditions (daytime hours for instance), increases management flexibility and guarantees a certain efficiency without requiring any transfer of work to the current members of the crew.

Deciding on the number of people in a work crew should logically depend of course on the type of work, its regular and transient requirements, but also on the actual work activity itself, including not only absences but also pauses, whether these be exceptional or regular.

To sum up, the workforce available means that it is either impossible or possible to decrease the human cost of alternating work times, to the extent that any adequate workforce may be:

- completely employed under certain circumstances (e.g. complex sub-tasks, holidays) and during certain shifts (e.g. at night when the functional capacity of the human organism is affected);
- or especially partially employed under other circumstances (e.g. simple, reduced or postponed sub-tasks) and during uncomfortable shifts (e.g. night shift, disturbed rotation).

Finally, independent of the workforce, a problem is raised by the number of crews working according to alternating work times. Thus an organisation based on a minimum number of crews will need fewer individuals working according to alternating times. But a corollary of this means that these individuals will be permanently subject to alternating times, and the problem of managing absenteeism remains as does the difficulty of calling meetings between members of crews outside of the work time planned in the rotation.

With six, seven or eight crews, the number of nights and weekends to be manned is reduced. On the other hand, more individuals will have to work at alternating times, even though it is possible to assign them irregularly. In fact it then becomes possible to combine periods of alternating work times with periods of ordinary working days. This combination makes it easier to call inter-crew meetings, as well as to replace absent workers who are on holiday, sick leave, or who have been seconded elsewhere, or who are being trained or participating in the professional, union, cultural or social activities of the industrial facility, ... More especially this limits the frequency of exposure to risks generated by alternating work times. In these terms it could be a preventive measure to prevent the intolerance threshold from being reached. This critical individual threshold is reached more quickly to the extent that the person is older.

Such a preventive measure might be considered as a preliminary stage for limiting the number of work years under alternating work time conditions, in other words as a preliminary stage for restoring workers to a daytime routine. This reclassification could be based on the age and seniority of these workers who so desired. Of course, given that individuals are unique, an identical measure should not be forced on all individuals.

12.5 Conclusion

This chapter deals with the occurrence of human failures involved in the triggering of incidents. It is only a necessary stage in understanding, and further study is required.

Our intention was to further research into possible alternative arrangements, without proposing any specific changes. We therefore tried to raise pertinent questions throughout.

To sum up, the results mainly emphasise the existence of potentially penalising moments in the different shifts, moments which are related to:

- normal nycthemeral fluctuations in the psychophysiological state of individuals;
- but also the beginning of shifts;
- or again to the end of shifts;
- as well as working at different periods according to the rotation plan;
- or even the progression within the normal week;
- and finally to the stage reached in the reactor's operating phase.

It is important always to try to understand the occurrence of failures leading to incidents, in order to learn the best way of managing the circumstances in which these failures take place, and the extent to which it is possible to avoid them. Highlighting the existence of failure risks can at least partially help guide efforts to reduce them, to find solutions to reliability problems in industrial facilities.

When certain conditions related to the running of facilities are taken into account, conditions which are a source of difficulty for human beings and lead to an increase in human cost, this may lead to alternative arrangements being chosen. The options chosen might involve, when necessary:

- training, but also the organisation of work;
- establishing work stations and working and rest areas, or again improving the physical environment;
- and finally the distribution of functions between man and the technical devices (involving the man–machine interface and work aids).

For example, we should remember the help provided by automatons for achieving certain complex sub-tasks, or memory aids consisting of lists of check points, or again aid which consists in attracting the operator's attention by indicating the potential difficulties intrinsic in a given work situation.

A change in the characteristics of work situations, in order to make them less uncomfortable, to carefully relieve man of certain tasks, and to encourage comfort for the purpose of efficiency, obviously facilitates the operator's task. Consequently it is probable that this would help to reduce problem-causing factors and limit the occurrence of human failures. Nevertheless it should be noted that any optional alternative arrangement will be efficient only if the nature and requirements of the work are taken into account and related to the variability of the actual activity underlying different work situations (in terms of the sub-tasks to be performed, the means and the methods) as well as to the unstable functional state of individuals.

This is why, as far as temporal organisation is concerned, any attempt to transform work conditions should, in particular, envisage several organisational modes. Indeed, given the variability in the actual work activity, both a common solution for all types of situation and a single way of implementing choices would be unsuitable.

It is precisely for this reason, when considering gradual changes within a facility, and in order to avoid contradicting improvement policies, that none of the modes of temporal organisation should be hidebound. All solutions, which by nature are compromises, as we have shown, should be provisional and, within reasonable limits, open to later criticism in relation for instance to negative effects on the health of individuals.

Finally, the personnel should be invited to take part in discussions for finding solutions, and also for implementing the chosen solutions. These are necessary conditions for a smooth transformation of work conditions, whatever these may be.

Acknowledgements

The French Ministry for Research and Technology and also the CNRS (National Centre for Scientific Research) funded this work which was done with the permission of the French Electricity Board. The author would like to express his heartfelt thanks for this co-operation and in particular to the management and personnel of the power stations, who welcomed him.

Notes

1 The obvious ignorance of the effect of the body's circadian rhythms on actual work activity, and the implicit reference to a model of the working man as a stable operating identity, were the determining reason for our interest in this relation.

2 When considering the operation of a company which relies on alternating work times, it is fundamentally important not to hide any consequences, whether these are of an economic or financial nature, but especially if they are of a socio-economic nature (Brunstein and Andlauer, 1988).

Gains and costs, which can be quantified in financial terms, can be identified. A few respective examples are increased production on the one hand, wages and equipment on the other hand. Some gains and costs are more difficult to quantify. This is respectively true for an increase in productivity, and for the results of training, of the fight against industrial accidents, against absenteeism, and against poor quality. Other gains and costs cannot be quantified, for instance motivation, and the internal communication and information system.

3 During the working time, a varying set of specific sub-tasks – sub-tasks which are either continuous or limited in time, but which are also either unique or repeated – contribute to the achievement of a task entrusted to a given member of the work group in the industrial plant.

4 Each of the three nuclear power stations houses two independent reactors as well as their corresponding system equipment.

5 References to a given standard and the observed value of a deviation are concepts which have been used by several authors to mean an error (e.g. Leplat and Pailhous, 1973–4; Rasmussen, 1982).

These notions themselves become a subject of discussion whenever 'it is necessary to determine on the one hand the reference standard and on the other the value of the deviation' (Cellier, 1990) in order to determine the error. It is easy to understand the difficulty of identifying an error in that, with a certain degree of freedom, the error appears in relation to a codified objective standard (e.g. Leplat, 1985), but also with respect to a subjective standard which is the required result (e.g. Reason, 1990).

6 The expression 'significant incident' refers to an incident which affects or which might affect the safety of the corresponding facility.

7 To the extent that it is possible to question the adjective 'natural' used to describe the work conditions to be found in an industrial facility, it refers explicitly here to actual work situations in nuclear power stations, and to their environment, neither of which could be as faithfully reproduced by a simulation.

8 This log contains an automatic printout of the successive states of the numerous items of equipment in the various systems of the facility.

9 The lack of action, just as much as an action carried out too late or too early, or partial neglect and confusion, is defined as a wrong action.

10 The documents used are the logbooks used respectively in the main control room and in the decentralised control room.

11 The stopping-phase for loading nuclear fuel into the reactor implies a different organisation of work and in particular a reorganisation of the actual activity of members of the various power station departments.

12 The other people responsible for failures do not work according to alternating times.

13 The disturbance in the roll of crew members present for duty is caused by the replacement of one or several members in a control crew by one or several members from other control crews. These replacements are required when leave is taken for training courses, but also for annual leave or again for sick leave.

14 Special attention should be paid to the way in which the latest information is taken into account by the relieving crew, and the mental updating of the states of the various equipment and facility systems, in comparison to the states known at the end of the previous work shift for this relieving crew. In addition to the time taken for the change over of crews itself – the time during which the two crews are both present is limited – the time needed for taking the latest information into account and performing a mental update of the situation may, in the case of an operator in the control room, require a time period of between a few minutes and the whole of the shift, or even several shifts.

Indeed the power station can either have been powered up, with no anomalies detected, for several days and weeks, or it may be in a starting-up stage following loading of nuclear fuel into the reactor. These states are moreover combined with either the regular daily presence of the operator, for instance for a week, or with a first return to his work at six in the morning, with no transition, when returning from an extended rest period of several days, or even from holiday leave of two, three or four weeks.

15 The fatigue of the retiring crew is related to the end of the shift.
16 (d) represents both so-called normal days and due rest days. Having available due days is the same as adding rest days to the rest days initially scheduled in the rotation plan.
17 The seven successive night shifts manned disappeared in 1984. A type of rotation characterised by quick alternation was then directly introduced. In comparison, during the same year, the seven successive night shifts manned, which were still in force in almost a third of the French steel industry, only disappeared when they were replaced by a 5–5–5-type rotation (Dorel *et al.*, 1984).
18 During certain rotation periods during the summer holidays, the needs for change become more permanent than temporary.
19 Care should be taken to check that the numbers of morning, afternoon and night shifts manned are equal during periods showing disturbances.
20 Self-control is an integral part of the action at the moment that the latter is being performed. It is part of the operator's awareness of what he or she is doing. This awareness, generally simultaneous with the action, functions to prevent inappropriate actions. Following the action, it reveals a deviation with respect to normal cognitive functioning which may be observed through operative procedures and methods for performing the task.

References

ARLAT, J., BLANQUART, J.-P., COSTES, A., CROUZET, Y., DESWARTE, Y., FABRE, J.-C., GUILLERMAIN, H., KAÂNICHE, M., KANOUN, K., LAPRIE, J.-C., MAZET, C., POWELL, D., RABÉJAC, C. & THÉVENOD, P. (1995) *Guide de la sûreté de fonctionnement* (Reliability, availability, maintainability and safety guide). Toulouse, France: Cépaduès-Editions.

BRUNSTEIN, I. & ANDLAUER, P. (1988) *Le travail posté chez nous et ailleurs* (Shift work assignments in France and elsewhere). Marseille, France: Les Editions Octarès/ Entreprises.

CELLIER, J.-M. (1990), L'erreur humaine dans le travail (Human error in work). In Leplat, J. and Terssac, G. de (eds), *Les facteurs humains de la fiabilité dans les systèmes complexes*, pp. 193–209. Marseille, France: Les Editions Octarès/Entreprises.

DOREL, M. (1978) Variation du comportement et horaire alternant de travail: Etude préliminaire de quelques indicateurs comportementaux dans le cadre d'un poste de contrôle d'un système automatique (Behaviour variations and alternating work times: A preliminary study of a few behavioural indices in the framework of a control station for an automatic system). *Psychologie et Education*, **3**(2), 51–61.

(1983) *Le travailleur à horaires alternés et le temps: Introduction à de simples choses observables* (The shift worker and time: Introduction to simple observations which can be made). Unpublished doctoral thesis, Université de Toulouse-Le Mirail, Toulouse, France.

DOREL, M. & QUÉINNEC, Y. (1980) Régulation individuelle et inter-individuelle en situation d'horaires alternants (Individual and inter-individual regulation in alternating work time situation). *Bulletin de Psychologie*, **33**, 465–71.

DOREL, M., TERSSAC, G. DE, THON, P. & QUÉINNEC, Y. (1984) *Nouveauté technique, restructuration et travail en équipes successives dans la sidérurgie française* (Technical innovations, restructuring and successive crew rotations in the French steel industry) (Research Report). Toulouse, France: Association pour le Développement des Etudes sur l'Amélioration des Conditions de Travail.

FAVERGE, J.-M. (1970) L'homme, agent de fiabilité et d'infiabilité (The human being, an agent of both reliability and unreliability). *Ergonomics*, **13**, 301–27.

LECKNER, J.-M., BOUILLON, P., DIEN, Y. & CERNES, A. (1987) Etude de conception détaillée du repérage en centrales nucléaires REP 900 MW (A detailed study of the design of marking in nuclear power stations with 900 MW PWR). *Le Travail Humain*, **50**, 157–63.

LEPLAT, J. (1985) *Erreur humaine, fiabilité humaine dans le travail* (Human error, human reliability in work). Paris: Armand Colin.

LEPLAT, J. & PAILHOUS, J. (1973–4) Quelques remarques sur l'origine des erreurs (A few comments on the causes of errors). *Bulletin de Psychologie*, **27**, 729–36.

MARQUIÉ, J.-C. & QUÉINNEC, Y. (1987) Vigilance et conditions de travail (Watchfulness and work conditions). *Neuro-Psy*, **2**(1), 27–37.

NEWELL, A. & SIMON, H. A. (1972) *Human Problem Solving*. Englewood Cliffs, NJ: Prentice-Hall.

POYET, C. (1990) L'homme, agent de fiabilité dans les systèmes automatisés (The human being, an agent of reliability in automated systems). In Leplat, J. and Terssac, G. de (eds), *Les facteurs humains de la fiabilité dans les systèmes complexes* (pp. 223–40). Marseille, France: Les Editions Octarès/Entreprises.

QUÉINNEC, Y. & TERSSAC, G. DE (1987) Chronobiological approach of human errors in complex systems. In Rasmussen, J., Duncan, K. and Leplat, J. (eds), *New Technology and Human Error*, pp. 223–33. Chichester: Wiley.

QUÉINNEC, Y., TEIGER, C. & TERSSAC, G. DE (1987) Travailler la nuit? Mais dans quelles conditions? (Night work? Yes, but under what conditions?). *Cahiers de Notes Documentaires* (Nd 1642-128-87), 429–45.

RASMUSSEN, J. (1982) Human errors: A taxonomy for describing human malfunction in industrial installations. *Journal of Occupational Accidents*, **4**, 311–33.

RASMUSSEN, J., DUNCAN, K. & LEPLAT, J. (eds) (1987) *New Technology and Human Error*. Chichester: Wiley.

REASON, J. (1987) The Chernobyl errors. *Bulletin of the British Psychological Society*, **40**, 201–6.

(1990) *Human Error*. Cambridge: Cambridge University Press.

REINBERG, A. (1971) Les rythmes biologiques (Biological rhythms), *La Recherche*, Mars, pp. 241–61.

(1979) Le temps, une dimension biologique et médicale (Time has a biological and a medical dimension). In Reinberg, A., Fraisse, P., Leroy, C., Montagner, H., Péquignot, H., Poulizac, H. and Vermeil, G. (eds), *L'homme malade du temps*, pp. 23–62, Paris: Stock.

SHANNON, C. E. & WEAVER, W. (1949) *The Mathematical Theory of Communication*. Urbana: University of Illinois Press.

TERSSAC, G. DE, QUÉINNEC, Y. & THON, P. (1983) Horaires de travail et organisation de l'activité de surveillance (Work times and organisation of supervising activity). *Le Travail Humain*, **46**, 65–79.

CHAPTER THIRTEEN

Work-related stress and control-room operations in nuclear power generation

TOM COX[1] and SUE COX[2]

[1]*Department of Psychology, University of Nottingham*
[2]*Centre for Hazard and Risk Management, Loughborough University of Technology*

During the UK Public enquiry into the construction of the Sizewell B nuclear power station, the appointed Inspector Sir Frank Layfield drew attention to the 'stress problem':

> (I) would be grateful for a clearer picture, in due course, of what one may call the stress problem – in the sense that I think we all recognise that if we take two quite different people with the same technical professional or any other sort of training, when the going becomes very difficult, the time when it really matters, one will perform well and the other may not (Sizewell, Day 197, 1984).

His question focused concern on individual differences in the response to work-related stress, and, in doing so, appears to assume the existence of stress in control-room operations in nuclear power generation. This review is primarily concerned with the nature of work-related stress and its effects on those involved in such operations. Attention is drawn to the social context to control-room operations and the role and functioning of the shift team. However, the scientific literature pertaining to the problem of control-room operations in nuclear power generation is relatively sparse compared to that focused on other work systems and organisations. As a consequence, this review has to make reference to studies on control operations other than those involved in the nuclear industry, and to draw on the general literature on work-related stress.

13.1 Background

The safe operation of a nuclear power plant depends on the precise interaction of many different control systems involving not only hardware and software but also people. Such systems cannot be adequately characterised as a community of man–machine interactions: rather they represent a socio-technical system in which the shift team is the functional unit. Should an event occur, the operator's role is to prevent or limit the development of both hazardous situations and equipment damage. In order to perform this role efficiently and safely, operators are selected, trained and formally qualified. Furthermore, they must have and use agreed procedures, they must be supported by their colleagues, the organisation and its management, and they must be provided with adequate and relevant information regarding plant status and behaviour.

In 1979, a major nuclear incident occurred at the Three Mile Island plant in the USA,[1] and some seven years later, an even worse incident occurred at Chernobyl in the Ukraine.[2] These incidents served to emphasise the importance of human performance in systems reliability. However, despite the general comments made on operator behaviour by the two inquiries into Three Mile Island, it remained unclear what role stress played in the aetiology and management of that incident. In none of the published accounts was there any specific mention of stress affecting operator behaviour (see Kemeny, 1979; Malone *et al.*, 1980; Patterson, 1983; Rubenstein and Mason, 1979). While it was apparent that the control-room operators were clearly under increasing stress during the course of the incident, there is no real indication that inappropriate actions (or inactions) were due to that stress (Malone *et al.*, 1980). It is also difficult in hindsight to identify the role that stress played, if any, in the course of the Chernobyl incident. The EPRI Report 1987 (see, for example, Moore and Diemich, 1987) stated that, although many argued that the incident showed a greater need for automation to improve the man–machine interface, Gary Vine of the Nuclear Safety Analysis Centre concluded:

> there were serious problems with the man, serious problems with the machine but very few obvious problems with the interface between them . . . the engineers did not lack good information, they lacked good judgement.

Could impaired judgement be the result of working under stress? Less dramatic, but nevertheless important, is the observation that about 30% of the events reviewed in the US nuclear industry's Licensee Event Reports (LERs) have involved operator behaviour relating to perception, decision-making, forgetting, the commission of incorrect responses or the omission of a correct response (Sabri and Husseiny, 1978). Furthermore, the US Nuclear Regulatory Commission's Reactor Safety Study (WASH-400, 1975) argued that human failure rates can be higher than those of major mechanical or electrical components in a nuclear power plant, and that under conditions of high stress, human failure rates may approach 100%. What then is the role of stress in determining operator behaviour and systems performance?

The authors of the Norwegian Halden Report (Baker and Marshall, 1987) concluded that no answer could yet be provided to the question 'are operators more likely to make increased errors under the stressful conditions which may arise in the control room?' In reaching this conclusion, they noted that generally speaking the available reports of nuclear power plant incidents seldom pointed to stress as a causal factor, even under difficult conditions. However, a closer reading of these reports combined with the somewhat scant scientific literature, suggests that stress could, in fact, have played some role in the development and management of such incidents. The special inquiry group for Three Mile Island attributed operator errors to factors generally outside the operator's control, including inadequate training, poor procedures, a lack of diagnostic skill on the part of management groups, misleading instrumentation, plant deficiencies and poor control-room design. Though stress was not mentioned *per se*, such factors are likely to have given rise to an experience of stress by the operators.

Mackie *et al.* (1985) surveyed the likely impact on the performance of sonar operators of 19 different stressors. Their study suggested that the different stressors affected performance in a variety of ways: the complexity of the relationship between work conditions and task design, on one hand, and the experience of stress, on the other, suggested that simple generalisations might have little value. It is therefore logical to suggest that every particular control situation needs to be analysed and assessed separately against existing knowledge of the effects of stress on performance. The following sections of this review therefore consider the range of factors which might give rise to the experience of stress in control-room operations and 'shape' operator performance. The review does not touch on the effects of being involved in the control of a major nuclear incident on operators' subsequent well-being, although there is evidence of such post-traumatic stress (Koscheyev *et al.*, 1993; Baum *et al.*, 1993).

The framework adopted in structuring this review, and offered for any subsequent analysis of control room operations is captured by the phrase 'the person in their job in the organization'. This framework has been explicated in more detail by the authors in their book *Systems, People and Safety* (Cox and Cox, 1996). The focus is on the working environments in which the operators act and on the social psychological conditions that reflect their roles as members of the shift team. Such a focus is shared with the dynamic reliability techniques for the analysis of human error in man–machine systems (DREAMS) of Cacciabue *et al.* (1993). Five different domains are explored: factors relating to the operators' role, factors relating to their job and the design of the tasks which comprise that job, factors relating to the physical and social environments, and, finally, those relating to the organisation of work.

13.2 Operator Factors

Operators appear to fulfil two different types of role in control-room operations: those demanded by their technical tasks and those which are more informal and social, and are determined within their work group, the shift team. These roles, and their interaction, largely determine operators' behaviour, although there is also scope for the expression of individual difference in that behaviour.

Traditionally, the operator has been said to fulfil two different task roles in relation to his overall job: passive monitor and active intervener (Johansson, 1989a, b). These two roles may require different behavioural styles and aptitudes, and stress may arise at times when it is unclear which is the most appropriate for the operator because there is uncertainty over the task. This area has been reviewed by Ainsworth (1989). Both these traditionally conceived roles relate to the management of information. However, control-room operations involve more than just the management of information; they also necessarily involve the management of operations, projects and events, and the management of people. The former includes the performance of routine operations and the management of unplanned events and non-routine situations. The management of people is largely set in the context of the shift team, but also involves interacting with staff on plant and people outside of the plant.

An unpublished study by the first author has suggested that operators evolve informal roles within their shift teams. Several particular roles were identified during that study, those of: technical 'guru', 'joker', and 'fixer' or 'doer'. The first role was occupied by the person who was recognised as having the greatest insight or experience in relation to the operation of the plant (this was not necessarily the most senior person); the second was the person who held the team together socially during shifts and who provided the 'crack'; the latter was the person who 'did things': organised tea and coffee, food, social events (etc). The possibility of role-related problems was recognised, in particular, conflict between informal team and task roles, and between the former and organisational roles.

This unpublished study also identified three dimensions to the behaviour style of the 'ideal' operator. These reflected both their task and informal team roles: (a) being able to handle the process information in a team context, (b) being able to work in a team and support team working, and (c) being able to effectively process information on one's own. The latter appeared somewhat less important than those relating to team working. The majority of operators in the study reported relying on the team in times of stress, and most denied that there were marked individual differences in coping with operations under stress. The differences which were reported to exist appeared to relate to how good a 'team player' the operator was. However, there is evidence in the literature of individual differences in coping with emergencies and stress.

The literature on individual differences in the response to stress suggests that, among other things, locus of control may moderate the effects of stress on well-being and performance at work (see Jackson, 1987; Cox and Ferguson, 1991). For example, Anderson (1977) showed that, when coping with disaster 'entrepreneurs' with an internal locus of control perceived less stress, employed more task-centred and less emotion-centred coping behaviours than those with external locus of control. Gertman and colleagues (Gertman and Haney, 1985; Gertman *et al.*, 1985a, b) have examined the effects of stress on decision making and problem solving in operators, and how such effects might interact with individual differences among operators, including locus of control. Their data suggested that the level of workload, the detail of available procedures, and individual differences, as defined in their studies, all had an impact on performance. They found evidence that impaired decision making was related to operator feelings of depersonalisation and emotional fatigue, and that operators who assumed responsibility for their own decisions, as identified by their internal locus of control scores, performed better under stress than those with an external locus of control. The responses of operators who had coped with stress in the past seemed to be superior to those who had not. Generally, the literature suggests that those with an internal locus of control report fewer symptoms of stress, are more adaptable in their choice of coping strategies and make better use of their social resources (Lefcourt 1983; Sandler and Lakey, 1982). The latter two findings are likely to be important in the context of control room operations and team working.

13.3 Task Design

The available literature, and operators' self-report, suggest several different sources of stress inherent in the design of the operator's task. Five particular task characteristics are briefly reviewed here: (a) uncertainty, (b) the interdependency of tasks, (c) work load, (d) time pressure, and (e) automation and proceduralisation.

13.3.1 Uncertainty

No matter how well defended systems are and how well prepared operators are events and incidents do occur. Most are unpredictable. Operators must work knowing that such events might occur, anticipating the unknown and holding themselves ready to respond to the unknown. Such a situation generates uncertainty and feelings of stress (Saint Jean *et al.*, 1986; Johansson and Sanden, 1989). Operators sometimes refer to the 'coiled spring syndrome'. However, uncertainty in control room operations is not narrowly confined to the scenario of unpredictable events. Saint Jean *et al.* (1986), on the basis of their studies, suggested three different types of uncertainty:

1 that related to problems generated by the organisation and particularly by interdepartmental relationships;
2 that relating to the inherent risks of operating the installation; and
3 that relating to the occurrence and management of unexpected events.

Interestingly, those authors concluded that stress is a possible, but not inevitable consequence, of these uncertainties.

13.3.2 Interdependency of Tasks

Saint Jean *et al.* (1986) also described how the tasks which are performed by the staff of various departments and services are heavily interdependent in the nuclear industry. In addition to the co-ordination required to complete any particular action itself, there usually has to be forward planning and consultation through a complex system of work requests. For such a system to work, there must be adequate communications and the continual (24 hour) availability of the information that is required by the operators. This is not always reported to be the case (Depigny and Gertman, 1986). As the control room is the 'hub' for plant-wide communications, operators receive a huge number of phone calls during a shift; this often poses a problem in itself.

13.3.3 Workload

Johansson (1989a, b) has offered a description of the psychological workload inherent in control tasks in terms of monitoring and intervening (see above). Monitoring under routine operations is usually a very monotonous and passive task, although it requires continuous attention. Monitoring stands in stark contrast to intervening. In failure or breakdown situations, it is associated with high demands on perception, information processing and decision making.

Activities at start up and shut down of the process also require the operator to carefully follow certain rules and procedures and take complex actions at precise moments in time. These procedures can last for several hours, even days. A characteristic feature of control work is, therefore, the rapid shifts from passive monitoring to active problem solving which can sometimes occur (see Otway and Misenta, 1980). A somewhat similar situation exists for sonar operators. Smith *et al.* (1985) have described how their routine workload can suddenly escalate: their routine appears to be punctuated unpredictably by periods of high demand. During these periods, the demands of information processing may be excessive, requiring the accurate perception and cross matching of visually and aurally displayed information, skilled control manipulation, extensive communication and complex decision making. From studies such as these, it is not unreasonable to conclude that the workload problems of operators logically arise for two reasons: first, as a reflection of the workload condition itself, or, second, as a result of the unpredictable and rapid transition from low to high workload.

Frankenhaeuser and Gardell (1976) described workload or demand (see Cox and Griffiths, 1996) as a continuum, from work overload to work underload, with maximal work performance occurring at some optimal midpoint. While high and low workloads might be defined against some agreed and independent standard, the conditions of 'overload' and 'underload' can only be defined against the workers' ability to cope with their tasks, and must necessarily vary both from person to person, and from task to task. Frankenhaeuser and Gardell (1976) argued that both overload and underload are associated with the experience of stress and with sub-optimal performance: this may be particularly so for underload when it gives rise to feelings of boredom or is associated with a need for sustained attention. It is interesting to compare low and high work loads in terms of their effects.

Gebhart *et al.* (1990) compared mental strain in operators, as defined by heart rate, working in monotonous conditions (low work load) and working with high density information flow (high work load). Greater mental strain occurred in the high workload condition. However, the situation is more complex than this study might suggest. Johansson and Sanden (1989) examined the mental workload and job satisfaction of control-room operators in a steel plant. Process operators performing passive monitoring tasks in a monotonous environment (low workload) were compared to operators in an active planning and production control environment characterised by high information load and time pressure (high workload). Those authors found that operators in the monotonous environment were more isolated and experienced fewer interruptions and disturbances, whereas active communication was vital for the other group. Sources of job satisfaction in these situations reflected a good social atmosphere at work, independence, interesting and responsible work allowing learning on the job, pride in being able to manage a complex set of tasks and a qualified job. Naturally the group involved with active planning and production control reported significantly higher feelings of job satisfaction than those involved with more monotonous process control. In terms of operator performance, Cox (1985) and Williams (1982, 1985) have both observed that prolonged inactivity or highly repetitious tasks with low mental workload can result in boredom, stress and an increase in operator error. Furthermore, Smith *et al.* (1985) identified boredom as a factor affecting sonar operators' vigilance and motivation, which can give rise to sub-optimal levels of arousal, increased lapses of attention, and increased variability in detection time.

Seminara *et al.* (1977) interviewed personnel at a number of power plants in the US. They reported that a majority of the operators and trainers spoken to (69.2%) saw the control room operator's job as one in which there was much boredom especially when the plant was operating normally. Such monotony and boredom may be most evident during long night shifts.

The link between work underload and the requirement for sustained attention is important. Johansson and Sanden (1989) reported that for control operators involved in passive monitoring, the experience of stress resulted not only from the monotony of their situation but also in the demand for sustained attention (and uncertainty about how well they would manage any major process disturbance or breakdown). It would appear that stress arose from demands for sustained attention and action readiness in an environment offering little sensory or cognitive stimulation and forcing low levels of arousal. The need to maintain action readiness in such an environment – the coiled spring syndrome – is part of the problem posed by unpredictable oscillations between underload and overload. Indeed, Otway and Misenta (1980) recommend that the design of control-room operations should ensure that operators are kept sufficiently aroused and alert during normal operations so as to prevent errors occurring and maintain readiness for active intervention (if necessary).

High workloads and high levels of demand have also been reported to be a source of stress, and, in particular, where they represent work overload (see above). A number of factors appear to moderate the relationships among high workload and level of demand, work overload and work-related stress; three are discussed below and relate to issues of multi-tasking, control and sustained performance requirements and sleep loss (see Figure 13.1).

While workers appear to be able to cope with a relatively high workload if it is generated by a single task or information source, if it is generated by several different tasks it is likely to prove more difficult to manage. A multi-tasking requirement may further increase the level of demand and stressfulness of high workloads. This is important in relation to the design of control-room operations as these generally require attention to many different and competing tasks and sources of information. Furthermore, if the worker is operating under time pressure, the extra demands that this imposes may serve to further exacerbate work overload problems.

Not unrelated, the level of control that workers have over their workload, or perceive themselves to have, is a moderator of the experience of stress and related outcomes (Ganster and Fusilier, 1989; Karasek, 1979; Cox and Griffiths, 1996). Karasek (1979) has argued that the level of job demand interacts with control over work in relation to their effects on worker health. It is possible that a similar argument might be applied to their effects on performance. Analysing data from the United States and Sweden, he found that workers in jobs perceived to have both low decision latitude and high job demands[3] were particularly likely to report poor health and low satisfaction. The lowest probabilities for illness and death were found among

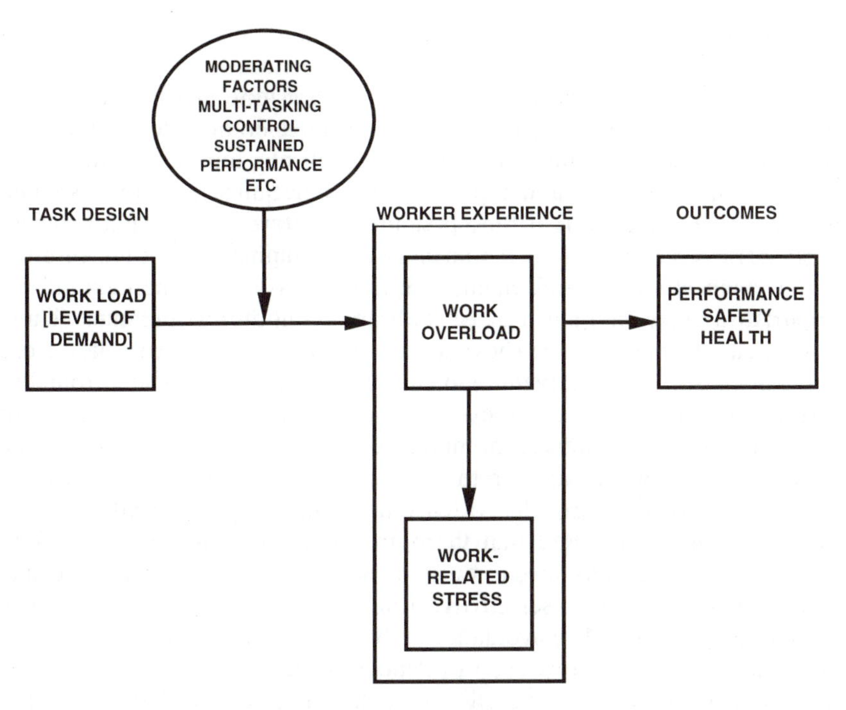

Figure 13.1 Workload, work overload and stress.

work groups with moderate workloads combined with high control over work conditions. Later studies appeared to confirm the theory (see Karasek and Theorell, 1990). The combined effect of these two work characteristics is often described as a true interaction but despite the strong popular appeal of this suggestion there is only weak evidence in its support (Warr, 1990). Simple additive combinations have been reported by a number of researchers, for example, Hurrell and McLaney (1989), Perrewe and Ganster (1989) and Spector (1987).

It is possible for individuals to cope with a high workload for a short period of time (Cox, 1978). However, such coping may not be sustainable. The effort after faultless performance may eventually be abandoned and lower criteria accepted: error rates may rise sharply (see Hamilton and Warburton, 1979). Operators may be forced to prioritise tasks and 'load shed', and such strategies will have an effect on their performance. Interestingly, they attempt to protect what is perceived (under pressure) to be the critical aspects of the overall task (Baddeley, 1972; Idzikowski and Baddeley, 1983). The problems that are posed by work overload in nuclear control operations may be made worse by the presence of other stressors, for example, sleep deprivation (see, for example, Broadbent, 1963; Monk and Folkard, 1983) and time pressure.

13.3.4 Time Pressure

Time pressure has been reported to be one of the most significant stressors which people experience during their working lives. Undoubtedly in emergency situations, and when managing events, operators are under increased time pressure, and report feeling under such pressure. Various researchers have argued that time pressure can impair human performance and give rise to error (including Usdansky and Chapman, 1960; Hutchinson, 1981; Keinan *et al.*, 1987; Edland, 1985). Usdansky and Chapman (1960) compared subjects on a sorting task with and without time pressure. Those working under time pressure showed an increase in error scores; they made more associative and irrelevant errors. Williams (1982) has stated that 'the shortage of time available for error detection and correction' is a major industrial problem. However, the nature of the task needs to be taken into account: while it would appear that the performance of complex tasks is impaired by time pressure, that of easy tasks might be improved.

Edland (1985) concluded that there are systematic changes in cognitive processes when decisions have to be made under time pressure. The changes seem to involve increased selectivity, a more frequent use of non-compensatory decision rules and an avoidance of the negative aspects of the situation. These changes may be preceded by a 'speeding up' of processing and occur when there is no further possibility of processing information faster. The person then has to filter information by increasing selectivity, focusing on a subset of the available information and basing the decisions or judgements on that. When time becomes extremely restricted, there may be a total change of strategy to handle the situation and reach a decision or judgement. Of course, avoidance (or denial) of the entire problem may occur when the pressure of time is so strong that there is no possibility of solving the problem.

13.3.5 Automation and Proceduralisation

Increased automation and engineering control, while being widely seen as enhancing safety, may also increase the passive monitoring role of operators at the expense of their active intervening role (see Johansson, 1989a, b). This may have a variety of effects but, in particular, reduce the operators' control over their work, and exacerbate two existing problems by increasing the underload component of their job and the difficulty of switching from their passive to their active role. There is, therefore, concern about the impact of automation of control-room operations on the operator (Bainbridge, 1983, 1987). A somewhat similar challenge to the discretion that control operators can exercise in their work has resulted from proceduralisation.

Procedures are written to provide supportive and consistent guidance on monitoring and reacting to changes in plant status, and to establish performance standards. However, they are often perceived by operators to be inaccessible and ambiguous, to make excessive demands on memory and often to provide little help with the actual diagnosis of plant problems. Operators may experience stress in situations where such procedures demand a different course of action from that they would otherwise choose to follow. Not surprisingly, on occasions, operators opt not to follow written procedures (see below).

Embrey (1989) has observed that a large number of accidents involve operators deliberately ignoring or transgressing known procedures. This appears to occur for one of four reasons:

1. procedures do not correspond to the way the job is actually done;
2. the information contained in the procedures, although correct, is not cast in a usable form;
3. procedures are not updated on the basis of operational experience; or
4. rules and procedures are not seen to apply to the operators or the situation in question.

The more 'proceduralised' sets of tasks are and the more experienced the operators are, the greater the potential both for stress and transgression. Ainsworth (1989) has discussed this aspect of proceduralisation.

Embrey (1989) has also described rule-fixated safety cultures and the risk that organisations will develop the belief that their system of safety rules is invulnerable to human error. It can become an assumption of such cultures that if the rules are followed then accidents are impossible. This is based on the belief that their rules cover every possible contingency and that operators will never be required to interpret them in 'new' situations. This belief is fallacious and, for that reason, dangerous to safety. In such cultures, plant management's investment in the application of procedures may also result in operators developing a lack of confidence in their own ability.

13.4 Working environment

The tasks that engage the operator are set in the wider contexts of their physical and social work environments. These are discussed below beginning with the physical work environment. Two possible sources of stress relating to the design of the physical work environment are discussed: noise and distraction. Several studies have identified both as problems in nuclear control rooms. The effects of noise are dependent on a range of factors, including the type and meaning of the noise, its intensity, the length of exposure, the characteristics of the task performed, and of the operator. Detailed discussions of such effects are available in Jones (1985) and in Hockey (1983). The literature on distraction is less extensive.

The problem posed by noise is complex and involves more than just the level of background noise, which can often be sustained at a high level, and includes the intrusiveness and implications of both telephone calls and auditory alarms. Furthermore, in addition to its more direct effects on performance through any changes in arousal and information processing that it effects, noise can act as a distractor, and also interfere with communication (see below). These effects may be particularly troublesome during complex interventions or periods of difficult information processing.

Noise seems to increase attentional selectivity (see, for example, Glass and Singer, 1972; Hockey 1970a, b, 1983; Jones, 1985). If a task requires high attention to all cues, noise may adversely affect performance. However, noise may improve performance if the attention can be usefully focused on the more relevant aspects of a task. In the control room environment, where there is an interplay of many complex tasks, high noise levels might impair performance through increased arousal, narrowing of attention, and less efficient working memory (Glass and Singer, 1972; Hockey 1970 a, b, 1983).

Operators often report that not only noise but also some peripheral activities in the control-room operations distract them from what they perceived to be their central task. Examples of such activities include: answering external telephone lines, issuing keys, and attending to visitors to the control room. There are, at least in the UK, a surprising number of 'visitors' present in control rooms over the course of a working week. In Johansson and Sanden's (1989) comparison of an active planning group of process control operators and a group of passive monitors, they also found that stress was associated with too many people in the control room.

13.5 Social Environment

It has already been argued that the shift team is the functional unit in control-room operations, and that such operations represent a sociotechnical system. Operators work as a team. One characteristic of team work is that the overall reliability of action is dependent on the interaction between team members so that (a) the probability of one member of the team making an error is related to the probability of another making an error, and (b) the probability of detecting and correcting such errors is determined at a team and not an individual level. These two points are important for the design and management of control-room operations.

When people work together in teams they bring with them different objectives and agenda, personalities and cultural backgrounds. In turn, teams both influence and are influenced by their general task and social and organisational environments. To a considerable extent that team's structure (reflected, for example, in its role system and communication channels) evolves to help the team cope with the demands both of its task and organisational environments, and of teamwork itself. Not only can the team act as a support for its members, both in terms of task completion and more personally, but it may also act as a potential source of stress for them.

Decision making may be enhanced by team work, but it may also become more risky (Stoner, 1961) and more fixed (group think: Janis, 1972). Often it is some sense of 'cohesion' and 'invulnerability' that leads teams to make riskier decisions (Hutchins, 1988). Individuals become committed to group decisions. Group think is most likely to occur when the decision-making unit displays high cohesion, when it is insulated from the advice of experts and when a strong leader actively promotes their solution to the problem.

Research has shown that social support from superiors, colleagues and family can reduce both perceived stress and its psychological effects (Caplan *et al.*, 1975; French *et al.*, 1974). Support from colleagues appears to produce the most consistent gains. There is, however, confusion over the precise mode of action of this support. Some argue that it has a direct or additive role and counterbalances the effects of stressful situations (La Rocco and Jones, 1978). Other researchers suggest a more interactive role, such that support can only mitigate or buffer the impact of stress. The evidence is inconclusive. However, it is clear that support from colleagues is likely to be important where the team is an integral feature of work. There is also some suggestion that those with an internal locus of control can make use of their social resources (Lefcourt 1983; Sandler and Lakey, 1982).

Working in a team may itself make various demands on the operator, ranging from the resolution of personality clashes and other interpersonal difficulties to the effective increase in workload associated with the need for team maintenance activities (see, for example, Otway and Misenta, 1980). Those working within teams are also subject to social pressure from others and a felt lack of privacy, especially in confined groups. The close confinement enforces socialisation to a greater degree than many crew members would prefer.

13.5.1 Group Reactions to Stress

Under stress, some groups break down, show maladaptive behaviour and begin to function as a collection of individuals: others respond more adaptively. Torrance (1954) suggested that, under extreme conditions in the Second World War, some US Air Force bomber crews that survived crash landings subsequently panicked, suffered internal communication breakdowns, and acted as competitive individuals. Other more cohesive bomber crews, under similar conditions, held together and problem-solved collectively. Survival rates were higher in the latter cases (Torrance, 1954). Some groups appear to become more cohesive under stress while others become less so. This observation might be explained by an inverted U function between stress and group cohesion (Stein, 1976). (In the same way there is also some evidence of an inverted U relationship between stress and group performance: total group performance increasing with stress until it becomes too high (Pepinsky *et al.*, 1960).) However, a study of behaviour during natural disasters such as earthquakes, storms and floods by Quarantelli and Dynes (1972) showed considerable reliance on teamwork in almost all cases.

The leadership of teams under stress appears to be a key issue determining the team's overall reaction. The various studies of teams and groups under stress show an increased need for authoritarian leadership (Mulder and Stemerding, 1963) and increased authoritarian behaviour by leaders (Fodor, 1976), a greater effectiveness of authoritarian rather than more participative leaders (Rosenbaum and Rosenbaum, 1971), and, on occasion, the replacement of leaders (Hamblin, 1958; Rosen, 1989). Team members expect their supervisors (leaders) to act as a buffer against frustration and to help minimise stress within the team. The supervisor who fails to perform these critical aspects of team maintenance often loses the team's confidence and support (Rosen, 1989). The issues around team leadership under stress are partly determined by the nature of the team; for example, is it a work group or a social group defined solely by the situation, and by its normal mode of operation and level of cohesiveness? It is not yet clear how far the existing literature on leadership explains group behaviour under stress, or to what extent conventional leadership training may enhance such behaviour or interfere with it.

13.6 Organisational Factors

Much of the work on operator stress, both inside and outside nuclear power generation, has largely focused on their immediate tasks, equipment and work environments. However, it is clear from talking with operators, and other staff, that many of the problems that they face relate to the very nature of their organisation, and to its management of their work (Cox *et al.*, 1990). Three issues are discussed here: conflicting commercial and safety pressures, communication and decision making, and the design and preparation for shift work.

13.6.1 Conflicting Pressures

Many operators world-wide report conflicting pressures on them to operate plant safely but, at the same time, to maintain the generation of electricity. Frequent variations in the demand for electricity increase the difficulty of the operator's task and increase the opportunity for an emergency shut down caused by an inappropriate manoeuvre (Saint Jean *et al.*, 1986). At the same time, there is the fear that decisions on safety may become conditioned by commercial considerations particularly in ambiguous and uncertain situations. Such decisions will also be influenced by other management and organisational factors. These various factors may interact to determine other forms of operator error. Reason (1987, 1989), for example, has argued that the greatest danger facing high hazard/low risk installations, such as nuclear power plants, comes from the insidious accumulation of latent failures emerging from a complex interaction between technical, social and organisational aspects of the overall system. In the same vein, Layton and Turnage (1980) have inferred from reports of various safety related events in plants, that management, supervisory and staff practices play an important part in determining safety. These authors have argued that operator error, initially considered as inadequacies on the part of the person involved, can, on reflection, be seen to have occurred because of inadequacies in management and supervisory practice and an organisational climate conducive to errors.

13.6.2 Communication and Decision Making

It has already been stated that the control room acts as the hub or nerve centre for the entire nuclear plant. As such it seeks and receives data, processes that data and then disseminates it to the rest of the plant. This requirement alone makes operator and team communication extremely important to systems performance, and it is clear that there are both technical and organisational constraints on operator communication which can act as a source of stress and an additional threat to both individual and systems performance. These have been referred to above in relation to associated problems of noise and distraction in the control room. First, each plant employee works in a complex communication network and generally has to communicate with a number of other people before accumulating sufficient information to accomplish the task in hand. Second, because of the multiplicity of the tasks involved, each operator is consulted or must consult on a wide range of unrelated topics, and the information exchanged in relation to each topic must be appropriate and precise. Finally, in nuclear power stations, certain key positions must be manned round the clock while others only function during normal office hours. For the plant to remain in continuous and safe operation, up-to-date information is required by the operators and must be available at all times: often it is not. Depigny and Gertman (1986) have described the programme established by EDF to resolve communications problems in French nuclear power stations. The project was launched following a series of incidents in which such problems were shown to be an important factor.

On analysing verbal and other auditory communications, EDF found that typically the control-room supervisor received an average of 23 calls an hour (Depigny and Gertman, 1986). This load varied with time and day. It was heaviest from 10.00 h to 15.00 h and towards the end of the week. The other control room staff experienced a similar load profile. More than 30% of these calls represented requests for information and came either from other members of the control crew or from outside callers. The majority of callers requested operating information from supervisors and a smaller proportion inquired as to plant status. 14% of calls consisted of requests for the supervisor to search for or page someone in the plant. Smith *et al.* (1985) report that 'command pressure' is strongly associated with pressures from superiors for operators to report information in minimal time. They felt that such pressure could give rise to increased arousal, response rate and selectivity of attention, but also reduce quality of performance, increase frustration and anxiety, increase distractability and reduce parallel information processing.

EDF underlined a fundamental principle upon which any action to enhance plant communications must be based (Depigny and Gertman, 1986). The information necessary to help the operator perform tasks must be readily available at all times and be of an appropriate quality. Enhancement programmes should involve: (a) improving information sources, (b) improving information quality and (c) diminishing the burden of information search on the persons requiring and delivering the data. Many technical and staff-oriented solutions have been suggested by Depigny and Gertman (1986). However, there are many obstacles to improving communication. For example, a familiar context breeds a tendency towards informality. Operators who have worked together for any length of time (and as a team) develop confidence in one another and acquire some kind of automatic response to each others' actions. They tend to develop a false sense of security in each other, and this may result in a tendency not to share information because it is judged to be self-evident.

It was argued by Depigny and Gertman (1986) that problems of communication can not only lead to operating incidents but also increase the workload for plant staff if they need to spend time overcoming those deficiencies. Feedback from EDF nuclear operators showed that communication breakdowns can result from: (a) poor design of communications equipment, and (b) inadequate communications practices. In some cases communications hardware itself may be of an unsatisfactory design: in others the problem may be linked to the way people attempt to communicate with each other.

Under conditions of crisis induced stress, there is a tendency for participation in decision making to be limited to a small number of individuals. It is possible that the central decision unit comes to consist of a tightly knit homogeneous group guided by a strong leader. The dynamics of such a group can contribute to error. Janis (1972) and Janis and Mann (1976) have suggested that during crises, group pressures in such a decision-making unit can bring about a deterioration of mental efficiency, reality testing and moral judgement. This may promote a condition called group think (see above: social environment). Individuals become committed to group decisions and, as a result, their own personal attitudes and models of reality shift to reflect those of the group. This happens as a result of their attempt to maintain consistency. Group think is most likely to occur when the decision-making unit displays high cohesion, when it is insulated from the advice of experts and when a strong leader actively promotes their solution to the problem. However, not all cohesive groups are prone to group think.

13.6.3 The Design and Preparation for Shift Working

Nuclear plants work on a shift basis. The available evidence suggests that, regardless of the technological and economic justification for shift work, it can cause problems for shift workers, having possible negative consequences for their social, physical or mental well-being as well as their productivity.

Many physiological functions, and possibly related psychological processes, appear to operate cyclically (see, for example, Monk and Folkard, 1983). These cycles appear to be driven by a number of internal clocks and synchronised against various external '*Zeitgeber*' (time markers). Many approximate to 24 h cycles and map onto the day–night pattern of activity. Social systems follow more complex patterns with daily cycles being imposed on a weekly cycle and that in turn on a seasonable pattern. Shift work superimposes an artificial pattern of activity onto these natural and social cycles. Problems arise when the person has to work when naturally they would be sleeping, and sleep when they would otherwise be working. Furthermore, the family and friends of shift workers may not share their pattern of work and non-work activities, and be active when the shift worker is trying to sleep, and be tired or sleeping when the shift worker is awake after work. Shift work can cause a disruption of the shift worker's social life.

There is evidence of negative effects of shift work on physical and mental health, and on social relationships and constructive use of leisure time. However, there are several different types of shift system available, including compressed work weeks, and different systems have different effects on health. Those that ameliorate health effects appear to disrupt social life, while those that allow social adjustment appear to preclude physiological adjustment. Shift working can also affect operator performance. For example, Glenville and Wilkinson (1979) found that computer operators' mean reaction time increased significantly on night as compared to day shifts. This adverse effect on the night shift was not apparent at first but developed over successive comparisons. The general finding of the research on the effects of night working seems to be a decrease in the speed of performance with accuracy only marginally affected if at all (Folkard, 1975, 1983; Monk and Folkard, 1983). Clearly there is a need, where possible, to further develop the design of shift-work systems. Where this is not possible, operators should be prepared for shift working through a programme of shift-work education.

13.7 Concluding Remarks

The literature reviewed here highlighted several different types of demand, associated with control-room operations in nuclear power generation, which might be experienced as stressful by operators. These sources of stress arise from the nature of the operator's tasks and their role in the overall system, the design of their working environments, and the organisation itself.

Several particular demands have been identified:

- work load, including work overload, and the need to switch between passive monitoring and active intervention;
- uncertainty over unplanned situations;
- the interdependency of tasks;
- time pressure;
- automation and proceduralisation;
- noise and distraction;
- lack of social support and group think;
- conflicting safety and commercial pressures;
- poor communication; and
- the design and preparation for shift work.

What is important to note is that none of these potential stressors can be treated in isolation from the others. The authors have attempted to make this clear throughout the review: the various sources of stress associated with control-room operations appear to interact both in the way in which they arise and in the effects they have on operators. The interaction of stressors, convincingly demonstrated in the early 1960s by Broadbent (1963), is an important phenomenon. There are two immediate implications: first, that the study of any one source of stress in isolation will not provide a good basis for generalising to the real-life situation, and second, interventions targeted on any one source of stress should carefully consider their possible impact on other problem areas. Failure to take the interaction of stressors into account in research and intervention may lead to misunderstanding and counter-productive effort.

Acknowledgements

The authors wish to acknowledge not only the contributions to their work made by their colleagues in their respective Centres, but also the support for their research provided by a large number of organisations, nuclear and non-nuclear. The first author wishes to acknowledge the general support of the World Health Organization (Regional Office for Europe). The views expressed here are the authors' and do not necessarily reflect those of any other person or organisation.

Notes

1. During March, 1979, there was a reactor core melt down at the Three Mile Island nuclear power plant (USA) resulting in the release of radioactive material into the atmosphere. The incorrect action of the plant's control room engineers, in switching off safety injection pumps, contributed to the melt down. The following shift crew correctly diagnosed the problem and took corrective action.

   Both of the independent investigations into the Three Mile Island incident, those of the Kemeny Commission (1979) and the Nuclear Regulatory Commission, reached similar conclusions. They were that the weakest element in operational reactor safety, the greatest area of risk, relates to human behaviour. The inquiries also concluded that the value of reducing this risk has been under-emphasised compared to the improvement of safety hardware, and the design and manufacture of engineering systems. Human factors specialists working for the NRC argued that:

   > there is indeed room for the improvement in design of control rooms to aid engineer performance of the job, particularly during the highly stressful minutes following emergencies. Personnel selection, training and education, procedural systems, control board design, diagnostic information systems, management structure, staff motivation, these and other aspects of the total system of interaction between humans and machines need to be examined.

   Many changes were recommended by these two inquiries, the most urgent of which focused on the training, qualification and overall competence of on-site operations staff and their management.

2. At 01.24 h on Saturday 26 April 1986, two explosions blew off the 1000 tonne concrete cap which sealed the Chernobyl-4 reactor (Ukraine) releasing molten core fragments into the immediate vicinity and fission products into the atmosphere. This became the worst accident in the history of nuclear power generation. It was allowed by a design which necessitated strict compliance with written instructions and procedures to maintain the reactor in a stable and safe state. These instructions and procedures were ignored.

   Maclachan (1986) reviewed the Soviet authorities' investigations into Chernobyl. They blamed the cause of the accident on a convergence of 'deliberate and wilful' violations of operating rules and procedures by the unit's engineers. They argued that superficially, at least, the engineers appeared to demonstrate 'kamikaze behaviour', driving a turbine test to completion while the reactor was dangerously unstable and after they had disconnected all safety and protection systems. However, it was also pointed out by the Soviet authorities that the accident occurred at a bad time psychologically: at the end of the working week and early in the morning. It was also just before the May Day holiday and the engineers were under pressure to complete an experiment which would otherwise have to have been postponed to the next shut-down one year later. The Soviet authorities suggested that these different aspects of the situation combined and predisposed the engineers to narrow down their view of the reactor process to the specific demands of the experiment. This may have led them to disregard overall plant safety status (Baker and Marshall, 1987).

Reason (1987) has pointed to the way in which these factors did combine, arguing against any assumption that an accident like Chernobyl could never happen in the UK. He believes that the ingredients for the disaster were present at many different levels. There was a society committed to the generation of energy through large-scale nuclear power plants. The overall system was complex and hazardous, and operated outside normal conditions. The management structure was remote and slow to respond. The engineers possessed only a limited understanding of the system they were controlling and were set tasks that made violations of safety procedures inevitable: an accident was waiting to happen.

3. Karasek (1979) defined 'decision latitude' as 'the working individual's potential control over his tasks and his conduct during the working day'. He defined 'job demands' as 'the psychological stressors involved in accomplishing the workload'.

References

AINSWORTH, L. (1989) *A Comparison Of Control Room Evaluation Techniques*, Vickers Shipbuilding and Engineering Limited, on behalf of HM Nuclear Installation Inspectorate.

ANDERSON, C. R. (1977) Locus of control, coping behaviours and performance in a stress setting: a longitudinal study, *Journal of Applied Psychology*, **62**, 446–51.

BADDELEY, A. D. (1972) Selective attention and performance in dangerous environments, *British Journal of Psychology*, **63**, 537–46.

BAINBRIDGE, L. (1983) Ironies of automation, *Automatica*, **19**, 775–9.

(1987) Ironies of automation. In Rasmussen, J., Duncan, K. & Leplat, J. (eds), *New Technology and Human Error*, Chichester: Wiley.

BAKER, E. & MARSHALL, S. (1987) *The Halden Report*, Halden, Norway.

BROADBENT, D. E. (1963) Differences and interactions between stresses, *Quarterly Journal of Experimental Psychology*, **15**, 205–11.

CACCIABUE, P. C., CARPIGNANO, A. & VIVALDA, C. (1993) A dynamic reliability technique for error assessment in man-machine systems, *International Journal of Man–Machine Studies*, **38**, 403–28.

CAPLAN, R. D., COBB, S., FRENCH, J. R. P., VAN HARRISON, R. & PINNEAU, S. R. (1975) *Job Demands and Worker Health*, Washington: National Institute for Occupational Safety and Health, HEW Publication No. (NOISH), 75–160.

COX, S. & COX, T. (1996) *Systems, People and Safety*, London: Butterworth-Heinemann.

COX, T. (1978) *Stress*, London: Macmillan.

(1985) Repetitive work: occupational stress and health, in Cooper, C. L. & Smith, M. J., *Job Stress and Blue Collar Work*, Chichester: Wiley.

COX, T. & FERGUSON, E. (1991) Individual differences, stress and coping. In Cooper, C. L. & Payne, R. (eds), *Personality and Stress: Individual Differences in the Stress Process*, Chichester: Wiley.

COX, T. & GRIFFITHS, A. (1996) The nature and measurement of work stress: theory and practice. In Wilson, J. & Corlett, N. (eds) *The Evaluation of Human Work: A Practical Ergonomics Methodology*, London: Taylor and Francis.

COX, T., LEATHER, P. & COX, S. (1990) Stress, health and organizations. *Occupational Health Review*, **23**, 13–18.

DEPIGNY, C. & GERTMAN, F. (1986) Improvement of communications in nuclear power plants, EDF Report, *Proceedings of the ANS/ENS International Topical Meeting on: Advances in Human Factors in Nuclear Power Systems*, Knoxville, Tenn.

EDLAND, A. (1985) *Attractiveness Judgements of Decision Alternatives Under Time Stress*. Reports from the Cognition and Decision Research Unit, Department of Psychology, University of Stockholm.

EMBREY, D. (1989) The management of risk arising from human error, *Conference Proceedings Human Reliability in Nuclear Power*, Cafe Royal, London, October.

FODOR, E. M. (1976) Group stress, authoritarian style of control and use of power, *Journal of Applied Psychology*, **61**, 313–18.

FOLKARD, S. (1975) The nature of diurnal variations in performance and their implications for shiftwork studies. In Coluhoun, W. P., Folkard, S., Knauth, P. & Rutenfranz, J. (eds), *Experimental Studies of Shift Work*, Proceedings of III International Symposium on Night and Shift Work, Dortmund, October.

(1983) Diurnal variation. In Hockey, J. (ed.), *Stress and Fatigue in Human Performance*, Chichester: Wiley.

FRANKENHAEUSER, M. & GARDELL, B. (1976) Underload and overload in working life: outline of a multi disciplinary approach, *Journal of Human Stress*, **2**, 35–46.

FRENCH, J., ROGERS, D. W. & COBB, S. (1974) Adjustment as person–environment fit. In Coelho, G. V., Haburg, D. A., & Adams, J. E. (eds), *Coping and Adaptation*, New York: Basic Books.

GANSTER, D. C. & FUSILIER, M. R. (1989), Control in the workplace. In Cooper, C. L. and Robertson, I. (eds), *International Review of Industrial and Organizational Psychology*, Chichester: Wiley.

GEBHART, J., KOZENY, J., PROCHAZKOVA, Z. & BOSCHEK, P. (1990) Heart rate response as an indicator of mental strain: interaction of personality and situational factors, *Activitas Nervosa Superior*, **32**, 174–8.

GERTMAN, D. I. & HANEY, L. N. (1985) Personality and stress: what impact on decision making? Paper presented to *Annual Ergonomics Conference*, University of Nottingham, March.

GERTMAN, D. I., HANEY, L. N., JENKINS, J. P. & BLACKMAN, H. S. (1985a) Operator Decision Making and Action Selection Under Psychological Stress in Nuclear Power Plants, NUREG/CR 4040 EGG 2387.

(1985b) Operator Decision Making Under Stress. Paper to: *II IFAC/IFIP/IFORS IEA Conference 'Analysis, Design and Evaluation of Man–Machine Systems'*, Varese, Italy, September.

GLASS, D. C. & SINGER, J. E. (1972) *Urban Stress: Experiments on Noise and Social Stressors*, New York: Academic Press.

GLENVILLE, M. & WILKINSON, R. T. (1979) Portable devices for measuring performance in the field. The effects of sleep deprivation and night shift on the performance of computer operators, *Ergonomics*, **22**, 927–34.

HAMBLIN, R. (1958) Group integration during a crisis, *Human Relations*, **11**, 67–76.

HAMILTON, V. & WARBURTON, D. M. (1979) *Human Stress and Cognition*, Chichester: Wiley.

HOCKEY, G. R. J. (1970a) Signal probability and spatial location as possible bases for increased selectivity in noise, *Quarterly Journal of Experimental Psychology*, **22**, 37–42.

(1970b) Changes in attention allocation in a multi-component task under loss of sleep, *British Journal of Psychology*, **61**, 473–80.

(1983) *Stress and Fatigue in Human Performance*, Chichester: Wiley.

HURRELL, J. J. & MCLANEY, M. A. (1989) Control, job demands and job satisfaction. In Sauter, S. L., Hurrell, J. J. & Cooper, C. L. (eds), *Job Control and Worker Health*, Chichester: Wiley.

HUTCHINS, E. (1988) The technology of team navigation. In Galegher, J., Kraut, P. & Egids, C. (eds), *Intellectual Teamwork: Social and Technical Bases of Collaborative Work*, Hillsdale, NJ: Lawrence Erlbaum Associates.

HUTCHINSON, R. D. (1981) *New Horizons for Human Factors in Design*, New York: McGraw-Hill.

IDZIKOWSKI, C. & BADDELEY, A. D. (1983) Fear and dangerous environments. In Hockey, G. R. J. (ed.), *Stress and Fatigue in Human Performance*, Chichester: Wiley.

JANIS, I. L. (1972) Victims of groupthink: a psychological study of foreign policy decisions and fiascos, Boston, MA: Houghton Mifflin.

JANIS, I. L. & MANN, L. (1976) Coping with decisional conflict, *American Scientist*, **64**, 657–67. Job demands and worker health. Washington: National Institute for Occupational Safety and Health. HEW Publication No. (NOISH), 75–160.

JOHANSSON, G. (1989a) Categories of human operator behaviour in fault management situations. In Goodstein, L. P., Anderson, H. B. & Olsen, S. E. (eds), *Tasks, Errors and Mental Models*, London: Taylor and Francis.

(1989b) Stress, autonomy and maintenance of skill in supervisory control of automated systems, *International Review of Applied Psychology*, **38**, 45–56.

JOHANSSON, G. & SANDEN, P. O. (1989) *Mental Load And Job Satisfaction Of Control Room Operators*, Reports from the Department of Psychology, University of Stockholm.

JONES, D. M. (1985) Noise. In Hockey, R. (ed.), *Stress and Fatigue in Human Performance*, pp. 61–95, Chichester: Wiley.

KARASEK, R. A. (1979) Job demands, job decision latitude and mental strain: implications for job redesign, *Administrative Science Quarterly*, **24**, 285–308.

KARASEK, R. & THEORELL, T. (1990) *Health Work: Stress, Productivity and the Reconstruction of Working Life*, New York: Basic Books.

KEINAN, G., FIEDLAND, N. & BEN-PRATH, Y. (1987) Decision making under stress: scanning of alternatives under physical threat, *Acta Psychologica*, **64**, 219–28.

KEMENY, J. (1979) *The Need for Change: The Legacy of TMI*, Report of the President's Commission on the Accident at Three Mile Island, Washington, DC: Government Printing Office.

LA ROCCO, B. & JONES, F. (1978) Co-worker and leader support, *Journal of Applied Psychology*, **63**, 629–34.

LAYTON, W. L. & TURNAGE, J. J. (1980) *Relation of management, supervision and personnel practices to nuclear power plant safety*, Proceedings of Topical Meeting on Thermal Reactor Safety, Knoxville, Tennessee.

LEFCOURT, H. (1983) The locus of control as a moderator variable: stress. In Lefcourt, H. (ed.), *Research with the Locus of Control Construct*, New York: Academic Press.

MACKIE, R. R., WYLIE, C. D. & SMITH, M. J. (1985) Comparative effects of 19 stressors on task performance: critical literature review, *Conference Proceedings, Human Factors Society, 29th Annual Meeting*.

MACLACHAN, A. (1986) What the experts learned from Chernobyl post accident review, *Nucleonics Week*, **27**, 36.

MALONE, T. B. *et al.* (1980) *Human Factors Evaluation of Control Room Design and Engineer Performance at Three Mile Island*, NUREG/CR-1270, 1, January.

MONK, T. H. & FOLKARD, S. (1983) Circadian rhythms and shiftwork. In Hockey, J. (ed.), *Stress and Fatigue in Human Performance*, Chichester: Wiley.

MOORE, T. & DIEMICH, D. (1987) Chernobyl and its legacy: A special report, *EPRI Journal*, **12**, 4–21.

MULDER, M. & STERMERDING, A. (1963) Threat, attraction to group, and need for strong leadership, *Human Relations*, **16**, 317–34.

OTWAY, H. J. & MISENTA, R. (1980) Some human performance paradoxes of nuclear operations, *Futures*, October, 340–57.

PATTERSON, W. C. (1983) *Nuclear Power*, Harmondsworth: Penguin.

PEPINSKY, P., PEPINSKY, H. B. & PAVLIK, W. P. (1960) The effects of task complexity and time pressure upon team productivity, *Journal of Applied Psychology*, **44**, 34–8.

PERREWE, P. & GANSTER, D. C. (1989) The impact of job demands and behavioural control on experienced job stress, *Journal of Organizational Behaviour*, **10**, 136–47.

QUARANTELLI, E. & DYNES, R. (1972) When disaster strikes, *Psychology Today*, **5**(9), 66–70.

REASON, J. (1987) The Chernobyl errors, *Bulletin of the British Psychological Society*, **40**, 201–6.

(1989) The contribution of latent failures to the breakdown of complex systems, Paper presented at the *Royal Society, Discussion Meeting, Human Factors in High Risk Situations*, London (June).

ROSEN, N. (1989) *Teamwork and The Bottom Line: Groups Make a Difference*, Hillsdale, NJ: Lawrence Erlbaum Associates.

ROSENBAUM, L. & ROSEBAUM, W. (1971) Morale and productivity consequences of group leadership, stress and type of task, *Journal of Applied Psychology*, **55**(4), 343–8.

RUBINSTEIN, T. & MASON, J. F. (1979) An analysis of Three Mile Island, *IEEE Spectrum*, November, 37–57.

SABRI, Z. A. & HUSSEINY, A. A. (1978) *Operator Error Summaries*, ERISAR-78007, Iowa State: Nuclear Safety Research Group.

SAINT JEAN, T., GHERTMAN, F., FABER, H. & MALINE, J. (1986) Inventory of stress situations in nuclear power plants, *Proceedings of ANS/ENS International Topical Meeting on 'Advances in Human Factors in Nuclear Power Systems'*, Knoxville, Tenn.

SANDLER, I. & LAKEY, B. (1982) Locus of control as a stress moderator: the role of control perceptions and social support, *American Journal of Community Psychology*, **10**, 65–79.

SEBILLOTTE, S. (1988) Hierarchical planning as a method for task analysis: the example of office task analysis, *Behaviour and Information Technology*, **7**, 275–93.

SEMINARA, J., GONZALEZ, W. & PARSONS, S. (1977) *Human factors review of nuclear power plant control room design*, EPRI NP-309. Palo Alto: Electric Power Research Institute.

SMITH, M. J., MACKIE, R. R. & WYLIE, C. D. (1985) Stress and sonar operations: concerns and research methodology, *Proceedings, Human Factors Society, 29th Annual Meeting*.

SPECTOR, P. E. (1982) Behaviour in organizations as a function of employees' locus of control, *Psychological Bulletin*, **91**, 482–97.

(1987) Interactive effects of perceived control and job stressors on affective reactions and health outcomes for clerical workers, *Work and Stress*, **1**, 155–62.

STEIN, A. A. (1976) Conflict and cohesion: a review of the literature, *Journal Conflict Resolution*, **20**, 143–72.

STONER, J. A. F. (1961) *A Comparison of Group and Individual Decisions Involving Risk*, Unpublished Masters thesis, Massachusetts Institute of Technology, Cambridge, MA.

TORRANCE, E. P. (1954) The behaviour of small groups under the stress conditions of survival, *American Sociological Review*, **19**, 751–5.

USDANSKY, G. & CHAPMAN, L. J. (1960) Schizophrenic like responses in normal subjects under time pressure, *Journal of Abnormal and Social Psychology*, **60**, 143–6.

WARR, P. B. (1990) Decision latitude, job demands and employee well-being, *Work and Stress*, **4**, 285–94.

WASH-400 (1975) *Reactor Safety Study: An Assessment of Accident Risks in US Community Nuclear Power Plants*, NUREG 75/014, October.

WILLIAMS, L. J. (1982) Cognitive load and the functional field of view, *Human Factors*, **24**, 683–92.

(1985) Tunnel vision induced by a foveal load manipulation, *Human Factors*, **27**, 221–7.

PART FOUR

Organisational Issues for Safety

CHAPTER FOURTEEN

Human reliability assessment in the UK nuclear power and reprocessing industries

BARRY KIRWAN

School of Manufacturing and Mechanical Engineering, University of Birmingham

14.1 Introduction

14.1.1 Background to the Review

The UK has a mature and moderately sized nuclear power programme, and a relatively large fuel reprocessing capability. Although the UK has had no nuclear power plant accidents to date, and only one major reprocessing accident (the Windscale fire in 1957: Breach, 1978, p. 91), as with other nuclear industries around the world, significant resources are spent on carrying out formal and highly technical risk assessments of all nuclear-related stations. Such assessments occur both during the design stages, and periodically throughout the installations' lifetimes, until the plants are fully decommissioned, dismantled, and removed.

Part of the decision-making process which decides whether these stations are acceptably safe to build and operate relies on the assessment of the risk of such installations. Risk from all manner of events and failures (hardware, software, human and environmental events or failures) is assessed via the approach of probabilistic safety assessment (PSA: see Cox and Tait, 1991). As the engineering design of these complex systems has improved dramatically over the last three decades of commercial nuclear power generation, so the human element has come to be a more significant risk factor in the nuclear risk equation. This is not only because humans can cause or aggravate developing problems, but also because the human operators are currently a major recovery mechanism available when dangerous events occur (e.g. seismic events; pipe breaks and leaks, etc.). The reliability of the human operators therefore warrants significant scrutiny when determining risk for nuclear power plants and reprocessing plants.

The assessment of the human error contribution to risk is carried out via the generalised approach called human reliability assessment (HRA). Human reliability assessment dates back to the early 1960s, but really came of age, in the sense of becoming seriously used world-wide, following the Three Mile Island incident in 1979. HRA currently can be seen as doing three things: identifying what the critical human involvements in a system are (and how they can fail); quantifying the probability of success (and failure: called human error probability or HEP) of these involvements; and determining how to improve human reliability if performance needs improvement (i.e. because PSA-calculated risk levels are too high, and if human performance or error is contributing significantly to those high risk levels). This chapter therefore reviews how these HRA functions are achieved and implemented in actual HRA practice in commercial UK NP&R systems. This in turn enables an evaluation of HRA practice and the identification of research and development needs to support such HRA work.

Whilst this chapter introduces and outlines the HRA framework, the chapter's detailed nature necessarily assumes some knowledge of HRA by the reader. However, the reader not well-versed in HRA will still gain insights into the state-of-the-art of UK NP&R HRA, though some of the detailed arguments may be a little opaque to such readers. This is unavoidable without greatly expanding the chapter. For the reader wishing more information on HRA, or for an introduction to the basic concepts and techniques of HRA see Kirwan (1990a). Major texts on practical HRA include the following: Swain and Guttmann (1983); Dhillon (1986); Park (1987); Kirwan *et al.* (1988); Dougherty and Fragola (1988); Swain (1989); Gertman and Blackman (1994); Embrey *et al* (1994); and Kirwan (1994). For more theoretical aspects of human error in NP&R-type situations see Senders and Moray (1991); Rasmussen *et al.* (1987); Reason (1990); Hollnagel (1993); and Woods *et al.* (1994).

14.1.2 Scope of the Chapter

This chapter attempts to outline current practices and issues in UK NP&R HRA. The introductory sections outline firstly the HRA process, which gives a framework within which to consider the various developments and issues. Then, the UK NP&R industries themselves are briefly sketched, defining the types of reactors and the major reprocessing installations. The major assessments that have been carried out or finalised in the past five years in UK NP&R industries are then outlined. The remainder of the chapter then considers each stage of the generic HRA process as carried out in these assessments, showing the state of the art of UK NP&R HRA practice, noting any trends and insights, and short-term research and development needs. The final section summarises the status of HRA in UK NP&R, and the main HRA issues requiring resolution.

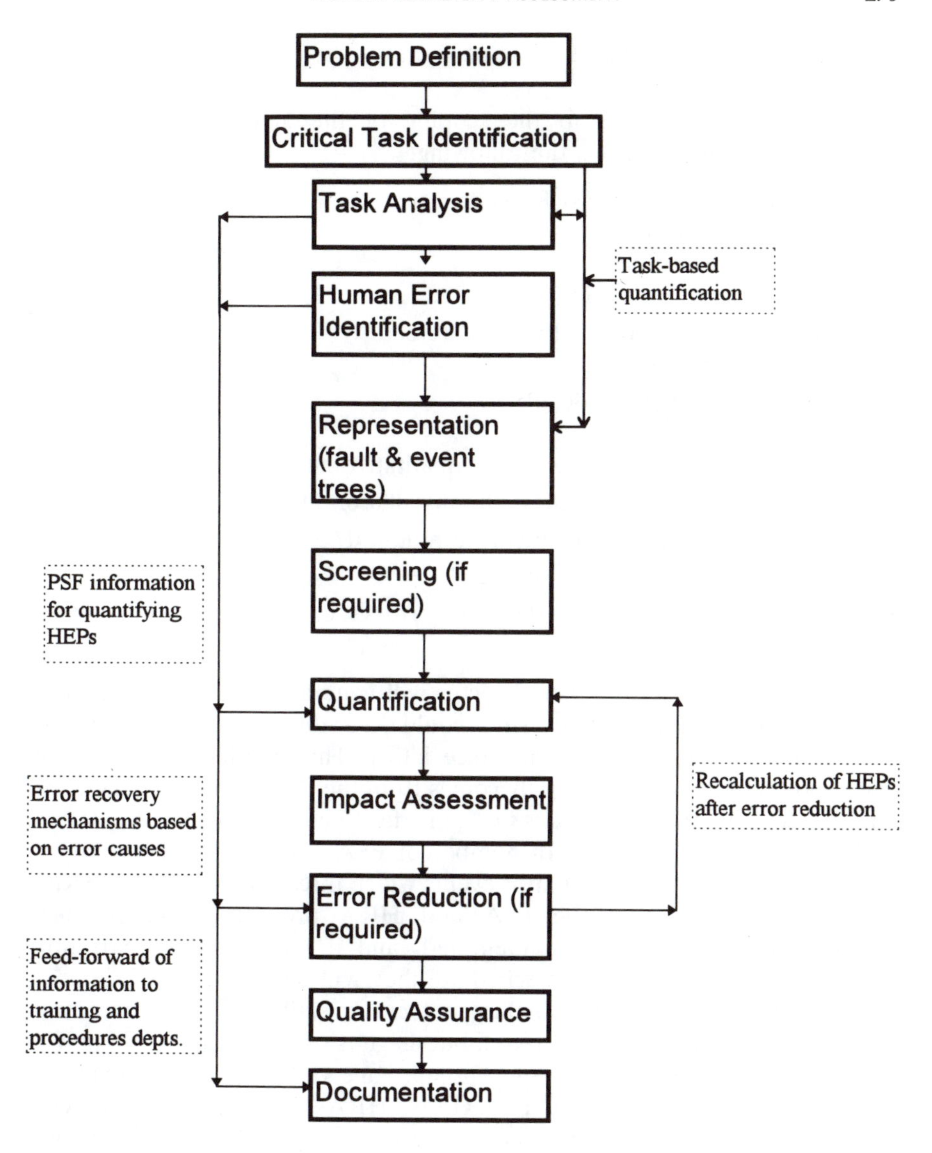

Figure 14.1 The HRA process (adapted from Kirwan, 1990a).

14.2 The HRA Process

The overall HRA process is shown in Figure 14.1 (adapted from Kirwan, 1990a). The principal components of this process are briefly outlined below.

14.2.1 Problem Definition

Problem definition refers to the scoping of the analysis, in particular answering questions such as the following:

- What tasks will the HRA focus on?
  - — emergency response;
  - — normal operations;
  - — maintenance operations.
- What types of errors are to be addressed?
  - — skill and rule-based;
  - — knowledge-based (diagnosis);
  - — rule violations.
- Is quantification required, or will qualitative analysis suffice?
- What stage of the life cycle is the installation at?
- Will error reduction assessment be required?
- What resources should be applied?
- Will the HRA identify detailed errors, or simply calculate task success/failure rates?

HRAs in the NP&R industrial sector will generally focus on emergency response, with some operations which could occur during normal operations, but maintenance treatment is rare (see later). The identification of which tasks should be analysed generally comes from the PSA preliminary systems analysis, wherein the major faults that can affect the system (called the system fault schedule) are compiled (a number of PSA and hazard identification methods are used to identify major events which threaten the safety integrity of the installation). As the PSA and HRA proceed, however, new human-related events may be identified, and if these are not currently 'bounded' by the existing fault schedule, they will be added to it.

NP&R HRAs will also usually largely focus on skill and rule-based errors or tasks (see Rasmussen *et al.*, 1981), but diagnosis and decision-making will be included in the PSA where appropriate. Rule violations may or may not be included by the individual PSA/HRA. HRAs in the UK commercial NP&R field are almost always quantitative since they feed into PSAs whose results are compared against quantitative risk criteria (see Kirwan, 1994; Cox and Tait, 1991). The life cycle stage of the system particularly influences the way in which data can be collected, but in practice has less effect on the actual choosing of techniques for the analysis. NP&R installations are assessed at numerous life-cycle stages until decommissioned and no longer hazardous.

Error reduction is increasingly being required as a result of HRAs/PSAs in the NP&R industries, and this in fact causes the HRAs to require more resources than would otherwise be the case. In general, the more novel and hazardous an installation, and the more human-dependent it is seen to be, the more resources will be allocated to the HRA.

Technically, PSA only requires task failure probabilities from HRA, i.e. that is all that is needed to calculate risk. In practice, though, there are certain benefits from modelling the actual errors that lead to task failure, particularly if error reduction is likely to be a requirement. However, detailed error modelling again will require more resources (see Kirwan, 1992a, b for a discussion of task-based versus error-based HRA).

The above outlines some of the major scoping or problem definition issues that are addressed in NP&R industry HRAs. Once the HRA has been scoped, the next phase is to analyse the tasks identified for assessment.

14.2.2 Task Analysis

Once the scenario being addressed is defined (e.g. dealing with a nuclear power plant accident sequence, such as a loss of coolant accident) task analysis will be used to define what human actions should occur in such an event, and what equipment and other 'interfaces' the operators should use. The task analysis will also (implicitly or explicitly) identify what training (skills and knowledge) and procedures (operating instructions) the operators should call upon during task execution. The task analysis phase of the study is critical in bounding the behaviours of interest in the study, whether concerned with maintenance, monitoring of process, carrying out control actions, or diagnosing a fault situation, etc. The task analysis is used to structure the operator tasks for further analysis, much as piping and instrument diagrams (P&IDs) and engineering flow diagrams (EFDs) are used by engineers to define the various states and operations of the process when designing and building a plant. Without some form of task analysis, the tasks under investigation will be vague and further HRA may not be reliable.

A large number of task analysis techniques exist (see Kirwan and Ainsworth, 1992), but certain task analysis approaches tend to be used in the NP&R industries, as follows:

- *hierarchical task analysis (HTA)* used to define the task goal and its sub-tasks, and operations, the task sequence, and the overall task structure;
- *tabular task analysis (TTA)* also called task decomposition – used to assess the degree to which the interface supports the operator during task execution.
- *timeline analysis (TLA)* used to define temporal characteristics of the task, and to determine if there are sufficient staff (via a basic workload measure), and to give an estimate of how long it will take to complete the task;
- *decision action diagrams (DADs)* a flowchart format used to map out the decision and diagnosis pathways and criteria during events.

The above are called task representation methods, and they allow analysis of the tasks, and are the bases for determining what can go wrong, and what factors in the task context will influence human reliability. In order to carry out such task analyses, other *data collection* task analysis methods must first be applied, such as incident review, observation, interviews, etc. (see Kirwan, 1994). The data collection task analysis techniques are strongly affected by the life cycle stage of the plant, since data collection is always far easier with an existing installation rather than one at an early design stage.

14.2.3 Human Error Analysis

Once task analysis has been completed, human error identification may then be applied to consider what can go wrong. At the least, this error identification process will consider the following types of error (adapted from Swain and Guttmann, 1983):

- *omission error* failing to carry out a required act;
- *action error* failing to carry out a required act adequately;
  - — act performed without required precision, or with too much/little force;
  - — act performed at wrong time;
  - — acts performed in wrong sequence;
- *extraneous act* unrequired act performed instead of or in addition to required act (also called Error of Commission, or EOC).

The human error identification phase can identify many errors, and there are many techniques available for identifying errors (see Kirwan, 1992a, b, 1995). Not all of these errors will be important for the study, as can be determined by reviewing their consequences on the system's performance. The ones which can contribute to a degraded system state, alone or in conjunction with other hardware/software failures and/or environmental events, must next be represented in the PSA risk analysis framework.

As noted under problem definition, however, some NP&R HRAs will not always model errors explicitly, instead simply modelling the human contribution to risk at the task-based level. For example, if the task involves manipulating a series of a dozen values, errors can be considered for each individual manipulation (error-based approach), or a single task failure can be considered (task-level approach).

14.2.4 Representation

Having defined what the operators should do (via task analysis) and what can go wrong (at a detailed error level or simply at the overall task execution level), the next step is to represent this information in a form in which quantitative evaluation of human error impact on the system, in conjunction with other failure events, can take place. It is usual that the human error impact be seen in the context of other potential contributions to system risk, such as hardware and software failures, and environmental events. This enables total risk to be calculated from all sources, and enables interactions between different sources of failure to be assessed. It also means that when risk is calculated, human error will be seen in its proper perspective, as one component factor affecting risk. Sometimes human error will be found to dominate risk, and sometimes it will have less importance than other failure types (e.g. hardware or equipment failures).

UK NP&R PSA typically uses logic trees, called fault and event trees to determine risk, and human errors and recoveries are usually embedded within such logical frameworks (see Henley and Kumamoto, 1981; Green, 1983; Cox and Tait, 1991; Kirwan, 1994). Fault trees look at how various failures and combinations of failures can lead to a 'top event' of interest (e.g. loss of core cooling). Event trees are typically more sequential, and look at how an event proceeds towards accidental circumstances (e.g. loss of cooling can eventually lead to core melt if certain hardware and human functions fail). Both fault trees and event trees can be very large and are usually evaluated using computerised methods.

A further representation issue is that of dependence between two or more errors or tasks, where for example, failure on one task will increase the likelihood of failure on a subsequent task (e.g. misdiagnosis may affect the successful outcome of a number of subsequent tasks). Dependence needs to be addressed whether quantifying at the task level or at the error-based level.

14.2.5 Human Error Quantification

Once the human error potential has been represented, the next step is to quantify the likelihood of the errors to determine the overall effect of human error on system safety or reliability. As much of applied and theoretical HRA has to date focussed on the quantification of human error potential, this area is somewhat more mature than some of the other stages of the HRA process (e.g. error identification).

Human reliability quantification techniques all quantify the human error probability (HEP), which is the metric of human reliability assessment. The HEP is defined as:

$$\text{HEP} = \frac{\text{number of errors occurred}}{\text{number of opportunities for error to occur}}$$

Thus, if when buying a cup of coffee from a vending machine, and on average one time in a hundred tea is accidentally purchased, the HEP is taken as 0.01 (it is somewhat educational to try and identify HEPs in everyday life with a value of less than once in a thousand opportunities, or even as low as once in ten thousand). In an ideal world, there would be many studies and experiments in which HEPs were recorded (e.g. operator fails to fully close a valve once every five thousand times this is required). In reality there are very few such recorded data. The ideal source of of human error 'data' would be from industrial studies of performance and accidents, but at least three reasons can be deduced for the lack of such data.

- difficulties in estimating the number of opportunities for error in realistically complex tasks (the so-called denominator problem);
- confidentiality and unwillingness to publish data on poor performance;
- lack of awareness of why it would be useful to collect in the first place (and hence lack of financial incentive for such data collection).

There are other potential reasons (see Kirwan *et al.*, 1990) but the net result in an extreme scarcity of HEP data. HRA therefore uses quantification techniques, which rely on either expert judgement or a mixture of data and psychologically based models which evaluate the effects of major influences on human performance.

Below are listed the major techniques in existence in the field of human reliability quantification, arguably the most developed field within Human Reliability Assessment today. These are categorised below into four classes, depending on their data sources, and mode of operation.

1 unstructured expert opinion techniques:
   (a) Absolute Probability Judgement (APJ: Seaver and Stillwell, 1983);
   (b) Paired Comparisons (PC: Hunns, 1982);
2 data-driven techniques:
   (a) Human Error Assessment and Reduction Technique (HEART: Williams, 1986; 1988a; 1992);
   (b) Technique for Human Error Rate Prediction (THERP: Swain and Guttmann, 1983);
   (c) Human Reliability Management System (HRMS: Kirwan and James, 1989; Kirwan, 1990b, 1994);
   (d) Justification of Human Error Data Information (JHEDI: Kirwan 1990b, 1994);
3 structured expert opinion techniques:
   (a) Success Likelihood Index Method using Multi Attribute Utility Decomposition (SLIM-MAUD: Embrey *et al.*, 1984);
   (b) Socio-Technical Approach to Assessing Human Reliability (STAHR: Phillips *et al.*, 1983);
4 accident sequence data driven techniques:
   (a) Human Cognitive Reliability (HCR: Hannaman *et al.*, 1984);
   (b) Accident Sequence Evaluation Program (ASEP: Swain, 1987).

All of these techniques generate human error probabilities. Swain (1989) and Kirwan *et al.* (1988) discuss the relative advantages and disadvantages of these techniques, and Kirwan (1988) gives the results of a comparative evaluation of five of them. Kirwan *et al.* (1988) and Kirwan (1994) also give selection guidance to help practitioners decide which one(s) to use for a particular assessment problem.

To give an indication of how data-driven techniques work (e.g. (2) above) the THERP technique uses a database of 'nominal' human error probabilities, e.g. failure to respond to a single annunciator alarm. Performance shaping factors (PSF) such as the quality of the interface design (e.g. whether alarms are prioritised, adequately colour-coded, near to the operator and in the normal viewing range, etc.), or time pressure, are then considered with respect to this error. If such factors are indeed evident in the scenario under investigation, then the nominal human error probability may be modified by the assessor (e.g. in this case increased to reflect poor quality of interface) by using an 'error factor' (EF) of, say, 10. Thus, if an initial nominal HEP is 10^{-3} (or 0.001), an EF of 10 can be used to increase the actual estimated HEP to a value of 10^{-2} (i.e. 0.01).

Expert judgement techniques, ((1) and (4) above) on the other hand, use personnel with relevant operational experience (e.g. more than ten years) to estimate HEPs, on the grounds that such personnel will have had significant opportunities for error and will have also committed certain errors (and seen others commit errors) and hence have information in their memories which can be used to generate HEPs. Such expert opinion methods may either ask 'experts' directly for such estimates, or may use more subtle and indirect methods, to avoid the various biases associated with human recall, which can occur (see Tversky and Kahneman, 1974).

14.2.6 Impact Assessment, Error Reduction Assessment, Quality Assurance and Documentation

Following such quantifications, and aggregation of these probabilities and those for other types of failures (hardware, software, environmental), that are included in the evaluation, risk will be calculated. It will then be determinable whether the installation's overall risk is acceptable. If not, it may be necessary or desirable to try to reduce the level of risk. If human error is contributing significantly to risk levels, then it will be desirable to attempt to find ways of reducing the human error impact. This may or may not involve the quantification technique in devising error reduction mechanisms (ERMs) which will reduce the system vulnerability to human error. If ERMs are derived, then the quantification technique will then be used again to re-calculate the HEPs, for the system as it would perform with ERMs in place. Following this stage (which can run through several iterations), the results will be documented and quality assurance systems should ensure that ERMs are effectively implemented, and that assumptions made during the analysis remain valid throughout the lifetime of the system (see Kirwan, 1994). This completes the HRA process. The next section gives a brief overview of the NP&R industries in the UK.

14.2.7 Background to the UK NP&R Industries

The UK Nuclear Power and Reprocessing (NP&R) industries fall firstly into two main categories, nuclear power utilities (Nuclear Electric and Scottish Nuclear) and the reprocessing industry (British Nuclear Fuels (BNFL)), although BNFL do possess a small number of their own power generation stations. The nuclear power generating stations are of three basic designs. The first and the oldest is the magnesium oxide cladded fuel (called Magnox) reactor design. The Magnox stations are one of the oldest designs world-wide, and use carbon dioxide gas to cool the reactor core. The advanced gas-cooled reactor (called the AGR), is an advance on the original Magnox design, but still uses carbon dioxide as the primary coolant. Both the Magnox and the AGR are unique to the UK. The third design, of which there is currently only one reactor in the UK, is based upon the (US) Westinghouse design, modified to meet UK requirements, and is a 1300 MW(e) single reactor, twin turbine pressurised water reactor (PWR). This uses water as the primary coolant, and falls under the generic category of the light water reactor (LWR). LWRs are the most prevalent design of nuclear power generating reactors world-wide, whether on land, or in use in nuclear powered submarines. The only existing British PWR is Sizewell B, recently operational (1995).

The reprocessing industry takes fuel which has been in a reactor (so-called 'spent fuel') and reprocesses it into reusable products (most of the fuel can be re-used once reprocessed as future fuel in reactors), and waste products (low, intermediate, and highly active waste products). Waste is treated and then stored for very long periods. Currently the highly active liquid waste is being vitrified into a glass-type (solid) form, so as to make it safer for long-term storage. The reprocessing takes place at (BNFL) Sellafield in Cumbria. Capenhurst (BNFL) prepares fuel compounds, and Springfields (BNFL) synthesises fuel elements for re-use in reactors, thus completing what is called the fuel cycle. The key element in the reprocessing cycle is the actual processing of the fuel elements once they have cooled down in cooling ponds. This used to take place in a plant called B205 at Sellafield, but now will largely take place in the thermal oxide reprocessing plant (THORP), at Sellafield, also recently operational (1995).

14.2.8 HRA Activities in Relation to UK NP&R Installations

The NP&R industry is regulated by Her Majesty's Nuclear Installations Inspectorate (HMNII, or more usually simply NII), which is itself part of the

UK Health & Safety Executive (HSE). Each station is granted a nuclear site licence by the NII, and as part of the licensing requirements, all stations require periodic risk assessments, as well as other audits of various types. These risk assessments currently must include assessment of the human contribution to risk, utilising human reliability assessment (HRA). In some cases the NII will also stipulate a requirement for formal human factors assessments, which may be linked to the HRA or may be an independent formal submission as part of the general safety case for the station. Each station generally has its own NII inspector or team of inspectors, responsible for checking the adequacy of the safety case submission. The NII also issue guidance to the nuclear utilities on safety assessment principles (SAPs) for industry, on how plants should be assessed. These SAPs currently include sections on ergonomics, task analysis, interface design, training and procedures, safety management systems, and human factors in PSAs (Whitfield, 1991, 1995).

Whilst there have been a number of individual periodic reviews of individual stations, several major HRA exercises have been recently carried out or finalised, which together give a representative picture of HRA in the UK NP&R industries:

- HRA for THORP – design stage assessment;
- HRA for Sizewell B – design stage assessment;
- continued operation PSAs (and HRAs) for all Magnox and AGR stations – existing station assessments.

These are briefly outlined below in terms of their content.

14.2.8.1 *HRA for THORP*

THORP is a very large plant, sometimes described as a number of conventionally sized plant modules joined together under one roof, and has a staff complement approaching 800 personnel. It has process complexity similar to that of a nuclear power plant. The HRA approach started in earnest in the detailed design stage for THORP, and was predicated upon a large human factors assessment exercise (Kirwan, 1988, 1989), amounting to approximately 15 person-years of effort. This assessment addressed the safety adequacy (from a human factors perspective) of the central control room, local control rooms, control and instrumentation panels local on plant, staffing and organisation issues, training, and emergency preparedness. The human factors reviews were used to determine the effect of performance-shaping factors on performance (e.g. the adequacy of the interface in the central control room during emergency conditions). Such information significantly influenced the two computerised quantification systems developed for the THORP HRA, namely the Human Reliability Management System (HRMS: Kirwan and James, 1989; Kirwan, 1990b) and the

Justification of Human Error Data Information system (JHEDI: Kirwan, 1990b).

JHEDI was designed to rapidly but conservatively assess all identified human involvements in the THORP risk assessment, and assessed over 800 errors. JHEDI requires a simplified task analysis, error identification via keyword prompting, quantification according to PSF, and noting of training or procedural implications to be fed forward to the respective THORP operations departments.

HRMS is a more intensive system, and is used for those errors that are found to be risk significant in the PSA (i.e. for the errors that have a potentially major risk impact on the THORP system: see Kirwan, 1990b). HRMS requires detailed task analysis, error identification and quantification, and computer-supported error reduction is also carried out based on the PSF assessment. Approximately twenty tasks were the subject of the more detailed HRMS assessment approach.

All assessments have been rigorously documented, and information arising out of the assessments is fed forward to the operational departments that are now running THORP and assessing its safety performance. JHEDI is still being applied to other plant designs at BNFL Risley, the design centre for BNFL.

14.2.8.2 *Sizewell B HRA*

The Sizewell B PSA and HRA (Whitworth, 1987; Whitfield, 1995) also were to an extent predicated upon extensive human factors assessment of the design of the interface and other systems, although the linkage between the human factors assessments and the HRA inputs was less formalised than for THORP. The HRA approach involved a very large amount of initial task analysis and error analysis. The human error assessment and reduction technique (HEART: Williams, 1986) was used as the main quantification tool, supplemented by the technique for human error rate prediction (THERP: Swain and Guttmann, 1983), and error reduction was carried out as required.

14.2.8.3 *Continued Operation PSAs/HRAs*

These PSAs and associated HRAs are part of a required programme of work to determine whether the ageing Magnox and AGR plants are safe to continue operating beyond their original sanctioned lifetimes (e.g. whether the Magnox stations can operate beyond 30 years). So far two Magnox plants have been shut down (due probably more to economic reasons than safety concerns), and the results of the continued operation of HRAs for the other stations are still being reviewed by the NII.

These HRAs have used a significant amount of task analysis, and have generally each followed a similar basic format: detailed task analysis for a small number of key scenarios, and less detailed analysis of the remainder of the scenarios. No error analysis is utilised, and task failure likelihood is calculated by using the HEART quantification method. Error reduction measures are identified either based on the HEART calculations or based on the task analysis. An example of the methodology from one of the Continued Operation HRAs is given in Kirwan *et al.*, (1995b).

Having defined the HRA process, outlined the UK NP&R industries and the major assessments carried out in the recent past, the next section considers UK practice within the framework of each individual stage of the HRA process.

14.3 Review of UK NP&R HRA Practice

14.3.1 Problem Definition

Problem definition broadly defines what is to be included in an HRA and what is not. It therefore scopes the HRA. Below are summarised generally (there will be exceptions) what tends to be included and excluded in the HRA:

- *Included in HRAs*:
  - — emergency response actions, actions during normal (pre-trip) operations;
  - — skill and rule based behaviour errors;
  - — mistakes (cognitive errors) – not necessarily quantified;
  - — some rule violations – not necessarily quantified;
  - — errors of commission – if identified, via informal methods – not quantified usually;
  - — recovery actions;
- *Excluded from HRAs*:
  - — maintenance actions;
  - — safety culture or organisational impacts on risk;
  - — socio-technical errors (EOCs due largely to emotional or affective factors – see Kirwan, 1993, 1994);
  - — specific errors related to software reliability.

Most assessments focus on required actions during developing accident sequences (so-called post-trip or post-fault scenarios), and most of the assessments are assessing operator performance in achieving an emergency recovery function. Far less frequently, rule violations, diagnostic errors, and errors of commission (unrequired and undesirable acts) are identified, though not always quantified (discussed later). Maintenance error impact tends to be assumed to be implicitly already in the PSA via using composite hardware failure data which implicitly includes maintenance errors already. However, such an assumption does not account for maintenance errors contributing to new failure paths, e.g. by an error on one sub-system affecting other normally unrelated sub-systems. Arguably incident frequency for initiating events (i.e. events which start off an accident sequence), based on historical records, will also already account for human errors. Such an argument is more easily defended in the case of existing plants with 30 years operational experience behind them than for novel plants being designed for the first time.

Safety culture and safety management impacts on risk (see Reason, 1990; Kirwan, 1994; Kennedy and Kirwan, 1995) are not generally modelled quantitatively in PSA or HRA at this stage in HRA's development (see however, Apostolakis *et al.*, 1992). Instead, qualitative audits are usually carried out to determine potential vulnerabilities from safety, cultural or organisational sources.

Since it is difficult to predict socio-technical errors (e.g. a communication failure of safety information due to two persons not being on speaking terms due to a disagreement), these are not generally identified in any HRAs. Errors when programming safety-related software are not assessed, instead software reliability is calculated using other software-reliability metrics.

Problem definition therefore generally focuses on the more straightforward issues in which the operators must perform critical actions in rare-event scenarios. More exotic error forms are less prevalent in safety cases.

14.3.2 Task Analysis

As task analysis has become increasingly important in NP&R HRA over the past five years, and as it represents the major qualitative input to HRA, a significant treatment is given of this part of the HRA process in this chapter.

14.3.2.1 Continued Operation HRA

The task analysis approaches generally fall into two categories, those associated with data collection, and those associated with the representation and analysis of operator performance as a function time, staffing and organisation, and interface adequacy. The overall HRA approach for one of the recent continued operation PSAs is shown in Figure 14.2 (Kirwan *et al.*, 1995b). The first third of this figure is concerned with data collection and task analysis approaches, and similar approaches have been used by the other Continued Operation PSAs. See also Umbers and Reiersen (1995) for a case study from a recent task analysis programme related to an existing Magnox station.

Data Collection Task Analysis Approaches The plant information gathering phase began with a number of approaches. Firstly documentation was reviewed in terms of operational procedures, training material, previous PSAs for other plants (in particular the PSA recently carried out for a similar Magnox plant), the previous (less detailed, and with little HRA content) PSA for the plant being assessed, and documentation of the NPP system itself.

Operational experience was reviewed both formally by accessing all previous incident data on the plant, and by discussions with operational and safety-responsible staff. Plant visits were held, to become familiar with the layout and general adequacy of ergonomics of the interfaces, both in the central control room (CCR), and in local plant locations. During visits a number of walk-throughs were run to help the ergonomics assessors more easily understand the operational details of the power production process and how it worked in practice. A number of interviews were also held concerning what tasks were believed by the operators to be both of risk concern, and particularly error-likely. It should be stressed that in practice the data collection stretched over a large part of the project, so that although 80% of data were collected during the first three months (also in connection with the task analyses), new data arose as the project progressed.

Task Analysis Representation Methods A global hierarchical task analysis (HTA: see Shepherd, 1989) was carried out for the plant, with the basic top goal of 'run plant', and this amounted to a top level description of normal, start-up and shutdown, and emergency operations and goal structures. This was followed by the detailed analysis of four key fault scenarios. These were chosen from a review of safety documentation, including the results from earlier PSAs at Magnox nuclear power plants (NPPs), as representative of operator actions required in the most demanding of the bounding scenarios identified in the PSA fault schedule. The scenarios were concerned with post-trip activities and the effects of degradation in the ability to achieve adequate post-trip cooling. In particular the effects of the following were analysed: loss of grid; loss of secondary cooling; loss of primary cooling; and loss of primary coolant. The failure of post-trip cooling places particularly severe demands on operations staff in terms of workload and difficulty of the required actions. Therefore, by choosing scenarios with the worst credible combinations of plant failures it was possible to argue that they would be at least representative of the most demanding fault scenarios that operations staff would possibly have to deal with.

The first main task analysis method employed, after carrying out a global HTA, was HTA for the four key scenarios. This involved two analysts interviewing a key member of staff in a conference room adjacent to the CCR. This conference room had all the procedures in it, and its location meant that the two analysts could clarify difficult points by simply going onto plant with the operating engineer to see whatever system or procedural operation was being discussed. The HTAs were verified by two independent CCR operators, and reviewed by the shift team, resulting in various refinements to the HTAs.

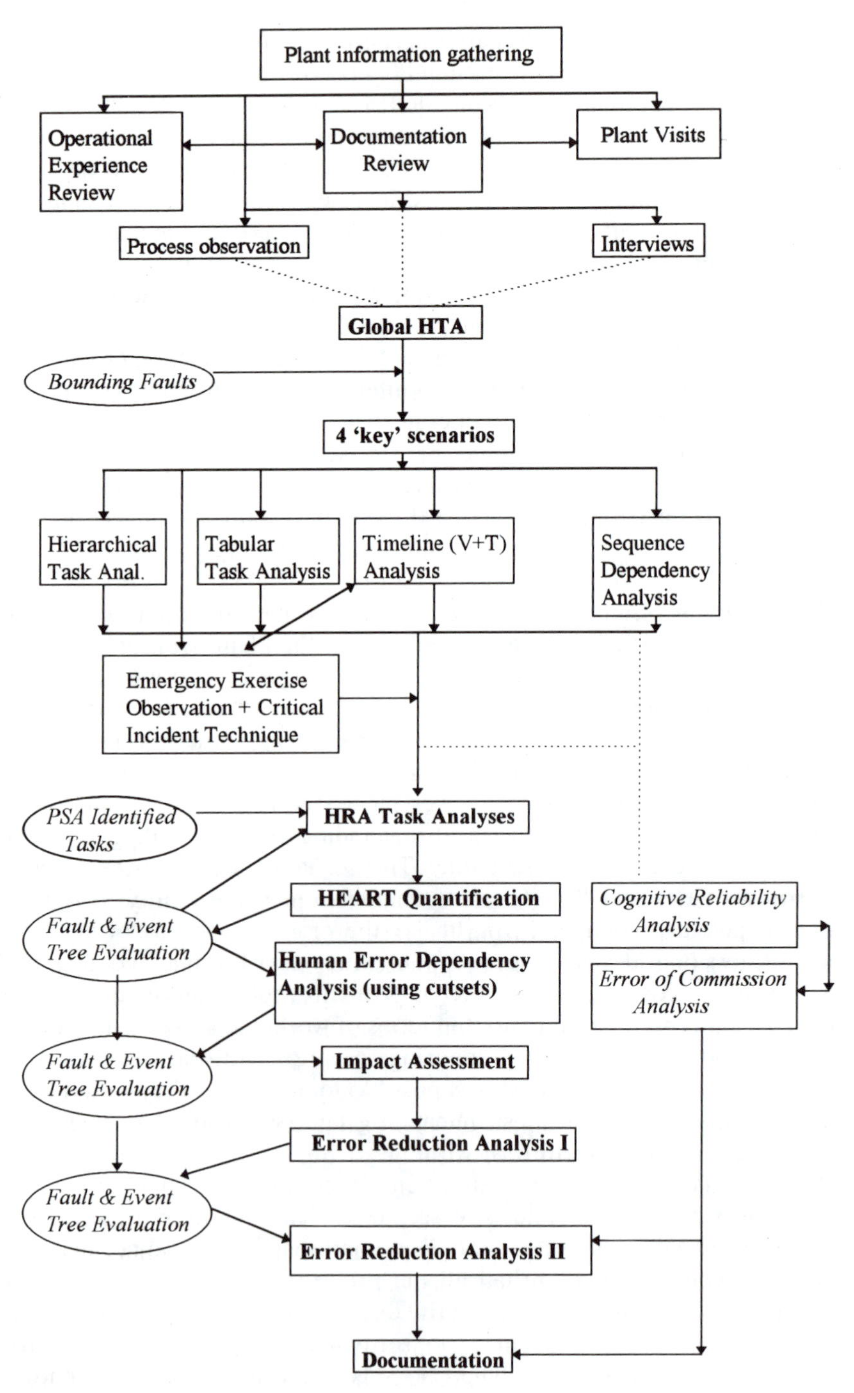

Figure 14.2 Continued operation HRA overview from a Magnox HRA (Kirwan *et al.*, 1995b).

The next stage was tabular task analyses (TTA) for the four scenarios. These were followed by two timeline analyses (TLAs), horizontal (to estimate the overall time to complete essential actions), and vertical (to model who would be doing what in each of the dynamic scenarios). These analyses resulted in a paper simulation of what the operators would try to do, what controls, displays and procedures they would use, when they would be reliant upon oral communication and co-ordination between different and disparate team members, and the detailed sequence of events. With respect to the sequence of operations, a simple sequence-dependency analysis was carried out. This entailed determining which tasks were critical for which other tasks (i.e. certain tasks could not be carried out before certain other tasks had been correctly executed), and led to the determination of critical tasks which could act as bottlenecks in the task sequence.

In order to help estimate some of the timings in the event of loss of coolant, a high-realism simulation exercise was observed by three members of the analysis team at the plant site. In this emergency exercise, forty members of plant staff took part in a simulated emergency (they did not know which emergency would be simulated), with a complete CCR shift occupying a simulated control room. The exercise was approximately three hours long, and was highly useful in gaining more realistic estimates of task durations. The number of communication activities had been underestimated, and although these were not critical to the performance of essential operator actions, they would represent a considerable secondary task during the early phase of the accident sequence. The critical incident technique was also employed related to a particular scenario, and again produced some highly useful information that had not been evident before. At this stage, approximately 9 months into the project, the first report was produced on the detailed task analyses. Already a number of preliminary error reduction recommendations had arisen from the task analyses, and in one scenario it was apparent that an essential operator action might not be possible within the timescales assumed by the PSA, if the plant was at the time being operated in a particular configuration. The NPP management decided not to operate the plant in the configuration that could leave the potential for such an (albeit extremely rare and unlikely) incident. The task analyses alone therefore already had an impact on plant safety and plant risk, without in this instance requiring error identification or quantification.

Less Detailed Task Analyses By this stage the PSA team had progressed to the full identification of tasks to be quantified in the PSA, many of which had come within the scope of the four key fault scenarios already analysed. This amounted to approximately 40 tasks to be included in the PSA fault trees and event trees, and the tasks were mostly (but not exclusively) post-trip. Each of these was then analysed in terms of how it would be performed, using a task analysis format specifically developed to support HRA (called simply HRA task analysis). An example of the format is shown in Figure 14.3. The information is judged to be the minimum required for conservative estimation of the task or error reliability (see also Kirwan, 1994).

Observations on Applied Task Analysis for HRA of Existing Installations Certain observations may be made concerning the task analysis phase for an existing installation. The ability to view actual equipment in its working environment was the only way to properly evaluate ergonomics aspects of operations: to see the equipment under its normal and emergency-lit state; to appreciate the thermal and noise aspects of the various rooms; and to fully appreciate the effectiveness of labelling, problems of reach or physical or visual access; etc. Use of walk-throughs was fundamental in identifying certain operator actions that would take some time, whereas if such equipment had been reviewed on paper, probably less conservative time estimates would have been produced. The walk-through during an emergency exercise was especially useful in giving reasonably robust time estimates for a number of emergency actions, since the operating crews are under some pressure in such exercises, and did not know (initially) the nature of the event in the exercise (the operators did successfully handle the simulated emergency event).

The critical incident technique (CIT) was applied formally with respect to one particular incident (loss of off-site power) and the operating team that had been involved. This one analysis was especially useful in adding context to the analysis of performance shaping factors, and gave the assessors more confidence in their assessment of the importance of such factors. The CIT was particularly useful in this instance as official records of the incident were also available for ratification purposes.

Hierarchical task analysis (HTA) differs in its benefits depending upon whether it is being applied to an existing or a design-stage installation. With the latter, there are many important and useful spin-offs, with the HTA formats derived often proving valuable to the operations departments who must generate training and procedural media. With existing installations, the HTA is more indirectly useful than directly useful. It firstly allows the assessors to build a structured understanding of the scenarios and general operation of the plant. Secondly it acts as the formal basis for the assessment of the tasks when the time comes for quantification. Perhaps more importantly it lays the foundation for the tabular and timeline analyses. This indirect rather than direct immediate benefit is not surprising, since the plant being assessed was an existing installation already equipped with procedures and training systems (two of the usual direct benefits of HTA). HTA was therefore necessary, but more specific insights on errors and vulnerabilities arose from the other task analyses.

Tabular task analysis can be directly useful in two main ways. Firstly it enables direct assessment of the ergonomics adequacy of the interface, where errors could occur, and how they could be recovered. This technique proved extensively useful in the THORP assessment (see Reed, 1992). Many of the detailed aspects of the interface and working environment which could lead

Operator Fails to reset pump after low suction trip

Scenario Background

In the event of a loss of main feed scenario, the emergency pump may trip on low suction. This event will require firstly that more feedwater is supplied to prevent further trips, and then that the Emergency Feed Pump (EFP) trip is reset. The pump may then be re-started.

a) Task steps

1. Control Room Supervisor (CRS) detects trip indication on VDU alarm display.
2. Control Room Supervisor instructs Turbine Hall Operator (THO) to reset pump.
3. THO goes to TH to reset pump.
   - 3.1 THO closes valve XV2357 EFP discharge valve.
   - 3.2 THO opens valve TV 6298 treated water supply valve.
   - 3.3 Demineralised water pump is started if not already in service.
   - 3.4 EFP 'Low Suction' trip device is reset in TH basement.
   - 3.5 THO communicates to CRS.
4. CRS restarts EFP from Control Room.

b) Personnel involved are the CRS and THO.

c) Location of the operation is in the control room, the TH, and the TH basement.

d) Duration of the operation is an estimated 6 minutes from detection (TH is adjacent to the control room).

e) Cues: EFP alarm in control room; Low suction pressure indicator light in TH; feed pressure and flow indicators in control room.

f) Procedures: Plant Operating Instruction POI 923A

g) Feedback: Alarm will reset when task completed; pump will start.

h) Potential errors in this task include:

- Go to wrong pump
- Valve sequence error
- Wrong valves operated upon
- Fail to switch pump back on
- Fail to supply additional feed

i) Diagnosis in this task is relatively straightforward.

j) Display and control interface adequacy appears adequate for the task.

Figure 14.3 HRA task analysis format (from Kirwan *et al.*, 1995b).

to task failure or task completion delay were identified during the TTA and supporting walk-through phases. For example, information which cued an operation to start or to stop was not always directly accessible to the operator, and might require access to a nearby location, or verification from a local operator, who would communicate the information back to the main control room. This would be problematical, however, in the event of an emergency requiring evacuation from local plant. Such an event would require the operators to have to infer indications from secondary and more indirect sources, increasing the risk of errors and misdiagnosis.

Secondly, and more fundamentally in terms of the shifting role of HRA towards error reduction, the TTA, walk-throughs, and general ergonomics assessments underpinning the TTA directly provide a high proportion of the error-reduction recommendations. The task analysis phase can therefore be to a real extent more responsible for the error-reduction recommendations than the application of an HEP quantification technique (discussed later), since the latter only considers major PSF at a fairly global level, and in a sense is a simpler or cruder model of the causes of task failure than a detailed task analysis (see Kirwan, 1996).

The timeline analysis can also be highly useful in determining likely human reliability in a number of ways. Firstly the vertical TLA shows when and where tasks could be delayed due to staff resource limitations (the PSA always assumes conservative staff resources, e.g. night shift etc., and due to staff having to evacuate certain areas or having to travel to certain areas on the NPP etc.). Secondly, the vertical TLA highlights where bottlenecks could occur (via the sequence dependency analysis), and was useful in pinpointing the critical-path tasks, i.e. those tasks which if unsuccessful would not allow completion of tasks further in the sequence. This is again important when considering error reduction, since such tasks would obviously be a priority in terms of achieving a high level of ergonomics adequacy.

Thirdly, the overall completion time estimates from the timelines gives a good impression of the general feasibility of the task, and the amount of time pressure that would be on the operators in the scenario, which is again important for the quantification phase (although time pressure is not always a dominant PSF, most HRA techniques use time pressure as a major factor influencing human reliability).

One approach which has not been particularly applied in UK HRAs is to extend the task analysis modelling into a network simulation, using Micro-SAINT (Laughery, 1984), for example. Such modelling, although initially resource-intensive, could yield more sophisticated information and insights about which operations are most time-dependent, or where bottlenecks are likely to be most damaging, etc.

Additionally, the more cognitive aspects of human performance are only superficially dealt with by most of the existing task analysis techniques. In particular, what makes certain tasks difficult is the rapidly evolving nature of the tasks due to compounding events occurring and worsening in the scenarios. What would be beneficial is a task analysis means of mapping the changing goal hierarchy and structures of the operating teams during emergency events, as a function of time and the operating team's perception of what is happening. This is therefore a potential research and development area.

In summary, task analysis in recent continued operation HRAs/PSAs has been extensive. This represents a significant shift in emphasis from purely quantitative analysis, seen in earlier HRAs, to a quantitative analysis firmly founded on significant (qualitative) task analysis using formal techniques.

14.3.2.2 *THORP Task Analysis*

The THORP Human Factors programme (Kirwan, 1989) utilised a large amount of task analysis, including hierarchical task analysis, tabular task and scenario analyses, timeline analysis, and usage of walk-throughs and talk-throughs, checklist evaluations, etc. (see Kirwan, 1992c for detail on the THORP task analysis programme). In parallel, the JHEDI system was utilised to document the basic task sequences for all identified critical tasks, and for very risk-sensitive tasks more comprehensive methods were utilised. The JHEDI task analysis format is a simplified form of HTA (see Kirwan, 1990b, 1994).

The HTAs for THORP, being carried out at the design stage, were in a sense far more useful for the plant, since they could be used as a major input to the development of training and procedural media. Similarly, the tabular and timeline analyses were able to produce recommendations on staffing and interface design which could be incorporated into the evolving design of the plant (see Kirwan, 1992c; Reed, 1992), usually without requiring quantification. One notable difficulty for task analysis at this stage of the life cycle of a plant however, is the much larger resources required. This is because data collection activities are significantly hindered by the unavailability of operators and existing workplaces to observe tasks being carried out, since THORP (like Sizewell B) was a novel plant design. Task analysis at the design stage therefore necessarily is more generative in nature, and sometimes is speculative and even creative.

14.3.2.3 *Sizewell B*

Sizewell B used task analysis extensively, both for the human factors analysis and for the operator reliability work. The task-analysis programme is described in Ainsworth and Pendlebury (1995) primarily from the human factors side, and in Whitworth (1987) more from the operator reliability (HRA) perspective. Fewins *et al.* (1992) also describe a task analysis case study for Sizewell B which, based on reliability considerations, led to the decision to automate a particular function in order to give more support to the operators in an emergency.

The task analysis support of Sizewell B took place in five main stages (Ainsworth and Pendlebury, 1995):

1 early task analysis of critical tasks;
2 more detailed task analysis of safety tasks;

3 preliminary walk-throughs and talk-throughs using a control room mock-up.
4 detailed validation of procedures via walk-throughs using a full-scope simulator;
5 task analysis of tasks outside the main control room.

A combination of approaches was utilised, including FAST (see Williams, 1988b), HTA, task decomposition (Kirwan and Ainsworth, 1992), and timeline analysis.

14.3.3 Human Error Identification

14.3.3.1 *Skill and Rule-based Errors and Tasks*

The Sizewell B and THORP PSAs both utilised detailed error identification approaches for identified accident sequences. THORP utilised the JHEDI system, with a checklist of basic error modes, and for more complex scenarios used the more substantiated HRMS error identification module (see Kirwan, 1994). Such sets of questions are used as a checklist by the assessors at BNFL, who are already familiar with error types, and apply these questions to the steps in their already-derived task analyses. The identified errors are then inserted into fault trees and quantified using JHEDI (see later sections).

The Sizewell B PSA use detailed task analysis, followed by an error identification phase in which basic error types (Whitworth, 1987) were identified for each task or task step. Notably, in the early Sizewell B HRA work reported by Whitworth (1987), attempts were made to also record unrequired acts, i.e. errors of commission (see later). Later, the Sizewell B PSA and HRA approach moved to a task-based rather than error-based approach, as utilised by all the Continued Operation HRAs, as discussed next.

14.3.3.2 *Task-based versus Error-based Modelling of Human Involvements in NP&R Systems*

The current PSAs reviewing the continued operation of the existing stations do not use formal human error identification methods (e.g. such as those defined in Kirwan, 1992a). Formal approaches such as SHERPA (Embrey, 1986) would utilise a bottom-up approach to error identification, i.e. starting from a detailed task analysis of operations, and analysing each step to see what errors would occur. The problem with this approach is that there are hundreds of procedures, and extensive task analysis and error identification would consume very large resources, even though many of the errors would probably themselves be of little ultimate significance in terms of operator risk. The alternative approach, therefore, adopted by the continued operation PSAs, is to use the PSA framework to identify the critical human interactions which are likely to have impact upon risk, and then to analyse these at the task level, rather than at the sub-operation level. The identification of critical operator interventions can be identified by the PSA's review of the safety systems and events which can affect the safety integrity of plant (e.g. coolant leaks; loss of power; seismic events, etc.), and by the operational experience review. Such analyses define those conditions in which the operator will have a critical role in restoring the safety of the plant.

These critical tasks, once identified, could be analysed in depth using task analysis, and then techniques like SHERPA or Human HAZOP (see Kirwan, 1994) could be utilised to identify detailed errors and their consequences. Such errors would then have their probabilities of occurrence estimated using human reliability quantification techniques such as the technique for human error rate prediction (THERP: Swain and Guttmann, 1983). However, an alternative paradigm is that of analysing the task (using task analysis techniques) but then quantifying likelihood of performance failure directly, at the task-based level, without decomposing the task failure into its elemental error probabilities. This approach makes two significant assumptions: firstly that there is a technique that can directly quantify at the task-based level, and secondly that this assessment approach will adequately address the human error contributions from the task in question. The first assumption is met via the availability of human error quantification techniques such as the Human Error Assessment and Reduction Technique (HEART: Williams, 1986) and the Success Likelihood Index Method (SLIM: Embrey *et al.*, 1984). Since HEART has recently been validated, such quantification at the task-based level appears to be currently acceptable (see Kirwan *et al.*, 1995a).

The second assumption is somewhat more involved. Assessing human reliability at the task failure level, there is an inevitable focus on the objectives of the task alone, and the failure of those specific objectives. However, errors within the task procedure which could affect other systems, or other systems' objectives, may be overlooked. For example, during a loss of coolant emergency, the operators might be required to close a number of emergency valves out on plant. The closure of these valves could be assessed at the task-based level, particularly if they are in the same physical location, etc., and if the valves will be closed in the same timeframe during the scenario. In such an instance, the task will probably be seen by the operators as a unified task. However, what if the operators also close another valve in the vicinity which is from another system, in fact one they will be utilising later in the scenario (which would constitute an error of commission) to restore feed? Will the task-based approach identify the event in the first place, and will its impact be treated in the PSA? If not, risk may be underestimated.

There is therefore a risk with adopting the task-based approach, of failing to identify all critical interactions. This risk is an unavoidable trade-off when carrying out real PSAs. There will always be limited resources available for any risk assessment, and task-based assessment is a cost-efficient approach to dealing with human involvements in a very large and complex system such as a nuclear power plant. Arguably, if detailed error identification were carried out for all identified tasks, this would either require significantly more resources (e.g. increasing the HRA resources by an order of magnitude), or else would result in detailed assessment of less scenarios than with the task-based approach. The reassuring aspect of the continued operation PSAs is that they have all used detailed task analysis as a back-up to quantification in the more important human-critical scenarios. When such task analysis is carried out, including, for example, tabular task analysis of the interface adequacy, then the assessors will be aware of errors which could occur during a particular task which could affect other tasks' effectiveness, or which could even compound the event or start a new one. To some extent, such concerns can also be addressed by error of commission analyses and cognitive error analyses, discussed below.

14.3.3.3 Errors of commission

THORP and Sizewell B As already noted, the human error analysis in the early HRA phase for Sizewell B did attempt to identify errors of commission, based on basic keywords and assessor experience and task analysis of the tasks. For THORP, such error potentials were more likely to be identified by the tabular task analyses which reviewed the interfaces related to the emergency tasks. Such analyses would often consider unrequired actions (e.g. 'right action on wrong object'), and would then make design changes to prevent such errors or ensure their prompt recovery. Some significant errors of commission were therefore identified and fed into the risk assessment and quantified accordingly. Errors of commission were also identified based on operational experience: although THORP was a novel design, several Sellafield plants bore close resemblance to certain THORP processes (e.g. fuel cooling and transfer ponds), and so incidents on these plants could be used to suggest credible EOCs.

Continued Operation HRAs Errors of commission were addressed to a limited extent in these HRAs. One HRA in particular used a HAZOP-based approach to identify credible EOCs – EOC Analysis or EOCA (Kirwan *et al.*, 1995b; Kirwan, 1994). The EOCs identified in this study were not fed into the quantitative PSA, but they were considered qualitatively, and error reduction mechanisms were accordingly identified.

14.3.3.4 Cognitive Errors

THORP and Sizewell B THORP identified requirements for diagnosis using the JHEDI and HRMS human error identification systems. HRMS in particular uses an algorithmic set of questions (see Kirwan, 1994) to determine if the task will require either knowledge-based or rule-based diagnostic behaviour (i.e. diagnosis: see Rasmussen *et al.*, 1981; Reason, 1990). If cognitive error potential is present, then the assessor proceeds to a set of cognitive error identification questions designed to elicit the type of cognitive error for that task (e.g. failure to consider special circumstances or side effects; misdiagnosis; etc.).

Sizewell B reviewed cognitive error potential firstly qualitatively, using an extensive and theoretically based cognitive error checklist, as well as examining diagnosis supporting interfaces. Then, based partly on procedures etc., decision-making (and diagnosis) tasks were identified in evolving plant emergency scenarios. Decision-making failures were then incorporated into the logic trees.

Continued Operation HRAs Decision-making failures or errors were identified in three HRAs, either as a function of failing to make a decision in time (e.g. when deciding how to achieve residual cooling after a loss of cooling event), or making the wrong decision. Some of the latter misdiagnoses in at least one HRA were identified using a confusion matrix analysis, based on a fault symptom matrix analysis (Potash *et al.*, 1981; Kirwan *et al.*, 1995b, Kirwan, 1994). In some cases cognitive errors were accounted for quantitatively in the PSA, and in others they were dealt with only qualitatively.

14.3.4 Representation

14.3.4.1 Logic Tree Usage and Dependence Treatment

Sizewell B and THORP The Sizewell B PSA used small event trees and very large fault trees, and this enabled identified human errors to be incorporated relatively easily into operator event trees. In the early phase of the analysis dependence was modelled using the five-stage dependency model that is part of the THERP technique (Swain and Guttmann, 1983). Modelling of dependence in event trees is relatively straightforward. In later analyses, when the HEART technique was utilised, dependence was modelled by inserting new events into the logic tree structures (see description of this approach for continued-operation PSAs).

The THORP PSA was predominantly fault-tree-based, so that all identified human errors were fed into fault trees rather than event trees, although a few event trees were used in special cases involving chronological event sequences. The THORP HRA did not utilise a THERP-type dependence model. Instead, firstly all JHEDIs are relatively conservative (i.e. pessimistic: see Kirwan *et al.*, 1995a). Secondly, where dependence was evident, conditional probabilities were utilised (e.g. first probability of an error = 0.01; probability of a second operator error checking the first = 0.1;

probability of third operator failing to check first = 1.0). Thirdly, human performance limiting values (HPLVs: Kirwan *et al.*, 1990) were utilised in all human error cutsets (see Kirwan, 1994). This prevents the type of optimism that can occur when dependence is not explicitly modelled.

Continued-operation HRAs Fault trees and event trees were utilised by the continued-operation PSAs, but generally small event trees were used to model consequences, with large fault trees being used to quantify the fault combinations. Therefore most of the human errors were modelled in fault tree format. Dependence was generally modelled in one of two ways. Firstly the THERP dependence model was used by some HRAs, and applied directly to HEPs in their respective cutsets (see Kirwan *et al.*, 1995b). Some other PSAs chose to model dependence by defining a task which was common to the two dependent tasks, and in fact was the mechanism of dependence between the two tasks (e.g. a diagnostic failure linking two otherwise independent errors). Once such an event was identified it would also be put into the fault tree and quantified directly. Such an approach has a benefit in that if the mechanism of dependence itself is identified, then if such a mechanism is contributing significantly to risk, then the error reduction can focus on the actual mechanism, rather than an arithmetical fix which accounts for the quantitative effects of dependence, but not the factors causing the dependence themselves.

14.3.5 Quantification

The following shows the techniques used by the various HRAs:

- THORP: JHEDI (all tasks) and HRMS (highly risk significant tasks);
- Sizewell 'B': THERP (early quantifications) and HEART (later quantifications);
- Continued-Operation PSAs: HEART (with occasional checks against THERP in at least one PSA).

The three techniques (THERP, HEART and JHEDI) have all recently been successfully validated (Kirwan *et al.*, 1995a), and hence are seen as acceptable by the industry and the regulatory authorities. Other techniques which would probably be seen as acceptable are the techniques SLIM (Embrey *et al.*, 1984), absolute probability judgement (APJ) or direct numerical estimation (Seaver and Stillwell, 1983), and possibly the paired comparisons technique (Hunns, 1982). These three are expert judgement techniques, with SLIM involving structured expert judgement based on performance shaping factors (PSF), and the other two involving more direct estimation of error likelihoods. If these techniques were utilised, the pedigree of the expert judges, and of the calibration data needed for SLIM and paired comparisons, would probably need to be justified.

The regulatory authorities, via the Advisory Committee for the Safety of Nuclear Installations (ACSNI), published a report in 1991 entitled 'HRA – a critical overview' (ACSNI, 1991). This review was critical of quantitative HRA in two main respects: firstly that there was insufficient validation of the techniques generally used to quantify HEPs; and secondly that ideally there should be a database of human error probabilities available both for calibration of techniques such as SLIM and Paired Comparisons, and for general reference purposes. The validation of the techniques has taken place and, as already noted, was successful. Furthermore, a 3-year data collection and compilation project was completed in 1995, culminating in the computerised operator reliability and error database (CORE-DATA: Taylor-Adams and Kirwan, 1995). This contains a reasonable amount of NP&R real data (i.e. data derived from incidents), as well as simulator data. The database will become accessible in 1996. Therefore, the two major concerns and actions arising out of the ACSNI UK review of quantitative HRA have been addressed.

Incidentally, the quantification of Human Error Probabilities, once the enveloping focus of concern of HRA practitioners and researchers alike, has become relatively routine, and concerns are now focused on other more qualitative aspects of the HRA process. However, this shift of concern away from quantification does not necessarily mean that these techniques are now entirely satisfactory. The further investigation of the validation exercise (Kirwan *et al.*, 1996; Kirwan, 1996) suggests that there is still an undesirably high degree of inconsistency in how different assessors utilise the techniques, which could affect, amongst other things, the appropriateness of the error reduction recommendations. Furthermore, it appears that the three techniques mainly in use in the NP&R industries are not particularly suited to particular types of errors, including errors of commission and cognitive errors, and hence it may be that there is a real and unique role for the expert judgement techniques, or techniques from further afield (e.g. INTENT, Gertman, 1991 and SNEAK, Hahn and de Vries, 1991). This is therefore an area for future research and development in HRA.

It appears more generally that assessors would benefit from more guidance/training in the practical application of the techniques THERP and HEART in particular, since THERP training has not been carried out by a direct source (e.g. Sandia National Laboratory's own (US) course run by Dr Swain and co-workers), and there are very few 'benchmark' examples of HEART applications, such that HEART therefore tends to be interpreted in different ways by different assessors. Such training and benchmarks are therefore a highly practical but principal research and development area for quantitative HRA.

14.3.6 Error Reduction Assessment

Error reduction has become more central to HRA in the UK NP&R industries. Both Sizewell B and THORP PSAs have been used as the impetus for changing aspects of operator support systems, such as emergency shutdown panels (THORP: Reed *et al.*, 1990), allocation of function (Sizewell B: Fewins *et al.*, 1992), interface design, training and procedures, and alarms (Magnox station: Kirwan *et al.*, 1995b). No longer is HRA simply a 'numbers game': it is instead often very much about influencing design and operational parameters and features, where such changes will better support the operator. This change is welcomed by, for example the Ergonomics and Human Factors community, and HRA is seen as a credible, though currently fragile, bridge between engineering-based risk assessment and the more behaviourally based sciences of ergonomics/human factors.

The error reduction assessments are notably relying more on the foundational analyses such as task and error analyses, than the quantitative metrics enshrouded in techniques such as THERP, HEART and JHEDI. This is also argued in Kirwan (1996) to be more appropriate, since quantification techniques often model performance-shaping factors at too crude a level to make robust and assured error-reduction recommendations (i.e. ones which will achieve a true reduction in error probability).

14.3.7 Quality Assurance (QA) and Documentation

In this current age of total quality management (TQM – e.g. see Holmes, 1992) there is inevitably much documentation associated with HRAs and PSAs, and much checking and signing of pieces of paper to say that a calculation has been checked, etc. The benefits of such processes are firstly that calculation errors in PSAs will themselves be relatively rare events, and secondly that more assumptions underlying assessments will be documented. In practice, the individual assessor has a good deal of flexibility when carrying out an HRA, and parts of the HRA process such as task analysis and error identification in particular, will be moderately assessor-dependent. In this sense HRA is still more an art than a science. However, TQM etc. is useful since individual assessors will be required to produce more information in terms of why particular errors were identified and others were ignored, etc. Such information is invaluable if the HRA information is ever to be used again, whether for operational purposes, or at a later periodic review stage. To the author's knowledge there are no true 'living HRAs', the logical correlate of living PSAs, which do exist (see Kafka and Zimmerman, 1995). But perhaps, with more detailed documentation, at least certain HRAs will be 'resurrectable', without too much new work, or at least without having to start again from the beginning.

A more short-term important QA aspect of real HRAs is associated with error reduction recommendations, particularly those which are assumed to be capable of reducing risk. The implementation of such measures is always somewhat more complicated than their initial identification, and it must be checked that ERMs have been properly implemented, without introducing new problems, and with the continuing support of the workforce. Otherwise, there may be a relatively short-lived improvement in reliability, before errors start to return to their previous probability levels. Such QA must operate in timescale terms of years rather than months.

14.4 Discussion of the Adequacy of UK NP&R HRA

HRA practice in UK NP&R appears to be relatively formalised and mature, at least with respect to quantification, and there has been a shift towards usage of a significant amount of task analysis to underpin the HRAs. The areas that would benefit from improvement, either because they are not being properly treated or because they are not being dealt with consistently, appear to be the following:

- task analysis of cognitive activities (diagnosis and decision-making);
- error of commission identification and quantification;
- cognitive error identification and quantification;
- rule violation identification and quantification;
- dependence treatment;
- task-based versus error-based HRA;
- provision of more guidance and benchmarks for certain HRA techniques;

The first, *cognitive task analysis for cognitive activities*, reflects the fact that current treatment of diagnosis and decision-making in HRAs is very 'black-box' in nature. Current UK HRA methods do not evaluate, in a detailed and explicit way, diagnosis as a function of operator knowledge and expectations in relation to the unfolding events and indications, except by using the operators as experts in fairly gross confusion matrix or other expert judgement sessions. HRA approaches also generally do not model the different and changing goal structures that actually drive a team, e.g. goals such as 'protect equipment', 'conserve power', 'stabilise plant', 'diagnose fault', 'rectify fault', 'shut down plant', 'maintain safety integrity', 'conserve back-up resources' etc. Such goals will interact with each other and the unfolding event nature, as a function of the perceived (by the operating team) danger as the event progresses (the failure to perceive danger, and therefore having non-safety-related goals as the driving ones, explains the Chernobyl operators' behaviour, and their failure to attempt to restore safety until danger became very evident, by which time it was too late). Essentially, the current modelling of cognitive error is both simplistic and gross, and warrants improvement: the treatment of cognitive error in UK NP&R HRAs needs to 'go cognitive'.

Error of commission and *rule violation identification* are difficult areas, but it is clear that these types of error do occur (see Kirwan, 1995; Kirwan *et al.*, 1995c). The difficulty is in identifying firstly credible EOCs and rule violations, and secondly of identifying ones that are not obvious. The EOCA approach appears to be able to do the former, and techniques such as SNEAK (Hahn and DeVries, 1991) purport to be able to do the latter (however, SNEAK appears to require significant resources). What is probably needed is a hybrid approach, based partly on analysis of interfaces etc., and partly using expert judgement. Certainly, it is difficult to identify rule violations in a plant that is at the design stage, since implicit in rule violation aetiology is a safety culture problem, difficult to identify and accept at the design stage. Quantification of EOCs and rule violations will probably be achieved by expert judgement techniques for the foreseeable future, backed up or calibrated by recorded data.

As noted above under task analysis needs, *cognitive error analysis* needs to be based more upon cognitive psychological concepts and approaches, if detailed cognitive errors are to be elicited. Some work has recently begun to try to explore the mechanisms of misdiagnosis as a function of depth of knowledge, complexity, and interface aspects (Folleso *et al.*, 1995). However, in the short term, improvements could be made to the existing techniques, to render them less superficial (e.g. considering a basic indication timeline and determining potential wrong decisions in each timeframe, as a function of indications, expectations, procedural guidance, and goal hierarchy). Quantification of cognitive errors, as for EOCs and rule violations, will probably be carried out using expert judgement approaches.

The treatment of *dependence* is currently achieved by three alternative methods:

1 use of the THERP dependence model;
2 identification of explicit dependence events;
3 use of conservative quantification plus the application of human performance limiting values.

Each of these three has its own strengths and weaknesses, and they are not necessarily mutually exclusive alternatives. What is therefore advisable is guidance on when to use each method, alone or in combination. Such guidance should arise from a detailed study of their effectiveness in actual PSAs.

The decision of whether to utilise a *task-based versus error-based HRA* approach is fundamental, and one that has not been properly resolved in HRA generally. What is required is a study which directly compares the two approaches based on the same set of scenarios. If the error-based approach is found to identify human contributions to risk which are both significant, and would not be included by the tasks in the task-based approach, then this would throw the validity of the task-based approach into question. What is most likely is that certain types of tasks may require more detailed error analysis or decomposition, and it would be useful therefore to determine what types of tasks these might be.

Although *quantification* techniques have been recently validated, the consistency of usage of the techniques warrants improvement. This is most likely achievable by the provision of NP&R examples (*benchmarks*), as well as possibly *training*, since there is very little formal training in the UK for HRA assessors. Ultimately this might lead to an accreditation system for UK assessors. Such steps would not be out of line with TQM developments in the NP&R and other industries, and would perhaps elevate the status of HRA practitioners.

Overall, HRA appears to be doing a reasonable job in a very difficult area, and has moved away from being a purely quantitative approach to a primarily qualitative one. The benefits of this are more face validity with the human factors world, and more tangible benefits in terms of being able to identify effective error reduction measures. There is still much room for improvement, however, and some suggestions for highly targeted research and development projects have been made.

14.5 Conclusions

UK NP&R HRA practice has been reviewed by consideration of recent high-profile NP&R HRAs. The approaches are relatively formalised, and there is a notable predominance of qualitative analysis underpinning all the HRAs, whether at the design stage or for existing installations. A second notable trend in UK HRA has been for the HRA to deliver credible and incisive error-reduction measures to improve human reliability.

Several areas have been highlighted for further development: cognitive task analysis, error identification, and representation issues (dependence and task vs. error-based HRA), as well as quantification of the more difficult human error types (cognitive errors and errors of commission), and the provision of more guidance to practitioners on the use of quantification techniques in NP&R.

Whilst HRA has moved to being more qualitative and hence has perhaps moved more towards human factors (from its original domain of reliability engineering), the next significant shift would logically be a move more towards cognitive psychology, since it is ultimately the minds of the operators that determines operator behaviour, and that is what HRA practice is all about.

14.6 Author's note

The author would like to take the unusual step of apologising for so many of the references being his own. This is due to the fact that he was responsible for THORP's HF&R programme for some years, and there are very few publications arising from the Sizewell B HRA programme, and almost none as yet (besides the author's own) from the continued operation PSAs. The opinions in this paper, and indeed the balance of perspectives in the paper, are therefore inevitably subject to the author's own biases, and should not be construed as necessarily representing those of any of the companies comprising the UK NP&R industries (NE, SN, BNFL, and NII/HSE).

References

ACSNI (1991) *Human Reliability Assessment – A Critical Overview*, Advisory Committee on the Safety of Nuclear Installations, Health and Safety Commission, London: HMSO.

AINSWORTH, L. & PENDLEBURY, G. (1995) Task-based contributions to the design and assessment of the man-machine interfaces for a pressurised water reactor, *Ergonomics*, **38**(3), 462–74.

APOSTOLAKIS, G., OKRENT, D., GRUSKY, O., WU, J. S., ADAMS, R., DAVOUDIAN, K. & XIONG, Y. (1992) Inclusion of organisational factors into probabilistic safety assessments of nuclear power plants, *IEEE Conference on Human Factors and Power Plants*, pp. 381–8, Monterey, CA, June 7–11.

BREACH, I. (1978) *Windscale Fallout*, Harmondsworth: Penguin.

COX, S. J. & TAIT, N. R. S. (1991) *Reliability, Safety and Risk Management*, Oxford: Butterworth-Heinemann.

DHILLON, B. S. (1986) *Human Reliability with Human Factors*, Oxford: Pergamon.

DOUGHERTY, E. M. & FRAGOLA, J. R. (1988) *Human Reliability Analysis: a Systems Engineering Approach with Nuclear Power Plant Applications*, New York: Wiley.

EMBREY, D. E. (1986) SHERPA – a systematic human error reduction and prediction approach, Paper presented at the *International Topical Meeting on Advances in Human Factors in Nuclear Power Systems*, Knoxville, Tenn.

EMBREY, D. E., HUMPHREYS, P. C., ROSA, E. A., KIRWAN, B. & REA, K. (1984) *SLIM-MAUD. An Approach to Assessing Human Error Probabilities Using Structured Expert Judgement*, NUREG/CR-3518, Vols 1 and 2, US Nuclear Regulatory Commission, Washington, DC 20555, 180 pages.

EMBREY, D. E., KONTOGIANNIS, T. & GREEN, M. (1994) *Preventing Human Error in Process Safety*, Centre for Chemical Process Safety (CCPS), American Institute of Chemical Engineers, New York: CCPS.

FEWINS, A., MITCHELL, K. & WILLIAMS, J. C. (1992) Balancing automation and human action through task analysis. In Kirwan, B. & Ainsworth, L. K. (eds), *A Guide to Task Analysis*, pp. 241–51, London: Taylor and Francis.

FOLLESOE, K., KAARSTAD, M., DROIVOLDSMO, A. & KIRWAN, B. (1995) Relations between task complexity, diagnostic strategies and performance in diagnosing process disturbances. Paper presented at the *Fifth Conference on Cognitive*

Science Approaches to Process Control (CSAPC '95), Helsinki (Espoo), 30 August–1 September.

GERTMAN, D. I. (1991) INTENT: a method for calculating HEP estimates for decision-based errors, *Proceedings of the Human Factors Society 35th Annual Meeting*, pp. 1090–4, San Francisco, CA, 2–6 September.

GERTMAN, D. I. & BLACKMAN, H. (1994) *Human Reliability and Safety Analysis Data Handbook*, Chichester: Wiley.

GREEN, A. E. (1983) *Safety Systems Reliability*, Chichester: Wiley.

HAHN, H. A. & DEVRIES, J. A. (1991) Identification of human errors of commission using Sneak Analysis, *Proceedings of the Human Factors Society 35th Annual Meeting*, pp. 1080–4, San Francisco, CA, 2–6 September.

HANNAMAN, G. W., SPURGIN, A. J. & LUKIC, Y. D. (1984) *Human Cognitive Reliability Model for PRA Analysis*, Report NUS-4531, Electric Power Research Institute, Palo Alto, CA.

HENLEY, E. J. & KUMAMOTO, H. (1981) *Reliability Engineering and Risk Assessment*, New Jersey: Prentice-Hall.

HOLLNAGEL, E. (1993) *Human Reliability Analysis: Context and Control*, London: Academic Press.

HOLMES, K. (1992) *Total Quality Management*, Leatherhead, Surrey: Pira International.

HUNNS, D. M. (1982) The method of paired comparisons. In Green, A. E. (ed.), *High Risk Safety Technology*, Chichester: Wiley.

KAFKA, P. & ZIMMERMAN, M. (1995) Operational risk management. In Watson, I. A. & Cottam, M. P. (eds), ESREL '95, pp. 123–37, London: Institute of Quality Assurance.

KENNEDY, R. & KIRWAN, B. (1995) The failure mechanisms of safety culture, *IAEA Conference on Safety Culture*, Vienna, April.

KIRWAN, B. (1988) Integrating human factors and reliability into the plant design and assessment process. In Megaw, E. D. (ed.), *Contemporary Ergonomics*, pp. 154–62, London: Taylor and Francis.

(1989) A human factors and reliability programme for the design of a large nuclear chemical plant, *Human Factors Annual Conference*, pp. 1009–13, Denver, CO, October.

(1990a) Human reliability assessment. In Wilson, J. R. & Corlett, N. E. (eds), *Evaluation of Human Work*, pp. 706–54, London: Taylor and Francis.

(1990b) A resources flexible approach to human reliability assessment for PRA, *Safety and Reliability Symposium*, pp. 114–35, Altrincham, September, London: Elsevier Applied Sciences.

(1992a) Human error identification in human reliability assessment, Part 1: overview of approaches, *Applied Ergonomics*, **23**(5), 299–318.

(1992b) Human error identification in HRA, Part 2: Detailed comparison of techniques, *Applied Ergonomics*, **23**(6), 371–81.

(1992c) A task analysis programme for THORP. In Kirwan, B. & Ainsworth, L. K. (eds), *A Guide to Task Analysis*, pp. 363–88, London: Taylor and Francis.

(1993) A human error analysis toolkit for complex systems, Paper presented at the *Fourth Cognitive Science Approaches to Process Control Conference*, 25–27 August, Copenhagen, Denmark, pp. 151–99.

(1994) *A Guide to Practical Human Reliability Assessment*, London: Taylor and Francis.

(1995) Current trends in human error analysis technique development. In Robertson, S. A. (ed.), *Contemporary Ergonomics*, London: Taylor and Francis, pp. 111–17.

(1996) (in press) Practical aspects of the usage of three human reliability quantification techniques: THERP, HEART and JHEDI, submitted to *Applied Ergonomics*.

KIRWAN, B. & AINSWORTH, L. K. (eds) (1992) *A Guide to Task Analysis*, London: Taylor and Francis.

KIRWAN, B. & JAMES, N. I. (1989) Development of a human reliability assessment system for the management of human error in complex systems, *Reliability '89*, Brighton, 14–16 June, pp. 5A/2/1–5A/2/11.

KIRWAN, B., EMBREY, D. E. & REA, K. (1988) *The Human Reliability Assessors Guide*, Report RTS 88/95Q, NCSR, UKAEA, Culcheth, Cheshire, 271 pages.

KIRWAN, B., MARTIN, B. R., RYCRAFT, H. & SMITH, A. (1990) Human error data collection and data generation, in *International Journal of Quality and Reliability Management*, 7.4, pp. 34–66.

KIRWAN, B., KENNEDY, R. & TAYLOR-ADAMS, S. (1995a) A validation study of three human reliability quantification techniques, *European Safety and Reliability Conference*, ESREL '95, Bournemouth, 26–28 June, London: Institute of Quality Assurance.

KIRWAN, B., SCANNALI, S. & ROBINSON, L. (1995b) Practical HRA in PSA – a case study, *European Safety and Reliability Conference*, ESREL '95, Bournemouth, 26–28 June, London: Institute of Quality Assurance.

KIRWAN, B., TAYLOR-ADAMS, S. & KENNEDY, R. (1995c) *Human Error Identification Techniques for PSA Application*, Report for HSE, Project HF/GNSR/22, Industrial Ergonomics Group, University of Birmingham, March.

KIRWAN, B., KENNEDY, R. & TAYLOR-ADAMS, S. (1996) (in press) A validation study of three human reliability quantification techniques: THERP, HEART and JHEDI, submitted to *Applied Ergonomics*.

LAUGHERY, K. R. (1984) Computer modelling of human performance on microcomputers, in the *Proceedings of the Human Factors Society Annual Conference*, pp. 884–8, Santa Monica, CA.

PARK, K. S. (1987) *Human Reliability: Analysis, Prediction, and Prevention of Human Errors*, Oxford: Elsevier.

PHILLIPS, L. D., HUMPHREYS, P. & EMBREY, D. E. (1983) *A Socio-Technical Approach to Assessing Human Reliability*, London School of Economics, Decision Analysis Unit, Technical Report 83-4.

POTASH, L. *et al.* (1981) Experience in integrating the operator contributions in the PRA of actual operating plants. In *Proceedings of the ANS/ENS Topical Meeting on PRA*, New York: American Nuclear Society.

RASMUSSEN, J., PEDERSEN, O. M., CARNINO, A., GRIFFON, M., MANCINI, C. & GAGNOLET, P. (1981) *Classification System for Reporting Events Involving Human Malfunctions*, RISO-M-2240, DK-4000, Riso National Laboratories, Roskilde, Denmark.

RASMUSSEN, J., DUNCAN, K. D. & LEPLAT, J. (1987) *New Technology and Human Error*, Chichester: Wiley.

REASON, J. T. (1990) *Human Error*, Cambridge: Cambridge University Press.

REED, J. (1992) *The Contribution of Human Factors to the THORP Project*, paper presented at SPECTRUM '92, Idaho, USA.

REED, J., VERLE, A. & KIRWAN, B. (1990) Design of emergency shutdown panels. In Lovesey, E. J. (ed.), *Contemporary Ergonomics*, pp. 393–8, London: Taylor and Francis.

SEAVER, D. A. & STILLWELL, W. G. (1983) *Procedures for Using Expert Judgement to Estimate Human Error Probabilities in Nuclear Power Plant Operations*, NUREG/CR-2743, Washington DC 20555.

SENDERS, J. W. & MORAY, N. P. (1991) *Human Error: Cause, Prediction and Reduction*, Hillsdale, NJ: Lawrence Erlbaum Associates.

SHEPHERD, A. (1989) Analysis and training of information technology tasks. In Diaper, D. (ed.), *Task Analysis for Human-Computer Interaction*, pp. 15–54, Chichester: Ellis Horwood.

SWAIN, A. D. (1987) *Accident Sequence Evaluation Program Human Reliability Analysis Procedure*, NUREG/CR-4722, Washington DC-20555: USNRC.

(1989) *Comparative Evaluation of Methods for Human Reliability Analysis*, Gesellschaft für Reaktorsicherheit, GRS-71. Schwertnergasse 1, 5000 Koln.

SWAIN, A. D. & GUTTMANN, H. E. (1983) *Human Reliability Analysis with Emphasis on Nuclear Power Plant Applications*, NUREG/CR-1278, USNRC, Washington, DC 20555.

TAYLOR-ADAMS, S. & KIRWAN, B. (1995) Human reliability data requirements, *International Journal of Quality and Reliability Management*, **12**(1), 24–46.

TVERSKY, A. & KAHNEMAN, D. (1974) Judgement under uncertainty: heuristics and biases, *Science*, **185**, 1124–31.

UMBERS, I. & REIERSEN, C. S. (1995) Task analysis in support of the design and development of a nuclear power plant safety system, *Ergonomics*, **38**(3), 443–54.

WHITFIELD, D. (1991) An overview of human factors principles for the development and support of nuclear power station personnel and their tasks, in the *Conference 'Quality management in the nuclear industry: the Human Factor'*, London: Institute of Mechnical Engineers.

(1995) Ergonomics in the design and operation of Sizewell B nuclear power station, *Ergonomics*, **38**(3), 455–61.

WHITWORTH, D. (1987) Application of operator error analysis in the design of Sizewell 'B'. In *Reliability '87*, NEC, Birmingham, pp 5A/1/1–5A/1/14, 14–16 April, London: Institute of Quality Assurance.

WILLIAMS, J. C. (1986) HEART – a proposed method for assessing and reducing human error. In *Proceedings of the 9th 'Advances in Reliability Technology' Symposium*, University of Bradford.

(1988a) A data-based method for assessing and reducing human error to improve operational performance. In *IEEE Conference on Human Factors in Power Plants*, pp. 436–50, Monterey, CA, 5–9 June.

(1988b) Human factors analysis of automation requirements. In Libberton, G. P. (ed.), *Proceedings of the 10th Advances in Reliability Technology Conference*, London: Elsevier Applied Sciences.

(1992) Toward an improved evaluation analysis tool for users of HEART, *Proceedings of the International Conference on Hazard Identification, Risk Analysis, Human Factors & Human Reliability in Process Safety*, Orlando, Fla.

WOODS, D. D. *et al.* (1994) *Behind Human Error: Cognitive Systems, Computers and Hindsight*, CSERIAC State of the art report, Ohio: Wright Patterson Air Force Base, October.

CHAPTER FIFTEEN

The promotion and measurement of a positive safety culture

RICHARD T. BOOTH
Aston University, Birmingham

15.1 Introduction

The objectives of the chapter are to seek to answer the following questions:

- What is safety culture, and how has the concept evolved?
- Why is safety culture a crucial element of health and safety management?
- How can organisations measure their safety culture, or at least its tangible manifestations?
- How can organisations plan and implement programmes to create and sustain a positive safety culture?

The content of the chapter follows very closely two principal sources: the report *Organising for safety* prepared by the Human Factors Study Group of the Advisory Committee for Safety in Nuclear Installations (Health and Safety Commission, 1993) and the preamble to the Institution of Occupational Safety and Health's *Policy on Health and Safety Culture* (IOSH, 1995). I am grateful to my colleagues on the ACSNI and IOSH working parties for their substantial contribution to the views expressed here.

The term 'health and safety culture', or simply 'safety culture', seeks to describe the characteristic shared attitudes, values, beliefs and practices of people at work concerning not only the magnitude of the risks that they encounter but also the necessity, practicality, and effectiveness of preventive measures. People's attitudes to safety are influenced by their attitudes and beliefs about the organisation for whom they work: safety culture is a sub-set of, or at least greatly influenced by, the overall culture of organisations. There is no standard definition of the term 'safety culture'. A number of definitions is given in the Glossary of Terms (Appendix 1). An organisation with a positive health and safety culture consists of competent people with strongly held safety values which they put into practice. In contrast, personnel in companies deemed to have a poor safety culture are likely to adopt safety arrangements with lukewarm acceptance at best, and at worst to ignore both statutory and company obligations.

The promotion of a positive culture is often advocated as one of the central objectives of health and safety management. The term is used increasingly as an indication of an organisation's determination and competence to control dangers at work (CBI, 1991). For example, Lord Cullen in the report on the Piper Alpha disaster (Department of Energy, 1991) observed that 'It is essential to create a corporate atmosphere or culture in which safety is understood to be, and is accepted as, the number one priority.' IOSH's Policy Statement on Safety Training (IOSH, 1992) recognised that 'training specifically designed to create a positive safety culture is . . . an essential part of company training . . . at all levels.'

The concept, and above all the ideas which underlie, safety culture are vitally important. These ideas should be considered by managers when planning any developments in the organisation, and by safety practitioners when reviewing and revising health and safety management systems. But there is a danger that the importance of the concept may be devalued by assumptions of two kinds. Firstly, managers may perceive that safety culture can be improved rapidly by prescription or decree: companies that have attempted to change their culture in this way have been unsuccessful (Beer *et al.*, 1990; Toft, 1992a). The essence of culture at any level in a society is its high resistance to change; indeed, attempts to change culture rapidly may reinforce pre-existing beliefs (Turner, 1992). Secondly, a case can be made that safety culture can be neither adequately defined nor measured, nor improved by direct management intervention alone.

This chapter seeks to steer a middle path between the expectations of some that safety culture can be improved by easily implemented measures, and the reservations of others that the behavioural complexity of people in organisations is likely to confound even well-considered attempts to promote a positive safety culture. In any event, the pursuit of a positive culture should not be seen as a substitute for a broadly based programme to improve health and safety performance. The rôle of safety culture in health and safety management can only usefully be discussed in the context of the key principles of proactive health and safety management.

15.2 Proactive Safety Management

Accident and ill-health prevention programmes must address the following distinctive elements of the accident causation process:

15.2.1 Multi-causality

Very few accidents, particularly in large organisations and complex technologies, are associated with a single cause. Rather accidents happen as a result of a chance concatenation of many distinct causative factors, each one necessary but not sufficient to cause a final breakdown (Reason, 1990). It follows that the coverage of prevention plans should seek to permeate all aspects of the organisation's activities.

15.2.2 Active and Latent Failures

Active failures are errors which have an immediate adverse effect. In contrast, latent failures lie dormant in an organisation for some time, only becoming evident when they combine with local triggers. The triggers are the active failures: unsafe acts, and unsafe conditions. The recognition of the importance of latent failures is useful because it emphasises the role of senior managers in causation, and draws attention to the scope of detecting latent failures in the system well before they are revealed by active failures.

15.2.3 Skill-, Rule- and Knowledge-based Errors, and Violations

The standard framework for classifying error is the skill-, rule-, knowledge-based model proposed by Rasmussen (1987) and described in Health and Safety Commission (1991).

Skill-based errors involve 'slips' or 'lapses' in highly practised and routine tasks. At a rather more complex level a person has to look at a situation and classify it into a familiar category as the basis for action; if it is mis-classified, this may be called a rule-based error, or mistake. Knowledge-based errors describe the most complex cases where people fail to create an adequate new rule to cope with a situation. Violations, sometimes referred to as 'risk taking', comprise a further category of error. Here, a person deliberately carries out an action that is contrary to a rule, such as an approved operating procedure.

The success of training programmes depends on an adequate diagnosis of the nature of the errors likely to be made. For example, task analysis and training that fail to consider violations may prove wholly ineffective.

15.2.4 Hazard Identification, Risk Assessment, and Preventive Action

The need to identify hazards, assess risks, and select, implement and monitor preventive actions is an essential foundation of safety management – the avoidance of latent failures. It is also the foundation for safe personal behaviour in the face of danger – the avoidance of active failures.

To create and maintain a safe working environment, and to work safely in a dangerous environment people must have the knowledge and skills and must know the rules, and be motivated, to (Hale and Glendon, 1987);

- identify hazards;
- assess accurately the priority and importance of the hazards (risk assessment);
- recognise and accept personal responsibility for dealing with the hazards in an appropriate way;
- have appropriate knowledge about what should be done (including specified rules);
- have the skills to carry out the appropriate necessary sequence of preventive actions, including monitoring the adequacy of the actions, and taking further corrective action.

The organisation should be aware of circumstances where managers, supervisors, and other personnel may:

- underestimate the magnitude of risks;
- overestimate their ability to assess and control risks;
- have an impaired ability to cope with risks.

15.2.4 The Aims of Safety Management

The primary aim of safety management is to intervene in the accident causation process and to break the causation chain. This involves preventing or detecting latent and active failures in the continuing process of hazard identification, risk assessment, control, and monitoring. However, the aim of safety management is not limited simply to hazard identification, control and monitoring. Employers must plan for safety. Decisions have to be made, for example, about priorities for resource allocation, about training needs, about the appropriate risk assessment methodologies to be adopted, about the need for human reliability assessment, and about the choice of tolerable risk criteria. Safety criteria should underpin every decision made by the enterprise. Safety must be considered as an integral part of day-to-day decision-making. Moreover, an employer must establish organisation and communications systems which facilitate the process of integrating safety within the management process, and which ensure that everyone in the organisation is at least fully informed of safety issues, and ideally has had an opportunity to contribute to the debate.

15.2.5 Key Functions of Safety Management

From the foregoing, the four key functions of the management of safety may be summarised as follows:

1 *policy and planning* determining safety goals, quantified objectives and priorities, and a programme of work designed to achieve the objectives, which is then subject to measurement and review;
2 *organisation and communication* establishing clear lines of responsibility and two-way communications at all levels;
3 *hazard management* ensuring that hazards are identified, risks assessed, and control measures determined, implemented, and subject to measurement and review;
4 *monitoring and review* establishing whether the above steps above are:
   (a) in place;
   (b) in use, and
   (c) work in practice.

The four key elements of safety management are underpinned by the requirements of the Management of Health and Safety at Work Regulations, 1992.

The report *Successful Health and Safety Management* (Health and Safety Executive, 1991) presents a truncated model of the above processes that seeks to distinguish between monitoring carried out by the management team and (external) auditing of the complete safety management system by third parties.

15.3 Safety Management and Safety Culture

The procedures and systems described above are necessary elements of an effective safety programme. But they are not the whole story. The danger exists that an organisation's safety policies, plans and monitoring arrangements, which appear on paper to be well-considered and comprehensive, may create an aura of respectability which disguises sullen scepticism or false perceptions among opinion-formers at management and shop-floor levels. The critical point is not so much the adequacy of the safety plans as the perceptions and beliefs that people hold about them. The next section focuses on the issues that determine whether the safety procedures just described are implemented with the full and enthusiastic support of the whole work force, or whether the procedures are, at best, put into practice grudgingly and without thought, or at worst are honoured in the breach.

15.4 The Concept of Safety Culture

15.4.1 Origins and Definition of Safety Culture

The concept of safety culture was introduced in a seminal paper by Zohar (1980), and to the nuclear safety debate by the International Nuclear Safety Advisory Group (1988) in their analysis of Chernobyl. The Agency has subsequently published an authoritative report which elaborates the concept in detail (International Nuclear Safety Advisory Group, 1991). The Agency (*ibid.*) has defined safety culture as:

> . . . that assembly of characteristics and attitudes in organisations and individuals which establishes that, as an overriding priority, nuclear plant safety issues receive the attention warranted by their significance. (IAEA, 1991).

Appendix 1 includes a range of definitions of safety culture. CBI (1990) describe the culture of an organisation as 'the mix of shared values, attitudes and patterns of behaviour that give the organisation its particular character. Put simply it is "the way we do things round here" '. They suggest that the 'safety culture of an organisation could be described as the ideas and beliefs that all members of the organisation share about risk, accidents and ill health.'

A possible shortcoming of the IAEA definition is that they use the term to describe only an ideal safety culture. The CBI's reference to shared ideas and beliefs does not make explicit the need for shared action. Neither definition quite captures the necessary elements of competency and proficiency. The Health and Safety Commission (1993) suggests the following as a working definition:

> the safety culture of an organisation is the product of individual and group values, attitudes, competencies, and patterns of behaviour that determine the commitment to, and the style and proficiency of, an organisation's health and safety programmes.
>
> Organisations with a positive safety culture are characterised by communications founded on mutual trust, by shared perceptions of the importance of safety, and by confidence in the efficacy of preventive measures. (Health and Safety Commission, 1993).

15.4.2 Characteristics of Organisations with a Positive Safety Culture

A positive safety culture implies that the whole is more than the sum of parts. The many separate practices interact to give added effect and, in particular, all the people involved share similar perceptions and adopt the same positive attitudes to safety: a collective commitment.

The synergy of a positive safety culture is mirrored by the negative synergy of organisations with a poor safety culture. Here the commitment to safety of some individuals is strangled by the cynicism of others. Here the whole is less than the sum of the parts. This is evident in organisations where a strong commitment to safety resides only in the safety department.

CBI (1990) have reported the results of a survey of 'how companies manage health and safety'. The idea of the culture of an organisation was incorporated in the report's title, *Developing a Safety Culture*. The dominant themes to emerge were:

- the crucial importance of leadership and the commitment of the chief executive;
- the executive safety role of line management;
- involvement of all employees;
- openness of communication; and
- demonstration of care and concern for all those affected by the business.

The objective of these and related organisational features is to cultivate a coherent set of perceptions and attitudes that accurately reflect the risks involved and which give high priority to safety as an integral part of shop floor and managerial performance. What is critical is not so much the apparent quality and comprehensiveness of health and safety policy and procedures. What matters is the perception of staff at all levels of their necessity and their effectiveness.

A constant theme of the discussion of safety culture is that it is a sub-set of, or at least profoundly influenced by, the overall culture of an organisation (see Appendix 1). It follows that the safety performance of organisations is greatly influenced by aspects of management that have traditionally not been 'part of safety'. This view has been supported by an extensive research programme carried out by the US Nuclear Regulatory Commission (Ryan, 1991). The expert judgement of the researchers who conducted the work believe that the key predictive indicators of safety performance in the US nuclear industry are, in rank order:

- effective communication, leading to commonly understood goals, and means to achieve the goals, at all levels in the organisation;
- good organisational learning, where organisations are tuned to identify, and respond to, incremental change;
- organisational focus; simply the attention devoted by the organisation to workplace safety and health;
- external factors, including the financial health of the parent organisation, or simply the economic climate within which the company is working, and the impact of regulatory bodies.

The point about these factors is that the first two points do not concern safety directly; they relate to all aspects of a company's culture. It follows that to make managers manage safety better it is necessary to make them better managers.

15.5 Promotion of a Positive Safety Culture

The conclusion drawn from the NUREG work and other studies, putting to one side the impact of external pressures and constraints, is that safety depends as much on organisational culture generally as on visible safety management activity. The best health and safety standards can arguably only be achieved by a programme which has a scope well beyond the traditional pattern of safety management functions. The ACSNI Human Factors Study Group (Health and Safety Commission, 1993) has advocated a 'gradualist' or step-by-step approach to the promotion of a positive safety culture (see next section). Continual variation in the approach to cultural development is likely to be needed. Culture is not a simple 'thing' that can be 'bolted on' to an organisation, nor a simple set of practices which can be implemented on a Monday morning after a weekend course (Turner *et al.*, 1989). Mutual trust and confidence between management and workforce are necessary for the development and maintenance of a strong safety culture (Health and Safety Commission, 1993). An important ingredient of company plans to promote a positive safety culture is 'organisational learning' (Toft, 1992a, b; Turner, 1992; Waring, 1992a). Organisational learning is the process whereby people in an organisation learn to change their characteristic ways of thinking and acting as a result of shared experience, addressing shared problems, coping with ambiguous situations and general immersion in the life of the organisation. Organisations do not learn merely by sending staff on training courses.

Culture has a powerful role in maintaining a sense of personal and professional identity for an organisation's members (Waring, 1992a). Attempts to change the culture may be perceived by some as a threat to their identity and interests, and this may result in resistance to, or limited co-operation with, the efforts to promote change. The critical role of power relations between different groups in an organisation and their effects on risk perception, decision-making and action is often not acknowledged (Waring, 1992b, 1993). The Royal Society report on risk (Royal Society, 1992) has emphasised that assumptions about risk acceptance and risk acceptability in organisations do not always recognise the variability in perceptions of risk between and among different groups.

It is difficult to determine and measure the complex mix of factors which together make up an organisation's culture (Johnson, 1992). A principal requirement is the development of means to use the tangible manifestations of safety culture to test what is underlying (International Nuclear Safety Advisory Group, 1991). Two matters need careful thought. First, what exactly is meant by the phrase 'tangible manifestations' and do they actually test what is underlying? Secondly, will the process of making measurements affect the validity of the measure itself?

One important issue is the extent to which the promotion of a positive health and safety culture should be an explicit goal of employers. A more rewarding approach might be for companies to develop health and safety programmes focusing on the separate factors which, taken together, constitute their health and safety culture (Turner, 1992). Managers need to take into account the effect their decisions and actions have (whether related directly to safety or not) on the perceptions that others in the organisation hold about the place of safety in the company. Equally, health and safety practitioners should consider that different approaches to the implementation of requirement to carry out risk assessments may lead to very different perceptions of the risks among managers and the rest of the workforce. The IOSH Policy Statement on Health and Safety Culture (IOSH, 1995) seeks to summarise these disparate issues. It is reproduced here in Appendix 2.

15.6 An Outline Plan to Promote a Positive Safety Culture

There are two barriers which might impede progress. Firstly, the advocates of a positive culture have proposed the simultaneous adoption of every conceivable measure which might lead to improvements. Secondly, the breadth of the concept may make the task of managing improvements appear both abstract and daunting. It must be emphasised therefore that while the outcome of well-conceived plans to improve the safety culture of an organisation may be revolutionary, the plans themselves should be evolutionary. A step-by-step approach is essential.

The first step should be to review the existing safety culture. An action plan may then be prepared on the basis of the findings of the initial review. Subsequent steps should be driven by analysis of the outcomes of each discrete stage. Some sections of proprietary safety audit systems may help in this evaluation.

Health and Safety Commission (1993) contains a detailed safety culture 'prompt list'. The list, drawn up in the context of nuclear safety, is relevant to safety culture evaluation in general. It is reproduced in here in the modified format presented by Booth and Lee (1993) in Appendix 3.

The distinctive assumption of the prompt list is that the organisation reviewing its culture already possesses an apparently impressive battery of safety and operating procedures and well-trained staff. The prompt list rather seeks to establish the adequacy of the steps that the organisation is taking to ensure that everyone in the organisation is genuinely committed to the successful implementation of the safety programme.

The foundations of the prompt list are:

- contemporary models of accident causation;

- the key functions of safety management;
- the ACSNI definition of safety culture (see above and Appendix 1);
- published research evidence.

Underlying plans to improve safety culture should be the goals that communications are founded on mutual trust, and all employees that the organisation's safety procedures:

- are founded on shared perceptions of hazards and risks;
- are necessary and workable;
- will succeed in preventing accidents;
- have been prepared following a consultation process with the participation of all employees;
- are subject to continuous review, involving all personnel.

15.7 Conclusions

Although a somewhat elusive and complex concept, the safety culture of an organisation is nonetheless 'real' for the organisation's members because it is real in its consequences. It is an inseparable part of the overall characteristic attitudes, values, beliefs and practices of people in the organisation. An organisation which has a positive safety culture consists of competent people with strongly-held safety values which they put into practice. A 'quick fix' approach to safety culture is unlikely to produce lasting benefits and employers need to make appropriate efforts to develop and maintain a positive safety culture. Health and safety practitioners have an important role to play in such efforts.

References

BEER, M. EISENSTAT, R. A. & SPECTOR, B. (1990) Why change programs don't produce change, *Harvard Business Review*, **68**(6), 158–66.

BOOTH, R. T. (1993) *Risk Assessment Workbook* (accompanies the training video *Where's the Harm in it?*) Monitor Films, 33 Market Place, Henley-on-Thames, Oxon RG9 2AA.

BOOTH, R. T. & LEE, T. R. (1993) *The Role and Human Factors and Safety Culture in Safety Management*, paper presented to the *Institution of Mechanical Engineers Conference 'Successful Management for Safety'*, 12–13 October, 1993, London: Mechanical Engineering Publication.

CBI (1990) *Developing a Safety Culture*, London: Confederation of British Industry.

DEPARTMENT OF ENERGY (1990) *The Public Inquiry into the Piper Alpha Disaster*, London: HMSO.

HALE, A. R. & GLENDON, A. I. (1987) *Individual Behaviour in the Control of Danger*, Amsterdam: Elsevier.

HEALTH AND SAFETY COMMISSION (1991) *Second report: Human Reliability Assessment – A Critical Overview*, ACSNI Study Group on Human Factors, London: HMSO.

(1991) *Successful Health and Safety Management*, Health and Safety Series booklet HS(G) 65, London: HMSO.

(1993) *ACSNI Human Factors Study Group Third Report: Organising for Safety*, London: HMSO.

INSTITUTION OF OCCUPATIONAL SAFETY AND HEALTH (1992) *Institution Policy Statement on Safety Training*, IOSH, 222 Uppingham Road, Leicester LE5 0QG.

INTERNATIONAL NUCLEAR SAFETY ADVISORY GROUP (1988) *Basic Safety Principles for Nuclear Power Plants*, Safety Series No. 75-INSAG-3 Vienna: International Atomic Energy Authority.

(1991) *Safety Culture*, Safety Series No. 75-INSAG-4, Vienna: International Atomic Energy Authority.

IOSH (1992) *Policy Statement on Safety Training*, Leicester: Institution of Occupational Safety and Health.

JOHNSON, G. (1992) Managing strategic change – strategy, culture and action, *Long Range Planning*, **25**(1), 28–36.

RASMUSSEN, J. (1987) Reasons, causes and human error. In Rasmussen, J., Duncan, K. D. & Leplat, J. (eds), *New Technology and Human Error*, Chichester: Wiley.

REASON, J. T. (1987) A framework for classifying errors. In Rasmussen, J., Duncan, K. D. & Leplat, J. (eds), *New Technology and Human Error*, Chichester: Wiley.

(1990) *Human Error*, Cambridge: Cambridge University Press.

ROYAL SOCIETY (1992) *Risk – Analysis, Perception and Management*, Report of a Royal Society Study Group, London.

RYAN, T. G. (1991) *Organisational Factors Regulatory Research Briefing to the ACSNI Study Group on Human Factors and Safety*, London, July (unpublished).

TOFT, B. (1992a) Changing a safety culture: decree, prescription or learning? paper presented at *IRST Conference on Risk, Management and Safety Culture*, London Business School, 9 April.

(1992b) Changing a safety culture: a holistic approach, paper presented at the *British Academy of Management 6th Annual Conference*, Bradford University, 14–16 September.

TURNER, B. A. (1992) Organisational learning and the management of risk, paper presented at the *British Academy of Management 6th Annual Conference*, Bradford University, 14–16 September.

TURNER, B. A., PIDGEON, N., BLOCKLEY, D. & TOFT, B. (1989) Safety culture: its importance in future risk management, position paper for the *Second World Bank Workshop on Safety Control and Risk Management*, Karlstad, Sweden, 6–9 November.

WARING, A. E. (1992a) Organisational culture, management and safety, paper presented at *British Academy of Management 6th Annual Conference*, Bradford University, 14–16 September.

(1992b) Organisations respond characteristically not ideally, paper presented at

BPP Conference on Risk Analysis and Crisis Management, London, 22–23 September.

(1993) Power and culture – their implications for safety cases and EER, paper presented at *1993 European Seminar on Human Factors in Offshore Safety*, Aberdeen, 29–30 September.

ZOHAR, D. (1980) Safety climate in industrial organisations: theoretical and applied implications, *Journal of Applied Psychology*, **65**(1), 96–102.

APPENDIX 1 Glossary of Terms

culture

(i) A set or system of unwritten and often un-admitted attitudes, behaviours, ideologies, beliefs, values, meanings, opinions, habitual responses, ways of doing things, language expression, rituals, quirks and other shared characteristics of a particular group of people. Cultures can be identified at different levels, e.g. nations, societies, organisations, departments, interest groups. Culture reinforces identity and behaviour.

(ii) A collective identity or world-view.

safety culture

(i) Those aspects of culture which affect safety.

(ii) The characteristic shared attitudes, values, beliefs and practices concerning the importance of health and safety and the necessity for effective controls.

(iii) '. . . that assembly of characteristics and attitudes in organisations and individuals which establishes that, as an overriding priority, nuclear plant safety issues receive the attention warranted by their significance.' (International Nuclear Safety Advisory Group, 1991).

(iv) 'The safety culture of an organisation is the product of individual and group values, attitudes, competencies, and patterns of behaviour that determine the commitment to, and the style and proficiency of, an organisation's health and safety programmes. Organisations with a positive safety culture are characterised by communications founded on mutual trust, by shared perceptions of the importance of safety, and by confidence in the efficacy of preventive measures' (Health and Safety Commission, 1993).

organisational learning

The process by which people in an organisation learn to change their characteristic ways of thinking and acting as a result of shared experience, addressing shared problems, coping with ambiguous situations and general immersion in the life of the organisation.

value
A belief, or set of beliefs, that is not testable by reference to objective standards, e.g. the belief that safety is more important than profit (or vice versa).

APPENDIX 2 Institution of Occupational Safety and Health Health and Safety Culture Policy Statement

1 Commentary

The institution recognises that:

1.1 The concept of health and safety culture is a useful shorthand description of the characteristic attitudes, values, beliefs and practices of people in an organisation concerning the importance of health and safety and the necessity for effective controls;

1.2 It is vital not to lose sight of the large number of factors which, when taken together, constitute a positive health and safety culture. Important indicators include:
- — the demonstrated commitment and leadership of directors and senior managers;
- — the acceptance among managers at all levels that health and safety is a line management responsibility;
- — participation in health and safety decisions by personnel at all levels;
- — training to promote competencies in health and safety;
- — shared perceptions of: the nature of hazards, the magnitude of risks, and the practicality and effectiveness of preventive plans.

A number of publications (CBI, 1990; Health and Safety Commission, 1993; Booth and Lee, 1993) contain checklists or 'promptlists' of factors which may influence the safety culture of an organisation. Guidance is also available (Booth, 1993) on the way risk assessment may be used to assist in the promotion of shared perceptions of risk and in the practicality of preventive plans.

1.3 Attempts to measure the health and safety culture of an organisation may prove complex and unrewarding. However, it is appropriate to seek to assess key indicators such as those listed in 1.2 above.

1.4 Attempts to change the health and safety culture of an organisation as a major company initiative may prove to be less successful than more tightly focused attempts to improve key indicators of the culture.

1.5 Plans to improve the safety culture of an organisation should take into account the fact that the attitudes, values and beliefs of personnel have evolved over an extended period, and that attempts to change such embedded characteristics may require a long-term and sustained programme.

1.6 The safety culture of an organisation is a part of the overall organisational culture. Employers' actions to change the safety culture may have much less impact on employees' perceptions than the signals given out by day-to-day management decisions and attitudes which may appear to have little to do with safety and health.

2 Recommendations

The Institution recommends that employers should:

2.1 Take steps to establish what their managers and employees actually believe about health and safety, in particular.
- — 'ownership' of health and safety responsibilities and benefits;
- — match between statements and actions;
- — attitudes towards risk taking and risk acceptance;
- — motivations to act appropriately;
- — conflicts of interest between different groups.

2.2 Codify what is expected of all their personnel in terms of values, beliefs, attitudes and practices concerning health and safety;

2.3 Consider the most appropriate way in which to address any differences between corporate expectations and the characteristics of the organisation's personnel concerning health and safety. In particular, account should be taken of:
- — the right balance between decree, prescription and 'organisational learning';
- — the likely timescale to achieve a permanent change in cultural characteristics.

2.4 The Institution further recommends that occupational health and safety practitioners should always consider health and safety cultural factors when:
- — planning, and assisting in, the completion of risk assessments;
- — drawing up plans for health and safety training;
- — preparing safe systems of work and permit-to-work procedures;
- — monitoring specific health and safety programmes;
- — auditing the implementation and effectiveness of the organisation's overall health and safety programme.

APPENDIX 3 Safety culture promplist

1 Review of Organisational Culture – Employers

Has the organisation evidence to demonstrate that:

1.1 Communications at all levels are founded on mutual trust?
1.2 All personnel understand, and agree with, corporate goals and the subordinate goals of their work group?
1.3 All personnel understand, and agree with, the means adopted to achieve corporate and work group goals?
1.4 The work practices of the organisation are under continuous review to ensure timely responses to changes in the internal or external environment?
1.5 Managers and supervisors demonstrate care and concern for everyone affected by the business?
1.6 Managers and supervisors take an interest in the personal, as well as the work, problems of their subordinates?
1.7 Managers and supervisors have been trained in leadership skills, and adopt a democratic and not an authoritarian leadership style?
1.8 Workforce participation in decision-making is not confined to peripheral issues?
1.9 Job satisfaction is maintained by, for example, verbal praise from supervisors and peers, equitable systems of promotion, minimisation of lay-offs, and the maintenance of a clean and comfortable working environment?
1.10 The organisation expects the highest standards of competence and commitment of all its employees, but retribution and blame are not seen as the purpose of investigations when things go wrong?
1.11 An appropriate distribution of both young, and more experienced socially mature employees is maintained in the workforce?
1.12 The organisation only recruits suitable personnel, but no automatic presumption is made that individuals are immediately competent to carry out the tasks assigned to them?

2 Review of Safety Culture – Employers

2.1 *Policy, Planning, Organisation and Communication*

Has the organisation evidence to demonstrate that:

2.1.1 The Chief Executive takes a personal and informed interest in safety?
2.1.2 The Chief Executive and the Board take explicit and continuing steps to ensure that their interest in, and commitment to, safety is known to all personnel?

2.1.3 A positive commitment to safety is visible throughout the management chain?

2.1.4 Safety is managed in a similar way to other aspects of the business, and is as much the responsibility of line management as any other function?

2.1.5 Safety practitioners have high professional status within the organisation with direct access to the Chief Executive or other appropriate Board member?

2.1.6 Safety committees have high status in the organisation, operate proactively, and publicise their work throughout the organisation?

2.1.7 Managers at all levels, and supervisors, spend time on the 'shop floor' discussing safety matters, and that steps are taken to ensure that all personnel hear of the visits and the matters discussed?

2.1.8 Managers and supervisors spend time commending safe behaviour as well as expressing concern if safety procedures are not being observed?

2.1.9 There are multiple channels for two-way communication on safety matters, including both formal and informal modes?

2.1.10 Safety representatives play a valued part in promoting a positive safety culture, and in particular contribute to the development of open communications?

2.1.11 Specially-convened discussion/focus groups are established to consider the safety aspects of new projects?

2.1.12 Everyone in the organisation talks about safety as a natural part of every-day conversation?

2.1.13 Everyone in the organisation recognises the futility of mere exhortation to think and act safely as a means of promoting good performance?

2.2 *Hazard Management*

Latent (decision) failures: has the organisation taken explicit steps to prevent and detect:

2.2.1 Cases where managers with responsibility for the development or implementation of safe operating procedures fail to:

- (a) search for, and identify, all relevant hazards?
- (b) assess risks accurately?
- (c) select workable and effective control solutions?
- (d) adopt appropriate methods to monitor and review the adequacy of the procedures?
- (e) determine whether foreseeable active failures are likely to be the result of errors at the skill-, or rule-, or knowledge-based levels, or the result of violations?
- (f) minimise or eliminate sources of conflict between production and safety?

(g) ensure that all relevant personnel have had an opportunity to comment on the procedures before finalisation or implementation?

(h) ensure that all personnel are adequately trained, instructed and motivated to follow safe operating procedures?

2.2.2 Cases where managers *personally* commit violations of safety procedures or professional good practice?

Active failures: has the organisation taken explicit steps to prevent and detect:

2.2.3 Personnel failing (as a consequence of errors and/or violations) to:

(a) search for and identify all relevant hazards?

(b) match their perception of risks to the actual risk magnitudes?

(c) accept responsibility for action?

(d) follow systems of work where specified, or otherwise adopt a safe method of work?

(e) continuously monitor and review the magnitude of risks to which they are exposed, and the effectiveness of the steps taken to keep the dangers under control?

Do the organisation's plans for preventing and detecting latent and active failures take explicit account of the following:

2.2.4 Managers, supervisors, and other personnel may tend to underestimate the magnitude of risks:

(a) with no significant potential (when dealing with major hazards)?

(b) where the consequences are delayed (for example, a long latent period between exposure and harm)?

(c) affecting people outside the immediate work group?

(d) where perceptions may not be adjusted sufficiently in the light of new information?

(e) where snap judgement are made on the basis of extrapolated information about other hazards?

2.2.5 Managers, supervisors, and other personnel may tend to overestimate their ability to assess and control risks:

(a) where the hazards have been encountered for long periods without apparent adverse effect?

(b) where the hazards present opportunities for ego enhancement (for example, public displays of daring (macho image); managers seeking to portray decisiveness)?

(c) where substantial benefits accrue?

(d) when the assessment is made by a volunteer?

2.2.6 Managers, supervisors, and other personnel may tend to have an impaired ability to cope with risks:
- (a) when affected by life-event stressors (for example, bereavement, divorce)?
- (b) when under stress as a result of a lack of confidence in the established procedures?
- (c) when they believe that they have no power to influence their own destiny or that of others (fatalism)?

Has the organisation adopted the following measures for improving people's perceptions of risks, and/or ability and commitment to control risks:

2.2.7 A scheme to identify managers, supervisors and other personnel who may:
- (a) be subject to life-event stressors?
- (b) lack confidence in the effectiveness of prevention?
- (c) harbour resentment or distrust of the organisation?
- (d) have an adventurous outlook on risks?

2.2.8 Steps to increase individual belief in their own ability to control events?

2.2.9 Steps to erode peer approval of risk taking?

2.2.10 Discussion groups to talk through individual perceptions of risks and preventive measures?

2.2.11 Safety training founded on:
- (a) a clear recognition and understanding of the likely distortions of people's perceptions of risk magnitudes and corrective measures?
- (b) the need for refresher training to counter people's changes in perceptions over time?
- (c) feedback of accident/near miss data?
- (d) explanations of not just how a job must be done, but why it must be done that way?
- (e) the need for team building?

2.3 *Monitoring and Review*

2.3.1 Has the organisation taken explicit steps to determine how its corporate goals compare with those of the local community and society at large?

2.3.2 Is the Board seen to receive regular safety reports, to review safety performance periodically, and to publicise the action it has taken?

2.3.3 Has the organisation:
- (a) a plan to review, and where necessary, improve its safety culture?
- (b) devised methods for selecting, quantifying and measuring (auditing) key indicators of safety culture?

(c) reviewed, and where necessary changed, its organisational structure to make manifest its commitment to safety?
(d) taken steps to ensure safety decisions are acted upon without delay?

2.3.4 Have members of the organisation been trained to:
(a) carry out a review of safety culture?
(b) devise and validate key indicators of safety culture?
(c) prioritise safety culture goals arising from a review?
(d) draw up an action plan to improve the safety culture of the organisation in priority areas?
(e) monitor the implementation and effectiveness of plans to improve the safety culture?

2.3.5 Has the organisation made arrangements to encourage reflection on, and elicit the views of all personnel about:
(a) the overall organisational culture?
(b) the safety culture of the organisation?
(c) their perceptions of the attitudes of others in the organisation about safety?
(d) their perceptions of risk?
(e) their perceptions of the effectiveness of preventive measures?
(f) themselves (self-assessment)?

2.3.6 Has the organisation introduced incident investigation procedures which take full account of:
(a) multi-causality?
(b) the need to explore the incidence of latent as well as active failures?
(c) the need to continue the investigation, even when an apparent cause has been found, to determine further causal factors?
(d) the importance of accepting that the ultimate responsibility lies with the organisation, rather than merely assigning blame to individuals?

3 Regulators & Safety Culture

Has the regulatory body:

3.1 Considered both the positive and potentially negative effects of regulatory intervention in the promotion of a positive safety culture?
3.2 Recognised that a positive safety culture demands that employers should have a sense of 'ownership' of safety?
3.3 Appreciated that key indicators of safety culture go beyond the conventional measures of safety performance?
3.4 Taken steps to ensure that employers' staff are adequately trained to carry out an objective review of their safety culture?

3.5 Taken steps to ensure that employers' staff are adequately trained in the skills necessary to develop an action plan for the promotion of a positive safety culture?

3.6 Recognised that employers' possible misperceptions of the quality of their safety culture may be shared by the regulator's own staff when the regulators have similar backgrounds and are part of the same social system?

3.7 Reviewed, in the context of safety culture, the qualifications, experience, and training of their own staff?

CHAPTER SIXTEEN

Key topics in nuclear safety

NEVILLE STANTON
Department of Psychology, University of Southampton

16.1 Introduction

This book contains 14 chapters on human factors aspects of nuclear safety. As far as possible, the main subject areas have been covered. As can be seen from the coverage of topics introduced by Welch and O'Hara (Chapter 2), who cite ten requirements in their human factor engineering program review model, all of these are covered. These ten elements together with pointers to the relevant chapters are shown in Table 16.1

Obviously, the enormity of the topic means that further reading will be necessary to gain an in-depth knowledge of the topic area. The references at the end of each chapter direct this reading. However, this book provides a good overview of human factors issues in nuclear safety. A content analysis of the chapters reveals seven key topics in the book, comprising: design and evaluation methodologies, safety, problems, selection and training, simulation, validation, and human performance. These topics may be related to the three main issues put forward in the introduction, organisational issues, personnel issues and design issues. The key topics are shown in Figure 16.1.

Figure 16.1 shows the authors and chapters numbers in the left-hand column, the topics across the rows at the top of the figure and the contributions to the topics by the authors within the matrix. The parts that the book is divided into are indicated on the right of the figure.

16.2 Key Topics

The contributions of the chapters to the seven key topics will be discussed in the remainder of this chapter.

Table 16.1 Pointers to chapters containing information related to HF review elements

Human factors review elements	Chapters
HFE program management	2, 3
Operating experience review	ALL
Functional requirements analysis and allocation	2
Task analysis	9, 10, 14
Staffing	9
Human reliability analysis	8, 12, 13, 14, 15
Human–system interface design	4, 5, 12, 13
Procedure development	6
Training program development	7, 10, 11
HF verification and validation	3

16.2.1 Design and Evaluation Methodologies

The coverage of design can be split into two principal parts: macro-design and micro-design issues. Both Welch and O'Hara (Chapter 2) and Dien (Chapter 3) provide a framework for the incorporation of human factors into the design process. The approach taken by Welch and O'Hara is to propose a development process that incorporates ten HF review elements, as introduced in Table 16.1. They suggest that this approach offers a structured and top-down approach to HF engineering, which should receive a high priority in the design process. Welch and O'Hara argue that HF needs to be proactive in nuclear power plant system development if it is to be more widely applied. Dien also argues that the control room needs to be tailored to the requirements of the users, and methodologies need to be developed to ensure that this occurs. In particular, he points out that different types of user may have different requirements from the system. Dien contrasts an ergonomic assessment with other types of assessment, to suggest that it needs to interface with both the functional/technical assessments and the verification assessments. It is proposed that a good deal of effort needs to be expended upon careful preparation of the assessment and methodical organisation is advised. Dien discusses a variety of methods of assessment including: expert judgement, mock-ups, prototyping and on-site evaluations. The general advice is to use a combination of approaches, some being more relevant at different points in the design process than others. To complement the Human Factors Engineering Program Review Model (Chapter 2), Dien proposes the main topics of assessment are: design of workstations, allocation or function, interface design, procedure design and organisational policy.

Within this overall framework of macro-design methodologies, approaches have been developed to answer specific questions. Methodologies presented within this book comprise: analysis of input devices (Chapter 4), analysis of

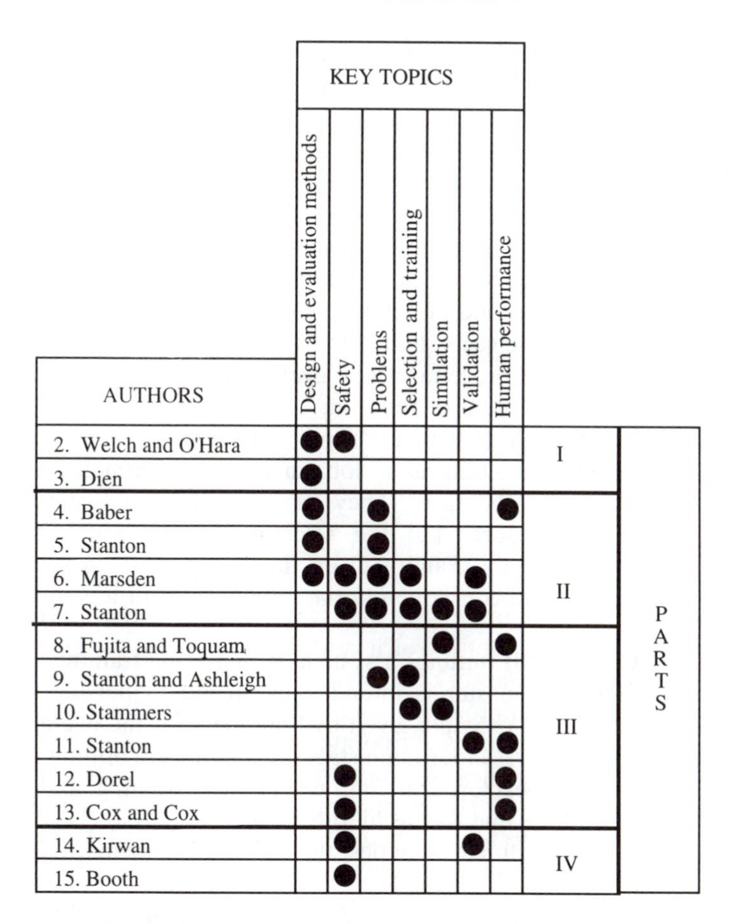

AUTHORS	Design and evaluation methods	Safety	Problems	Selection and training	Simulation	Validation	Human performance	PARTS
2. Welch and O'Hara	●	●						I
3. Dien	●							I
4. Baber	●		●				●	II
5. Stanton	●		●					II
6. Marsden	●	●	●	●		●		II
7. Stanton		●	●	●	●	●		II
8. Fujita and Toquam					●		●	III
9. Stanton and Ashleigh			●	●				III
10. Stammers				●	●			III
11. Stanton						●	●	III
12. Dorel		●					●	III
13. Cox and Cox		●					●	III
14. Kirwan		●				●		IV
15. Booth		●						IV

Figure 16.1 A content analysis of the chapters.

alarm systems (Chapter 5), design of procedures (Chapter 6), personnel selection procedures (Chapter 9), management of training (Chapter 10), human reliability analysis (Chapter 14) and measurement of safety culture (Chapter 15).

In a discussion of the ergonomic issues surrounding the choice of input devices for the control room, Baber (Chapter 4) suggests that four main criteria are important for selection of the appropriate device(s). These criteria are: compatibility (movement, spatial, operational and conceptual), performance (speed, accuracy and error), operational characteristics (space requirements and durability) and task characteristics (range of activity, physiological demands, strategy and process feel).

Stanton (Chapter 5) describes an alarm handling questionnaire which has been used to evaluate the reactions of operators toward their alarm system. The questionnaire has six main sections, comprising definition of alarm system, alarm identification, critical incidents, alarm responses, alarm diagnosis and problems with alarm systems. In a full alarm system review, this would be only one of the approaches, which would include: observations, static assessments and presentation of scenarios (Stanton, 1994).

Marsden (Chapter 6) presents two methodologies in the chapter on procedures. The first relates to the design of procedures and contains the following steps: analysis of task, select format for procedure, draft procedure, quality assure, approve procedure, monitor effectiveness and intermittently review procedure. The second method refers to procedure maintenance and integration. Marsden suggests that procedures need to be continually monitored and no set of procedures can ever be a finalised document, but rather it is in a constant state of revision.

Stanton and Ashleigh (Chapter 9) put forward a general personnel selection procedure that comprises: job analysis, recruitment, selection, decision making and validation. They argue that a systematic procedure is required to ensure that the selection of personnel is related to job-relevant criteria only. Failure to do this could result in sub-optimal selection procedures, and consequently the employment of less appropriate personnel.

Stammers (Chapter 10) outlines a system of managing training resources called Taskmaster. Taskmaster is intended as an aid to persons who are developing training. The example given by Stammers for the overall task of managing training comprises four main stages: promote and optimise training function, design and develop training, install new training and provide training delivery. In the same way as hierarchical task analysis (HTA) breaks down any task into subordinate activities, the task of training may be similarly structured.

Kirwan (Chapter 14) presents a generic human reliability analysis (HRA) methodology that comprises eleven main stages: problem definition, critical task identification, task analysis, human error identification, representation, screening, quantification, impact assessment, error reduction, quality assurance and documentation. Kirwan categorises quantification techniques under four main classes: unstructured expert opinion, structured expert opinion, data-driven and accident sequence data-driven. Kirwan points out that different techniques derive different data, so a combination of approaches may be appropriate.

Booth (Chapter 15) presents a detailed safety culture checklist to assist in promoting a positive safety culture. The checklist prompts the user to consider four main areas: policy and planning (determining safety goals and objectives), organisation and communication (establishing clear lines of responsibility and communication), hazard management (identifying and controlling hazards), monitoring and review (establishing the effectiveness of the organisation's attempts at: policy and planning, organisation, communication and hazard management). This sort of approach will assist the organisation in developing a proactive safety management programme.

16.2.2 Safety

As the title of the book indicates, safety is a central theme to this text. The theme is introduced by Welch and O'Hara (Chapter 2), who point out that the human element is an important link in the safety of nuclear power plants. They suggest that safety concerns and technological developments have made human factors an essential discipline to the nuclear industry. Stanton (Chapter 5) argues that alarm design is a critical safety feature in interface design, given the intended role that alarm systems are meant to play, i.e. communicating information that is of crucial importance to overall system integrity. This is all the more important when one considers that the operator's role is one of recovering the system from disturbances (Kirwan, Chapter 14). There may be procedures that assist the operator in plant recovery (Marsden, Chapter 6) and simulators (Stanton, Chapter 7) may be used to keep trained personnel refreshed (Stammers, Chapter 10). The experience and training of the operator appears to be the main determinant of successful management of incidents (Fujita and Toquam, Chapter 8). Stress in the control room leads to higher rates of human failure, and, under varying levels of stress, these rates may approach 100% (Cox and Cox, Chapter 13). Dorel (Chapter 12) suggests that increases in human failures at night compared to day shifts, irrespective of the differences in worker activity. This has important consequences for shift cycles. Unsafe work practices are likely to become apparent in human reliability assessment (Kirwan, Chapter 14). Safe working may be encouraged through developing a positive safety culture (Booth, Chapter 15).

16.2.3 Problems

Some HF problems have been identified in the individual chapters. Baber (Chapter 4) points out that SCADA technology may not necessarily remove the ergonomic problems associated with traditional control rooms. Indeed, the new technology may compound existing problems still further. Baber argues that many of the general ergonomic problems noted in the report on Three Mile Island were irrespective of the interface medium. New technology does not solve the problem of the operator navigating around wall panels, they still have to navigate round CRT screens which may be even more difficult. Advances in ergonomics will only come about through a greater understanding of human performance, not through technological development. Similarly, Stanton (Chapter 5) argues that problems with exiting alarm media cannot necessarily be solved by using a new medium for the presentation of alarm information. Problems with alarms in the reported study mainly related to the presence of too many alarms (on a CRT alarm display). There would seem to be two principal, and interrelated, difficulties. The first of these is the design of the alarm data to be presented and the second is the medium chosen to display the alarms. Stanton suggests that through a better understanding of the alarm handling task, we may be more equipped to design the alarm information and choose the most appropriate medium.

Marsden (Chapter 6) noted several problems with the design and use of procedures in the nuclear industry. By far the most important problem was the lack of formal processes for the development of procedures. This failing has led to incomplete, inaccurate and non-existent procedures, which has, in turn, contributed to a high proportion of nuclear power plant incidents. Marsden suggests a formal procedure development process that should reduce the percentage of incidents that are attributable to this cause. However, even well-designed procedures do not guarantee that they will be used. In an analysis of non-compliance in procedure use, Marsden argues that operators may not always perceive the procedures as useful. For example, they may be looked upon as there to protect the company. Also, they may make the job less rewarding or more difficult and unworkable. Marsden suggests user-centred design methods would overcome many of these difficulties.

Booth (Chapter 15) suggests that the term 'safety culture' is difficult to define, but that should not stop us from trying, nor should it stop us from trying to measure it. Booth offers some critique of the International Atomic Energy Authority (IAEA) definition of safety culture by suggesting that it only discusses the positive aspects and the Confederation of British Industry (CBI) definition in that it ignores the need for shared action. Instead, Booth proffers the Health and Safety Commission (HSC) definition:

> The safety culture of an organisation is the product of individual and group values, attitudes, competencies and patterns of behaviour that determine the commitment to, and the style and proficiency of, an organisation's health and safety programmes. Organisations with a positive safety culture are characterised by communications founded on mutual trust, by shared perceptions of the importance of safety and by confidence in the efficacy of preventative measures.

It would seem essential that 'safety culture' is clearly defined before it can be measured, changed or developed. It would also seem reasonable that 'safety culture' can be positive, negative or indifferent. Booth argues that although it is difficult to define, it nevertheless has important consequences for the organisation, and that development of a positive safety culture can only be brought about through commitment from the core of the organisation.

16.2.4 Selection and Training

According to the ACNSI reports outlined in the introduction (Chapter 1) selection and training of personnel is one of the utmost importance in the nuclear industry. Stanton and Ashleigh (Chapter 9) suggest that the selection process and the methods employed in selection could be improved. They also suggest a trade-off between selection, training and work design, e.g. resources can be expended on selecting more appropriate personnel or they can be put into training or making the task simpler. Obviously the cost-to-benefit ratio would determine how to best share out the finite resources. In a survey of selection methods used in the nuclear industry, Stanton and Ashleigh (Chapter 9) found that there is a heavy reliance upon traditional approaches, despite the poor performance of these approaches. They suggest that the assessment centre approach ought to be explored and that the industry should expend more effort on formally validating its selection process. Stammers (Chapter 10) argues that technological development makes greater demands upon training as they have created a complex task for control room operators. This problem is further exacerbated by the infrequent use of some of the skills. Skill loss can be quite significant after only a short time of non-practice. However, Stammers suggests that overlearning (practice beyond criterion performance) and rehearsal can help reduce skill loss, and thus assist in preparing the readiness of the operator. Marsden (Chapter 6) suggests that the procedures infrastructure need to be explicitly linked to the training programme.

16.2.5 Simulation

A good deal of investment has been expended on realistic simulation control room environments in the nuclear industry, and this has been put to good use. Examples include the collection of data about human performance (Chapter 8) and training (Chapter 10). Fujita and Toquam (Chapter 8) use a full-scope control-room simulator to good effect when investigating crew response to a number of emergency and non-emergency scenarios. Stammers (Chapter 10) and Stanton (Chapter 7) question whether this very high level of fidelity is always necessary. It has been suggested that the investment in high fidelity of simulators is largely as a result of ignorance about the benefits of their lower-fidelity counterparts. In particular, there is a reluctance of people to believe that unless the device mimics the real situation in every aspect, it cannot hope to produce behaviour that is transferable. However, there is a growing body of evidence to suggest that many of the benefits of the full-scope simulators could be realised for a fraction of the cost. This is not to say that low-fidelity simulators will always be appropriate, nor that fidelity is a unidimensional concept. Stammers (Chapter 10) argues that at least as much effort should go into making sure that they are effective training devices. For example, he suggests that the simulators should measure performance of the operators, provide part-task facilities and have graded difficulty. It is surprising how many simulators do not possess such facilities.

Table 16.2 Distinction between three types of validity

	Type of validity		
Topics	Predictive	Face	Theoretical
Selection	Correlates positively with future job performance	Looks as though it measures some aspect of the job	Based upon a psychological theory, e.g. theory of persona, such as Cattell *et al.* (1970)
Training	Correlates positively with future job performance	Looks as though it trains some aspect of the job	Based upon a theory of learning, e.g. Gagné (1965)
Simulation	Correlates positively with future job performance	Looks like the actual work equipment and task	Based upon a theory of simulator design, e.g. Stammers & Patrick (1975)
HRA	Correlates positively with future errors	Looks as though it captures relevant errors	Based upon a theory of human error, e.g. Reason (1990)

16.2.6 Validation

Validation is an important concept, and particularly when related to HF methods, and has been a recurrent theme throughout the book. The three basic types of validity are: predictive (the predictor is related to some future criterion of performance), face (the predictor looks as though it measures some aspect of performance), theoretical (the predictor is based upon some theoretical construct). These concepts are discussed with regard to selection of personnel (Chapter 9), training (Chapter 10), simulation (Chapter 7) and human reliability assessment (Chapter 14). Table 16.2 illustrates the differences between the types of validity.

In a practical sense, *predictive* validity is by far the most important, and it is possible that a method may have proven predictive validity independently of *face* or *theoretical* validity.

16.2.7 Human Performance

Ultimately, HF is concerned with human performance in technological domains. Therefore most of the chapters address performance-related issues to some extent. Baber (Chapter 4) discusses performance differences (both speed and errors) for input devices for two main types of task: cursor movement and text/data entry.

Performance is affected by task type, expertise of the user and the context within which the device is used. Baber reports that, in general, cursor keys are disliked and for a small number of objects the function keyboard is fast. The touchscreen seems to be the fastest input device for uncluttered screens for selecting objects. For selecting and dragging objects the mouse appears to have superior performance, whereas for entering text the keyboard is superior.

In an analysis of performance shaping factors, Fujita and Toquam (Chapter 8) attempted to identify which factors made the greatest difference to performance, using studies of crew performance on a full-scope simulator under a variety of scenarios. The crew consisted of a supervisor (who was responsible for operational decisions), a reactor operator (who was responsible for the primary system) and a turbine operator (who was responsible for the secondary system). Measures were taken on 6 generic PSFs: cognitive ability, personality, stress and coping, leader behaviour, experience and group interactions. Fujita and Toquam found that previous job and training experience accounted for 20% of the variance. Some other PSF measures (e.g. perceptual speed, attention, job knowledge, personality traits, social desirability and personal stress-coping mechanisms) were also significant contributors.

Differences in performance, such as those found by Fujita and Toquam, are of obvious importance in the selection of personnel and should be taken into account when developing job relevant selection criteria (Chapter 9). Stanton (Chapter 11) reviewed the major factors affecting team performance, comprising communication, co-ordination, co-operation and control. Stanton argues that team performance can be superior to individual performance, because it draws upon multiple skills, resources, knowledge and experience. However, under certain circumstances, group pathologies (such as 'groupthink') can lead to detrimental group performance. Some precautions, if implemented, should help prevent 'groupthink' from occurring.

The studies by Dorel (Chapter 12) show that demands of shiftworking can exceed the functional capabilities of human workers. He suggests that tasks are typically designed for day working, with little heed to psychophysiological changes in workers on night shifts. This failure to design work systems specifically for night shifts has been indicated as a contributory causal factor in some nuclear incidents. Dorel shows the effect of shift pattern and shift cycle on human failure. It is perhaps not surprising that there are more failures on night shifts, as humans are not normally nocturnal beings. Dorel suggests that as long as 24 h shift working remains, there is a need to consider ways of adapting night work to the capacities and limitations of workers. He suggests a number of strategies, including the provision of extra support, more workers and additional rest periods.

Cox and Cox (Chapter 13) suggest that groups adapt differently to stress; some groups break down whilst others appear to become more cohesive. The review of operator stress in the control room is consistent with the DREAMS framework (dynamic reliability technique for the analysis of human error in man–machine systems). Factors that give rise to stress can be identified at the level of the individual (e.g. inadequate training (see Stammers, Chapter 10), poor procedures (see Marsden, Chapter 6), misleading instrumentation (see Stanton, Chapter 5), plant deficiencies and poor control room design (see Part 2: Interface Design Issues)) and at the level of the organisation (e.g. conflicting commercial and safety pressures (see Booth, Chapter 15), communication and decision making (see Stanton, Chapter 11), design and preparation for shiftwork (see Dorel, Chapter 12)). Cox and Cox identify that extremes in workload demand are a common feature of control-room operation. A very quiet period requiring passive monitoring of equipment may suddenly be changed into an extremely busy period requiring hands-on problem solving. These transitions from low to high workload can, under some circumstances, be unpredictable and major sources of stress in the control room. Operators may be able to cope with high levels of workload for short periods, but they are likely to shed some tasks in order to sustain performance on perceived high-priority tasks. Cox and Cox question whether the decisions about load shedding are likely to be optimal under such conditions.

16.3 Future Developments

Some of the chapters indicate future developments in HF. For instance, developments in input devices (Chapter 4), alarm design (Chapter 5), training (Chapter 10), human reliability assessment (Chapter 13). Baber (Chapter 4) considers future possibilities of input devices in the control room, such as mobile communications, gesture-based systems (e.g. glove and pen devices) and speech. Mobile communications would permit the operator to have freedom of movement in the control room and beyond. Gesture systems could introduce a new dimension of communication into the control room. Speech (both recognition and synthesis) could free the operator from the desk as well as exploiting another communication channel. Baber is cautious in recommending these 'new' media, suggesting that at best, the research evidence is equivocal. Stanton (Chapter 5) proposes that future design of alarm systems should be based upon an improved understanding of human alarm handling. It is suggested that the traditional three-stage model of alarm handling (i.e. detect – diagnosis – compensation) is rather too simplistic and an inadequate description. A more representative description is provided by the process model of alarm-initiated activities and the taxonomy of human alarm handling. Stanton argues that the latter approach would lead designers to reconceive the design of alarm systems. Kirwan (Chapter 14) proposes that despite the wide use and acceptance of human reliability assessment (HRA), there are seven aspects in which improvements could be made. These improvements include analysis of cognitive activities and standards for development of methods. The current status of HRA techniques is well developed in the prediction of errors based upon physical activity, but not so well developed for the prediction of errors based upon cognitive activity. Kirwan proposes that effort for future development of HRA techniques should be concentrated in this direction. He also proposes that the development of HRA techniques could be improved through the provision of benchmark standards and better documentation and training.

There is likely to be an increasing amount of interest in the concept of a *safety culture* (Chapter 15) in the future. As our understanding improves and measures are developed to assess the tangible aspects of this phenomenon, we are likely to witness organisations becoming more proactive in developing positive safety cultures. Booth proposes that health and safety professionals have an important role to play in this future development.

Above all, the chapters in this book have emphasised *user-centred* approaches in the design, maintenance, operation and support of systems. For example, considering user performance with input devices (Chapter 4); listening to users' views and involving them in the design or procedures (Chapter 6); conducting experiments to increase our understanding of team performance (Chapter 8); involving employees at all levels of the organisation in the development of a positive safety culture (Chapter 15). This message cannot be over-emphasised. 'User-centred' does not just mean designing for users (e.g. designing the system to be used by the representative user groups), but it also means involving representative users in the design process (Chapters 2 and 3). Returning to the initial definitions of HF presented in the introduction (Chapter 1), the ultimate goal of HF is to match work systems to the capabilities and limitations of people with the objective of enhancing effectiveness and efficiency of the systems' performance whilst also improving safety, comfort, job satisfaction and quality of life of the individuals within the system. Application of the ideas and principles put forward within this book will help organisations meet these objectives.

References

CATTELL, R. B., EBER, H. W. & TATSUOKA, M. (1970) *Handbook for the Sixteen Personality Factor Questionnaire*, Champaign, Ill.: IPAT.

GAGNÉ, R. M. (1970) *The Conditions of Learning*, New York: Holt, Rinehart & Wilson.

REASON, J. (1990) *Human Error*, Cambridge: Cambridge University Press.

STAMMERS, R. B. & PATRICK, J. (1975) *The Psychology of Training*, London: Methuen.

STANTON, N. A. (1994) *Human Factors in Alarm Design*, London: Taylor and Francis.

Index

Note: HF is used as an abbreviation for human factors